Praise for
Grant and Sherman

"Charles Flood's *Grant and Sherman* is the story of two remarkable men, an extraordinary friendship, and a partnership that won the Civil War. Shedding fresh light on these two men, it is by turns evocative, charming, and often absorbing."

—JAY WINIK, author of
April 1865: The Month That Saved America

"Charles Bracelen Flood studies the friendship between Grant and Sherman and rightly concludes that it had a major impact on the results of the Civil War. Thoroughly researched and written with verve, this book is easy reading and provides rewarding insight into a friendship that influenced the lives of two significant individuals and the war in which they played such crucial roles."

—JOHN MARSZALEK, professor emeritus of history, Mississippi State University, and author of *Sherman: A Soldier's Passion for Order*

"For those who think there is nothing new to be said about the Civil War, *Grant and Sherman* will be a pleasant surprise. The book is a unique blend of emotional power and historical insight. A must-read."

—THOMAS FLEMING, author of
Liberty! The American Revolution

"*Grant and Sherman* is a profound study of the relationship of the generals who endured and determined the tide of the victory in our nation's most divisive and bloody war. This masterpiece ranges from vivid battlefield reports to intimate sketches of Grant's and Sherman's marriages to subtle cameo sketches of the officers and politicians who harassed or supported them. Fusing his talents as a resourceful scholar and distinguished novelist with a touch of the poet, Flood has achieved a moving and inspired classic of American history."

—SIDNEY OFFIT, president of
the Authors Guild Foundation

"This book describes with force, clarity, and admirable terseness the forging in the field of the historic leadership team that was essential to Union victory. Civil War scholars and general readers alike will profit from its insights."

—CHARLES P. ROLAND, author of
An American Iliad: The Story of the Civil War

"Ulysses S. Grant and William T. Sherman forged a superb partnership during the Civil War, a team of opposites drawn together to end a horrendous conflict without excessive suffering and bloodshed. Ironically, they succeeded by rendering war too terrible for the South to continue. Both Grant and Sherman have attracted more than their share of biographers, but never before has an author been audacious enough to tackle both at once. Their subtle and complex relationship deserves attention from a sophisticated and experienced writer; Flood is up to the task."

—JOHN Y. SIMON, editor of
The Papers of Ulysses S. Grant

Stephen Bates ©

About the Author

CHARLES BRACELEN FLOOD is the author of *Lee: The Last Years*, *Hitler: The Path to Power*, and *Rise, and Fight Again: Perilous Times Along the Road to Independence*, which won an American Revolution Round Table Award. He is a past president of the PEN American Center and lives in Richmond, Kentucky.

GRANT AND SHERMAN

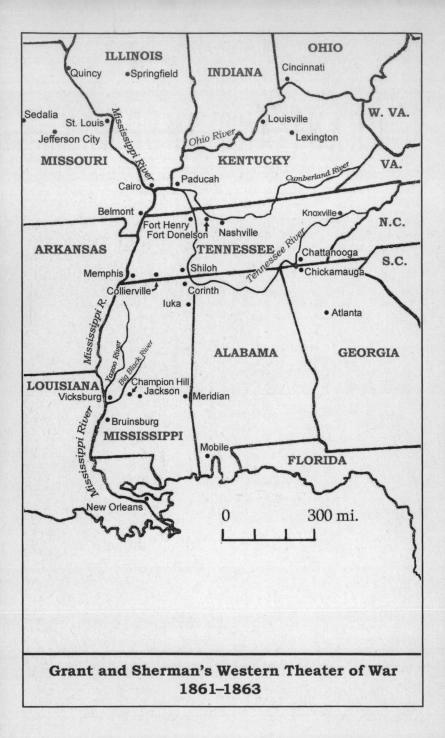

**Grant and Sherman's Western Theater of War
1861–1863**

GRANT AND SHERMAN

THE FRIENDSHIP THAT WON

THE CIVIL WAR

CHARLES BRACELEN FLOOD

HARPER PERENNIAL

NEW YORK • LONDON • TORONTO • SYDNEY

HARPER ● PERENNIAL

A hardcover edition of this book was published in 2005 by Farrar, Straus and Giroux.

GRANT AND SHERMAN. Copyright © 2005 by Charles Bracelen Flood. Maps copyright © by Edward Carrol Hale. All rights reserved. Printed in the United States of America. No part of this book may be used or reproduced in any manner whatsoever without written permission except in the case of brief quotations embodied in critical articles and reviews. For information address HarperCollins Publishers, 10 East 53rd Street, New York, NY 10022.

HarperCollins books may be purchased for educational, business, or sales promotional use. For information please write: Special Markets Department, HarperCollins Publishers, 10 East 53rd Street, New York, NY 10022.

First Harper Perennial edition published 2006.

Designed by Jonathan D. Lippincott

Library of Congress Cataloging-in-Publication Data is available upon request.

ISBN-10: 0-06-114871-7 (pbk.)
ISBN-13: 978-0-06-114871-2 (pbk.)

06 07 08 09 10 ❖/RRD 10 9 8 7 6 5 4 3 2

To my wife, Katherine Burnam Flood,

and to our children, Caperton, Lucy, and Curtis

As soon as real war begins, new men, heretofore unheard of, will emerge from obscurity, equal to any occasion.

—*William Tecumseh Sherman, six weeks before Bull Run*

I knew wherever I was that you thought of me, and if I got in a tight place you would come if alive.

—*Sherman to Grant, March 10, 1864, summing up their successful Western campaigns*

But what next? I suppose it will be safe if I leave General Grant and yourself to decide.

—*Abraham Lincoln to Sherman, after congratulating him on his capture of Savannah, Christmas 1864*

He stood by me when I was crazy and I stood by him when he was drunk, and now, sir, we stand by each other always.

—*Sherman, speaking of Grant*

I know him well as one of the greatest purest and best of men. He is poor and always will be, but he is great and magnanimous.

—*Grant, praising Sherman in a letter to Jesse Grant, his father*

We were as brothers, I the older man in years, he the higher in rank.

—*Sherman, summing up their friendship*

CONTENTS

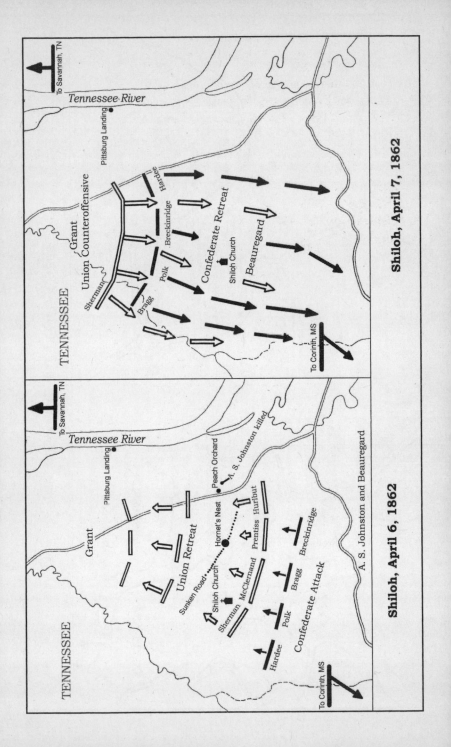

Shiloh, April 7, 1862

To Savannah, TN

Tennessee River

Pittsburg Landing

TENNESSEE

Grant
Union Counteroffensive

Sherman

Hardee

Breckinridge

Polk

Bragg

Confederate Retreat

Shiloh Church

Beauregard

To Corinth, MS

Shiloh, April 6, 1862

To Savannah, TN

Tennessee River

Pittsburg Landing

TENNESSEE

A. S. Johnston killed

Peach Orchard

Grant

Union Retreat

Hornet's Nest

Hurlbut

Prentiss

Breckinridge

Sunken Road

Shiloh Church

McClernand

Bragg

Sherman

Polk

Hardee

Confederate Attack

A. S. Johnston and Beauregard

To Corinth, MS

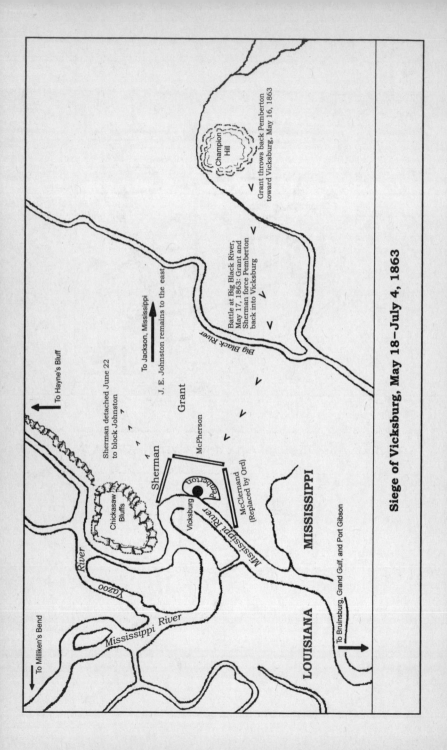

Siege of Vicksburg, May 18–July 4, 1863

To Hayne's Bluff

To Milliken's Bend

Yazoo River

Mississippi River

Chickasaw Bluffs

Sherman detached June 22 to block Johnston

To Jackson, Mississippi

J. E. Johnston remains to the east

Grant

Sherman

McPherson

Vicksburg

Pemberton

McClernand (Replaced by Ord)

Mississippi River

Big Black River

Battle at Big Black River, May 17, 1863: Grant and Sherman force Pemberton back into Vicksburg

Champion Hill

Grant throws back Pemberton toward Vicksburg, May 16, 1863

MISSISSIPPI

LOUISIANA

To Bruinsburg, Grand Gulf, and Port Gibson

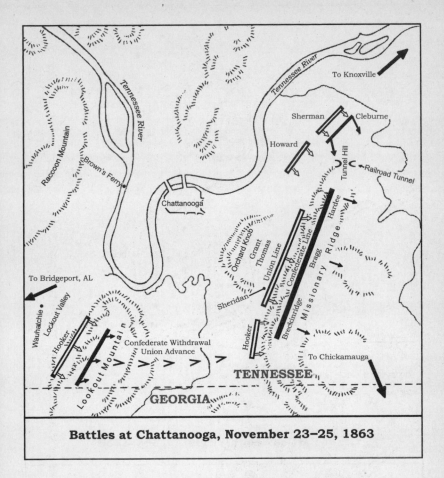

Battles at Chattanooga, November 23–25, 1863

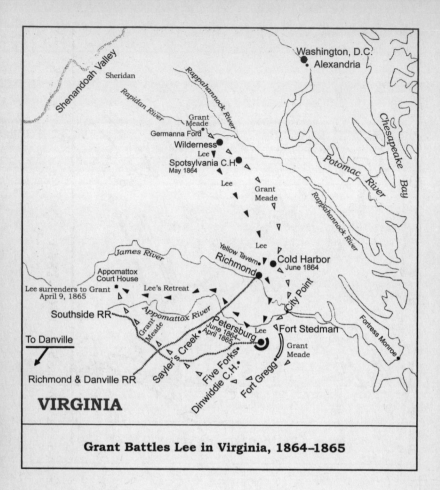

Grant Battles Lee in Virginia, 1864–1865

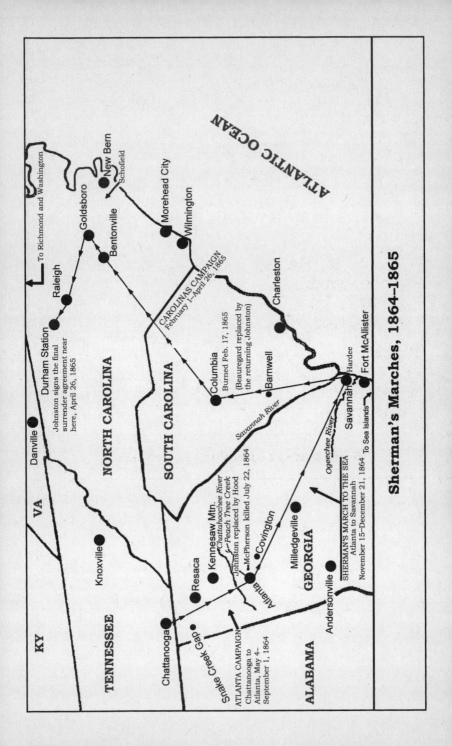

Sherman's Marches, 1864–1865

GRANT AND SHERMAN

PROLOGUE

———⟫●⟪———

In the early hours of April 7, 1862, after the terrible first day of the Battle of Shiloh, Brigadier General William Tecumseh Sherman came through the darkness to where his superior, Major General Ulysses S. Grant, stood in the rain. Sherman had reached the conclusion that the Union forces under Grant's command could not endure another day like the one just ended. When the massive Confederate surprise attack on the vast federal encampment beside the Tennessee River began at dawn on April 6, Grant's command had numbered thirty-seven thousand men. Now seven thousand of those were killed or wounded, another three thousand were captured, and more than five thousand were huddled along the bank of the river, demoralized and useless as soldiers. Sherman, who had been wounded in the hand earlier in the battle, was coming to tell Grant that he thought they should use the transport vessels near them at Pittsburg Landing to evacuate their forces so that they could "put the river between us and the enemy, and recuperate."

Sherman found Grant alone, under a tree. Hurt in a fall from a horse on a muddy road a few days before, Grant was leaning on a crutch and held a lantern. He had a lit cigar clenched in his teeth, and rain dripped from the brim of his hat. Looking at the determined expression on Grant's bearded face, Sherman found himself "moved by some wise and sudden instinct" not to mention retreat and used a more tentative approach. "Well, Grant," he said, "we've had the devil's own day of it, haven't we?"

"Yes," Grant said quietly in the rainy darkness, and drew on his cigar. "Lick 'em tomorrow though."

That was the end of any thought of retreat. At first light, Grant threw his entire force at the Confederates under General P.G.T. Beauregard, and after a second bloody day, Grant, with Sherman right beside him, had won the biggest Northern victory of the Civil War's first year. The author and Confederate soldier George Washington Cable wrote, "The South never smiled after Shiloh."

Shiloh was a great victory in itself, but that meeting in the rain symbolizes something more. Enormous military and political results flowed from the friendship between Ulysses S. Grant and William Tecumseh Sherman, two men who had been obscure failures before the Civil War. Their relationship as superior and subordinate began when they moved toward the Battle of Shiloh, which took place ten months into the conflict. At Shiloh they came together on the field, and here Grant and Sherman took each other's measure under fire and began two years of successful cooperation and friendship. They separated in the final year of the war to lead armies in different areas, but though their headquarters were hundreds of miles apart, they remained in virtually constant contact by what was then known as the "magnetic telegraph." Throughout the war, each supported the other's efforts in every way; each furthered and on occasion saved the other's career.

In some ways the two men were different. Grant, whom a fellow officer described as "plain as an old stove," was reserved in manner and worked with decisive inner power. A man who knew Sherman described his torrential energy: "He is never quiet. His fingers nervously twitch his whiskers . . . One moment his legs are crossed, and the next both are on the floor. He sits a moment, then paces the floor."

Sherman was an intellectual, widely read in military history and theory. Early in the war, Sherman, greatly talented but insecure, asked President Abraham Lincoln to agree that he would remain as second in command in a specific assignment and not have to lead it. By contrast, Grant operated on military intuition, thinking boldly and acting with quiet confidence: another officer said that Grant looked "as if he had determined to drive his head through a brick wall, and was about to do it." (As Grant advanced into Confederate territory, Abraham Lincoln said of him, "When Grant once gets possession of a place, he holds on to it as if he had inherited it.")

Grant needed a gifted and effective subordinate, and at first Sherman needed a man to give him orders and then stand by him, no matter what. And each needed a friend. They worked together for twenty-three months, planning, consuming countless cigars, learning the lessons taught them by their battles and campaigns.

At that point, in March of 1864, Lincoln summoned Grant east to assume command of all the Union armies and to oppose Robert E. Lee's Army of Northern Virginia during the final year of the war. Before they parted, Grant and Sherman agreed on what each had to do next. Grant would attack Lee in northern Virginia, working to outflank Lee until he could break through Lee's extended and continually thinning lines. Sherman would march southeast from Chattanooga, Tennessee, disemboweling the South.

Turning that strategy into action, Grant's forces and Sherman's Army of the West supported each other as effectively as if the two men had remained together. By then the two leaders thought alike, and any differences they had were quickly resolved.

After Grant came east to take the Union supreme command, he and Sherman did not meet again for a year. When they did, it was Sherman who traveled north on a swift courier vessel from his successful Carolina campaigns to meet Grant at City Point, Virginia, prior to a conference with Lincoln concerning what all three knew would be the closing scenes of the war. As Grant walked down the dock to where Sherman was coming ashore, one of Grant's staff witnessed this:

> In a moment, they stood upon the steps, with their hands locked in a cordial grasp, uttering words of familiar greeting. Their encounter was more like that of two school-boys coming together after a vacation than the meeting of the chief actors in a great war tragedy.

Soon after that conference at City Point, Grant forced Lee's final defeat at Appomattox Court House, and in North Carolina Sherman brought to an end the resistance of the South's other remaining large army under Joseph E. Johnston.

Grant and Sherman learned the lessons that led to the final victory during many desperate hours in dramatic campaigns. Those who believe that the North's greater industrial strength and manpower guaranteed the South's eventual defeat forget that those well-equipped Union columns had to be led by generals. The North had other good generals besides Grant and Sherman, as well as many that Lincoln tried in various areas who failed, but the partnership between these two leaders was unique. Grant and Sherman's way to victory required intelligence, luck, and brave soldiers, but it was built on the mutual trust that their friendship inspired.

1

TWO FAILED MEN WITH GREAT POTENTIAL

In December of 1860, five months before the Civil War began, two men who had resigned from the United States Army earlier in their lives reviewed their respective situations.

From Galena, Illinois, a small city of fourteen thousand, four miles east of the Mississippi River and just south of the Wisconsin border, the first of these men, former captain Ulysses S. Grant, wrote a friend, "In my new employment I have become pretty conversant . . . I hope to be made a partner soon, and am sanguine that a competency at least can be made out of the business."

A man who had graduated seventeen years before from the United States Military Academy at West Point, something that in itself conferred a certain prestige and social status, Grant was now a clerk in his stern father's small company, which operated a tannery as well as leather goods stores in several towns. Just six years previous, after four years as a cadet and eleven as an officer, including brave and efficient service during the Mexican War, his military career had come to a bad end. Stationed at remote posts in California without his wife and two children, Grant became bored and lonely. During the long separation from his wife, Julia, a highly intelligent, lively, affectionate woman who adored him as he adored her, he began to drink. In 1854, when Grant was thirty-two, his regimental commander forced him to resign from the army for being drunk while handing out money to troops on a payday.

Returning to Missouri, Grant struggled for four years to support his little family by farming land near St. Louis that belonged to his wife and father-in-law. Despite working hard, avoiding alcohol, and remaining optimistic—at one point he wrote to his father, who then lived in Kentucky, "Every day I like farming better and I do not doubt money is to be made at it"—events worked against him. A combination of weather-ruined crops and falling commodity prices left him with with one slim chance to get by. Hiring two slaves from their owners, and borrowing from his father-in-law a slave whom he later bought and set free, Grant and his new field hands began cutting down trees on the farm, sawing them into logs, and taking them to St. Louis to sell as firewood. Sometimes Grant brought his logs to houses whose owners arranged for deliveries, and on other days he peddled them on the street.

Wearing his faded old blue army overcoat, from which he had removed the insignia, he sometimes encountered officers who knew him from the past. Brigadier General William S. Harney, "resplendent in a new uniform" as he passed through St. Louis to campaign against the Sioux, saw Grant handling the reins of a team of horses pulling a wagon stacked with logs. Harney exclaimed, "Why, Grant, what in blazes are you doing here?" Grant answered, "Well, General, I'm hauling firewood." On another day, an old comrade looking for Grant's farm asked directions of a nondescript man driving a load into the city, only to realize that he was speaking to Grant. In response to his startled, "Great God, Grant, what are you doing?" he received the laconic reply, "I'm solving the problem of poverty."

On December 23, 1857, Ulysses S. Grant pawned his gold watch for twenty-two dollars to buy Christmas presents for Julia, who was seven months pregnant, and their three children. Nothing improved: bad weather destroyed most of the crops Grant planted in the spring of 1858, and a freak freeze on June 5 finished off the rest. During the summer, the Grants' ten-year-old son Fred nearly died of typhoid. In early September Grant wrote to his sister Mary that "Julia and I are both sick with chills and fever."

The end had come for Grant as a farmer. In the autumn of 1858, an auctioneer sold off his remaining animals, crops, and equipment. He, Julia, and their four children moved into St. Louis, where a cousin of Julia's had been persuaded to make him a partner in his real estate firm. Grant's job was to collect rents and sell houses, but even in a sharply rising real estate market, he could not make money. After nine months he was told that the partnership had been dissolved: he was unemployed. Next, after being turned down for the position of county engineer for

lack of the right political connections, he found a job in the federal cus-
tomshouse but was replaced after a month, again a victim of political pa-
tronage. Heavily in debt and behind in his rent, Grant could not support
his family. A friend who saw him walking the streets looking for work described a man "shabbily dressed . . . his face anxious," sunk in "profound
discouragement." Finally Grant turned in desperation to his austere father, who had earlier rejected his appeal for a substantial loan, and the
elder Grant created a job for him as a clerk at the leather goods store in
Galena. A man who ran a jewelry store across the street recalled this,
from the time when Grant was describing himself as "pretty conversant"
with his new job. "Grant was a very poor businessman, and never liked to
wait on customers . . . [He] would go behind the counter, very reluctantly, and drag down whatever was wanted; but hardly ever knew the
price of it, and, in nine cases out of ten, he charged either too much or
too little."

That was Grant as he lived in Galena on the eve of the Civil War—
an ordinary-looking man of thirty-eight, five feet eight inches tall and
weighing 135 pounds, somewhat stooped and with a short brown beard,
a quiet man who smoked a pipe and by then had some false teeth. He
had never wanted a military career: he went to West Point only because
his autocratic father, who had gotten him a congressional appointment
to the academy without consulting him, insisted that he go. While he
was there, Congress debated whether to close the nation's military
school, and Grant kept hoping that would happen. In studies, he said, "I
rarely read over a lesson a second time," but he devoured the library's
stock of novels, including the works of James Fenimore Cooper, Sir Walter Scott, and Washington Irving, and demonstrated skill and sensitivity
in the paintings and pen-and-ink sketches he executed in a drawing
course.

When Grant arrived as a plebe, seventeen years old, another cadet, a
big, swaggering youth named Jack Lindsay who was the son of a colonel
looked at this quiet and unassuming boy who then stood only five foot
one and weighed 117 pounds, and mistook Grant's politeness for weakness. Lindsay disdainfully shoved Grant out of line during a squad drill.
Grant asked him to stop. Lindsay did it again—and learned a lot about
Ulysses S. Grant when this little plebe knocked him to the ground with
one punch.

The incident may not of itself have ensured his acceptance and pop-

ularity, but Grant became a member of a secretive group known as the
T.I.O., standing for Twelve in One, a dozen classmates who pledged
eternal friendship and wore rings bearing a symbol whose significance
only they knew. In the evenings, he and his friends sometimes played a
card game called Brag. His classmate Daniel Frost, who was destined to
become a Confederate general, described him:

> His hair was reddish brown and his eyes grey-blue. We all liked him, and he
> took rank soon as a good mathematician and engineer . . . He had no bad
> habits whatever, and was a great favorite, though not a brilliant fellow.
>
> He couldn't, or wouldn't, dance. He had no facility in conversation with
> the ladies, a total absence of elegance, and naturally showed off badly in
> contrast with the young Southern men, who prided themselves on being fin-
> ished in the ways of the world.

In one area only, Grant stood first, in the entire corps of cadets: horse-
manship. From childhood on, he always had this intuitive relationship
with horses. At home, he broke them for their owners, trained them, and
rode them masterfully. The villain of the West Point stables was a big,
strong sorrel named York, who terrorized any cadet assigned to ride him
by rearing in the air and then tumbling backward onto the rider. Grant
asked the dismayed riding master for permission to work with York; when
that was granted, Grant hit the horse on the top of the head twice with
the butt of a pistol and began patiently showing the animal what he expected of him. A candidate for admission to West Point who was walking
around the academy described the eventual results of Grant's long work
with York, which he saw when he happened upon the part of the graduating class's final exercises that took place in the riding hall. After various
mounted drills performed for the audience of parents and dignitaries and
other guests,

> the class, still mounted, was formed in line through the center of the hall.
> The riding master placed the leaping bar higher than a man's head and
> called out "Cadet Grant!" A clean-faced, slender, blue-eyed young fellow,
> weighing about one hundred and twenty pounds, dashed from the ranks on
> a powerfully-built chestnut sorrel horse, and galloped down the opposite side
> of the hall. As he turned at the farther end and came into the stretch at
> which the bar was placed, the horse increased his pace and measuring his

stride for the great leap before him, bounded into the air and cleared the bar, carrying his rider as if man and beast had been welded together. The spectators were breathless.

During Grant's four years at West Point, some cadets, when they had a free hour, would go to the riding hall just to watch Grant school York and the other horses. In one event, Grant and York cleared a bar placed so high that their performance set an academy record that stood for twenty-five years.

Grant's roommate in his last year at West Point was Frederick Dent, a cadet from St. Louis. When newly commissioned Lieutenant Grant was assigned to the Fourth Infantry Regiment stationed at Jefferson Barracks, a few miles south of St. Louis, his friend Dent urged him to call on his family at White Haven, the large nearby farm to which the prosperous Dents annually moved from their winter house in St. Louis to spend much of the rest of the year. White Haven was not one of the great Southern plantations, but it had twelve hundred fertile acres situated on the broad Gravois Creek. In addition to the white-painted main house with its traditional big porches running along both the ground floor and the bedroom floor above it, all covered with honeysuckle and other vines, there were eighteen cabins in which the Dent family's slaves lived. The Dents' daughter Emma later described the place:

> The farm of White Haven was even prettier than its name, for the pebbly shining Gravois ran through it, and there were beautiful groves growing all over it, and acres upon acres of grassy meadows where the cows used to stand knee-deep in blue grass and clover . . . The house we lived in stood in the centre of a long sweep of wooded valley and the creek ran through the trees not far below it . . . Through the grove of locust trees a walk led from a low porch to an old-fashioned stile gate, about fifty yards from the house.

Emma was six years old when Lieutenant Grant came to call at this rural scene on a day when her eighteen-year-old sister Julia was away on a long visit to St. Louis. She described their first meeting: "I was nearing my seventh birthday, that bright spring afternoon in 1843 when, with my four little darky playmates, Henrietta, Sue, Ann, and Jeff, I went out hunting for birds' nests. They were my slaves as well as my chums, for father had given them to me at birth, and as we were all of about an age, we used to have some good times together. This day, I remember, we were out in front of the turnstile and I had my arms full of birds' nests and was clutching a tiny unfledged birdling in one hand when a young stranger rode blithely up to the stile."

In answer to this man on horseback's "How do you do? Does Mister Dent live here?" Emma was speechless. "I thought him the handsomest person I had ever seen in my life, this strange young man. He was riding a splendid horse, and, oh, he sat it so gracefully! The whole picture of him and his sleek, prancing steed was so good to look upon that I could do nothing but stare at it—so forgetting the poor little thing crying in my hand that I nearly crushed it to death. Of course, I knew he was a soldier from the barracks, because he had on a beautiful blue suit with gold buttons down the front, but he looked too young to be an officer."

When Emma recovered herself enough to answer "Yes, sir," after the lieutenant asked for the second time if this was the Dents' house, this scene ensued:

We children followed him up to the porch, trailing in his wake and close to his feet like a troop of little black-and-tan puppies . . . At the porch we heard him introduce himself to my father as Lieutenant Grant. Then my mother and sister Nellie came out to meet him . . . My own contribution to the entertainment of the stranger was one continuous stare up at his face . . . His cheeks were round and plump and rosy; his hair was fine and brown, very thick and wavy. His eyes were a clear blue, and always full of light. His features were regular, pleasingly molded and attractive, and his figure so slender, well formed, and graceful that it was like that of a young prince to my eye . . . When he rode up to White Haven that bright day in the spring of 1843 he was pretty as a doll.

Grant came to call several times, always urged to stay for supper by Mrs. Dent, who liked him immediately. Of the slender lieutenant's quiet political discussions with her husband, she commented, "That young man explains politics so clearly that I can understand the situation perfectly." Emma and her fifteen-year-old sister Nellie began to regard him as a gift that had somehow been bestowed upon them. Then their vivacious older sister Julia, who had recently turned nineteen, came back from St. Louis. "She was not exactly a beauty," Emma said, mentioning that one of Julia's eyes would go out of focus in a condition known as strabismus, "but she was possessed of a lively and pleasing countenance." Grant suddenly began to ride over from the barracks every other day. "It did not take Nell and myself long to see that we were no longer the attractions at White Haven," Emma noted. Having grown up on a big farm with three older brothers as well as her three younger sisters, Julia loved the outdoors. "He and she frequently went fishing along the banks of the creek, and many a fine mess of perch I've seen them catch together." Julia's impression of her new friend Lieutenant Grant was that he was "a darling little lieutenant."

An excellent rider, Julia had a spirited Kentucky mare. According to Emma, "Lieutenant Grant was one of the best horsemen I ever saw, and he rode a fine blooded animal . . . Many a sharp race they used to have in the fine mornings before breakfast or through the sunset and twilight after supper."

White-haired Colonel Dent—a courtesy title by which many men of his station in life were then known in the South, regardless of military experience—could be a peevish man, given to sitting by himself on the porch reading a newspaper and puffing on a long reed-stemmed pipe, but he, like his wife, believed in having many young guests. Grant was encouraged to bring his brother officers with him. There were picnics and dances around the countryside; one of the young officers always included was a handsome giant named James Longstreet, a cousin of Colonel Dent's who had been known at West Point as Pete.

When Julia's pet canary died, Grant organized a funeral for the bird. Julia remembered that "he was kind enough to make a little coffin for my canary bird and he painted it yellow. About eight officers attended the funeral of my little pet."

It seemed not to occur to this young couple that they were falling in love. At a time when Grant was home on leave visiting his family in Ohio, his regiment was ordered to Louisiana to become part of the Army of Observation during the annexation of Texas, in the confrontation that would lead to the Mexican War. An officer friend told Julia that if Grant did not appear at White Haven by the following Saturday, it would mean that he had gone straight on down the Mississippi from Ohio to catch up to his regiment and "would not be at the Barracks again." Julia later wrote: "Saturday came and no Lieutenant. I felt very restless and, ordering my horse, rode alone towards the Barracks . . . I halted my horse and waited and listened, but he did not come. The beating of my own heart was the only sound I heard. So I rode slowly and sadly home."

Grant was in fact hastening toward St. Louis from Ohio, where, on learning that he was about to be sent far from Julia for a long time, he discovered that there was something "serious the matter with me." When he arrived at Jefferson Barracks, the post was virtually deserted, with his friend Lieutenant Richard Ewell finishing the last of the departed regiment's paperwork before following the unit down the Mississippi. Ewell readily wrote out a few days' extension of Grant's leave, and Grant found a horse and set out for White Haven that evening. Normally the Gravois Creek was shallow, but a placid stream was not what he encountered that night: "On this occasion it had been raining heavily, and when the creek was reached, I found the banks full to overflowing, and the current rapid. I looked at it a moment to consider what to do. One of my superstitions has always been when I started to go anywhere, or to do anything, not to turn back, or stop until the thing intended was accomplished . . . So I struck into the stream, and in an instant the horse was swimming and I being carried down by the current." With Grant hanging on to the horse's mane as the animal swam through the foaming water in the darkness, both horse and man reached the opposite bank. When he arrived at White Haven, drenched and dripping, little sister Emma was right there, and her memory for such matters later enabled her to render this account of that moment:

We all enjoyed heartily the sight of his ridiculous figure with his clothes flopping like wet rags around his limbs, and none laughed more heartily than my sister Julia. Lieutenant Grant took it all good humoredly enough, but there was a sturdy seriousness in his usually twinkling eyes that must have suggested, perhaps, to Julia that he had come on more serious business, for the teasing did not last long. [Older brother] John carried him off to find some dry clothes, and when he returned the usually natty soldier looked scarcely more like himself . . . John was taller and larger than Grant, and his clothes did not fit the Lieutenant "soonenough." Of course, this roused more laughter, which the soldier took in the same good part, but those rosy telltale cheeks of his reddened, as usual with him when the inward state of his feelings did not agree with his outward composure.

Grant held his fire until a day soon thereafter when several of the Dents set off to attend a friend's wedding, with Grant included in the group. He arranged matters so that he and Julia were alone in a buggy, with him at the reins. As they approached a little wooden bridge across the still-turbulent Gravois, which had a torrent of water roaring just beneath the wooden planks, Julia began to worry about the safety of crossing.

> I noticed, too, that Lieutenant Grant was very quiet, and that and the high water bothered me . . . He assured me, in his brief way, that it was perfectly safe, and in my heart I relied upon him. Just as we reached the old bridge I said, "Now, if anything happens, remember I shall cling to you, no matter what you say to the contrary." He simply said "All right" and we were over the planks in less than a minute. Then his mood changed.

As Julia put it, he used her statement about clinging to him to ask her to cling to him forever. Grant's only recorded comment on his proposal was, "Before I returned I mustered up the courage to make known, in the most awkward manner imaginable, the discovery I had made on learning that the 4th infantry had been ordered away from Jefferson Barracks . . . Before separating it was definitely understood that at a convenient time we would join our fortunes, and not let the removal of a regiment trouble us." Julia told her "Ulys," as she had taken to calling him, not to ask her father for her hand in marriage just then; she was his, but she did not want an engagement to be announced.

During the next four years, the couple saw each other only once, when he returned from Louisiana on a brief leave before his regiment was sent into the Mexican War. On that visit, he received Colonel Dent's permission to marry his daughter, despite the colonel's dislike of what he knew of the conditions that army wives often encountered. From Mexico, Grant sent Julia letters that expressed great longing for her. Telling her of the American victory at Matamoros early in the war, he wrote, "In the thickest of it I thought of Julia. How much I should love to see you." His short, clear descriptions of the battles in which he fought mentioned little of his own part in them. In fact, young Lieutenant Grant participated in most of the major engagements. At the Battle of Monterrey, as his regiment advanced through city streets in house-to-house fighting, Grant's commander realized that his men were running out of ammunition. Someone had to ride back through streets swept from one side by enemy fire with the order that more ammunition be brought forward immediately. Believing that whoever tried to carry this message would probably be killed, the colonel asked for a volunteer. Grant swung up on a gray mare named Nellie and put his arm around the horse's neck and a foot over the hind part of the saddle. Then, hanging down along the side of the horse away from the enemy, he galloped back through the street crossings, which, he said, "I crossed at such a flying rate that generally I was past and under the cover of the next block before the enemy fired."

Serving at times as regimental quartermaster, a position calling for attention to supplies and transportation at the rear of the fighting, Grant did all that in superior fashion and still repeatedly fought at the front. He was to say, "I never went into a battle willingly or with enthusiasm . . . was always glad when a battle was over," but his friend Longstreet saw a different picture: "You could not keep him out of battle . . . Grant was everywhere on the field. He was always cool, swift, and unhurried . . . as unconcerned, apparently, as if it were a hail-storm instead of a storm of bullets." Among the officers in the thick of the fighting was Julia's brother Fred, Grant's West Point roommate. At Molino del Rey, Grant came upon Fred, who was in another regiment, minutes after his future brother-in-law was wounded in the thigh by a musket ball. He was soon able to write Julia that Fred would recover quickly.

On the day before General Winfield Scott's victorious entry into Mexico City, Grant distinguished himself in the attacks made along the aqueduct road leading toward the complex of buildings and defenses on the city's outskirts known as the Garita [city gate] San Cosme. He repeatedly took different kinds of initiatives. Grant moved forward on his own, actually working his way around an enemy breastworks until he was behind the Mexicans who were firing at the Americans. Returning to the American lines, he asked for volunteers and got twelve men. Grant led them, and an American company he came upon that was just entering the battle, back around to the side of the enemy position, attacked the Mexicans on their unprotected flank, and forced a retreat. When the numerically superior enemy reoccupied the breastworks later in the day, Grant's Fourth Infantry led the American counterattack that finally won it back: another lieutenant reported that he and Grant were "the first two persons to gain it."

As if that were not enough for the last afternoon of the war, Grant, scouting on his own again, "found a church off to the south of the road, which looked to me as if the belfry would command the ground back of the Garita San Cosme." Once again, Grant trusted his instincts: he rounded up an officer in command of a mountain howitzer and its crew, and directed them as they wrestled the small cannon up the steps of the bell tower. When they opened fire on the Mexican soldiers who were behind walls where they thought they could not be seen, "The shots from our little gun dropped in upon the enemy and created great confusion."

The commander of this wing of the American attack, Brigadier General William Worth, was studying the enemy position through a spyglass. This sudden successful development surprised him as much as it did the Mexicans. He sent his aide Lieutenant John C. Pemberton to bring Grant to him. Telling the gunners to keep up the fire, Grant reluctantly left the bell tower and reported to General Worth, who congratulated him, saying that "every shot was effective." He ordered Grant to take a captain with another mountain howitzer and its crew back with him, and get it up in the tower to double the fire.

"I could not tell the General," Grant said of that moment in which he combined military obedience and common sense in the middle of a hard-fought battle, "that there was not room enough in the steeple for another gun, because he probably would have looked upon such a statement as a contradiction from a second lieutenant. I took the captain with me, but did not use his gun."

By nine the next morning, General Scott entered the center of Mexico City and walked into the National Palace accompanied by a group of his officers. After a last uprising and enemy effort to reenter the city that was quelled within twenty-four hours, the fighting in the Mexican War came to an end. Grant wrote Julia that the highly professional American forces had won "astonishing victories," but added, "dearly have they paid for it! The loss of officers and men is frightful." Twenty-one officers of the Fourth Infantry Regiment, many of them guests at the merry picnics and dances held near White Haven, and some who had attended the funeral Grant organized for Julia's pet canary, had been sent from Jefferson Barracks to Mexico. Seventeen of them died there. In all, 78,718 American soldiers served in the Mexican War; 13,283 died, a higher percentage than in any other conflict in which the United States has been engaged. As for Grant's view of the war, he later termed it "one of the most unjust wars ever waged by a stronger against a weaker nation." That was Grant, in essence: he might disagree with his nation's policy, but he had sworn to carry it out.

So, after varying lengths of time serving in the army of occupation, the men who would have to choose between fighting for the North or South thirteen years later began coming home. There were officers clearly marked for future high command, such as Robert E. Lee, whom Winfield Scott called "the very best soldier that I ever saw in the field," and the two unrelated Johnstons, Joseph E. and Albert Sidney, who would also side with the South. George Gordon Meade, Winfield Scott Hancock, and John Sedgwick had gained important experience that they would use in fighting for the Union. Other men received notice: at the climactic moment of the Battle of Chapultepec, Lieutenant George Pickett, the future Confederate general whose division would be slaughtered as part of the doomed Pickett's Charge at Gettysburg, pulled down the Mexican flag that had been flying over the bravely defended Military College and hoisted an American flag. In the same battle, an artillery lieutenant from Virginia named Thomas Jonathan Jackson, later known as "Stonewall," managed to advance his cannon so far up Grasshoppers Hill in the face of intense musket fire that his gunners finally left the gun and hid behind some rocks. The tall young warrior with blazing gray eyes strode back and forth as bullets cracked past him, shouting to his crouching men, "There's no danger! See, I'm not hit!" In the final house-to-house fighting in Mexico City, Lieutenant George B. McClellan, who early in the Civil War would be the Union Army's general in chief, saw a Mexican kill an American sergeant; grabbing the sergeant's musket, McClellan killed the Mexican.

The talk in the army right after the Mexican War was that, of the junior officers, the one destined to rise highest was Don Carlos Buell, followed by George H. Thomas and Braxton Bragg. No one mentioned Ulysses S. Grant, although he had not only learned battlefield tactics during the bloody and demanding campaigns, but, fully as important, had mastered his regiment's complicated problems of supply and transportation.

When at the end of the war Grant returned from Mexico to claim his bride, his face, as Julia's sister Emma saw it, "was more bronzed from the sun, and he wore his captain's double-barred shoulder straps with a little more dignity than he had worn the old one[s], perhaps. His shoulders had broadened some, and his body was stouter, and it may be that he had grown a little more reserved in manner." They were married at the Dents' house in St. Louis, with Julia's cousin, Grant's West Point class-mate Lieutenant James Longstreet, acting as best man.

Grant's parents were not there: simple, spartan folk, they would have felt uncomfortable at this small but elegant wedding; besides that, Grant's father, Jesse, was opposed to the concept and practice of slavery, and his son was marrying into a family that owned slaves. (Grant seems to have been indifferent to the issue at this stage in his life, and there is no record of letters or discussions on the subject between father and son at just that time.)

Ulysses and Julia spent the beginning of their honeymoon aboard what Julia called "one of those beautiful great steamboats," going to Ohio so that Julia and his parents could meet. "Our honeymoon was a delight," Julia recalled. "We had waited four long years for this event and we adjusted to one another like hand to glove." Julia had never traveled outside the vicinity of St. Louis and had never been on a passenger ves-sel. "I enjoyed sitting alone with Ulys . . . He asked me to sing to him, something low and sweet, and I did as he requested. I do not remember any of the passengers on that trip. It was like a dream to me." Their visit with Grant's parents was a success: insofar as Jesse and Hannah Grant could be charmed by anyone, Julia succeeded.

After nearly four years of happy married life, during which the young Grants lived at army posts in upstate New York and in Michigan, in the spring of 1852 his Fourth Infantry Regiment was ordered to California. At that time Ulysses and Julia had a two-year-old son, Frederick Dent Grant, and she was due to have another baby in July. The route the Fourth Infantry was to take involved boarding a ship in New York to make the voyage to Panama, and at that time, long before the Panama Canal was built, this had to be followed by an overland trip across the of-ten disease-ridden isthmus to the Pacific, with the final long leg on an-other ship to San Francisco. Despite their deep desire to stay together, the Grants decided that the risks to Julia and their son and unborn child were too great, and that he must start serving this lengthy tour of duty alone.

On his regiment's harsh journey to California, Grant's hard-won knowledge of logistics, developed as a supply officer during the Mexican

War, briefly made him an unsung hero. His position as regimental quartermaster gave him the responsibility of drawing up and executing the plans for moving seven hundred soldiers, plus a hundred of their wives and children, across the Isthmus of Panama at a time when there was a cholera epidemic; it killed nearly a third of them. The toll would have been higher had it not been for Grant's energy and willingness to take the initiative. The army had authorized Grant an allowance of sixteen dollars, per mule, to rent the beasts of burden to carry women, children, equipment, and what had been assumed would be a few sick persons. Finding no mules for rent at that price during a mounting crisis, Grant cast aside the bureaucracy's rules, hired mules at double the price, and, burying along the way the men, women, and twenty young children who died, got the survivors to Panama City. There the sick were separated and sent to a vessel, anchored well out in the harbor, that Grant leased for a hospital ship. For two weeks, as more died all around him, Grant remained aboard, arranging for food, medicine, and care. A witness to his efforts said that Grant emerged as "a man of iron, so far as endurance went, seldom sleeping . . . His work was always done, his supplies ample and on hand . . . He was like a ministering angel to us all."

During his first few months in California, Grant received none of the letters Julia sent him and felt their separation deeply. When a letter came, in which Julia had traced the outspread hand of their new son, Ulysses, whom he had never seen, Grant proudly showed the drawing to a sergeant at his post and then, as he turned away, began silently shaking, tears in his eyes. The sergeant said of him, "He seemed always to be sad." After eighteen lonely months, he applied for orders that would take him back east. On February 6, 1854, writing Julia from Fort Humboldt, a remote post 250 miles north of San Francisco, he said, "A mail came in this evening but brought me no news from you nor nothing in reply to my application for orders to go home . . . The state of suspense I am in is scarcely bearable." Grant had already begun to drink. One of his fellow officers observed:

He was in the habit of drinking in a peculiar way. He held his little finger just even with the . . . heavy glass bottom of the tumbler, then lying his three fingers above the little one, filled in whiskey to the top of his first finger and drank it off without mixing water with it. This he would do more or less frequently each day.

Others described him as a man who seldom took alcohol, but went on "sprees" when he did. In any event, his drinking led to the payday when he was drunk while handing out money to the troops. Grant's colonel offered him the choice of resigning from the army without further explanation, or facing court-martial charges of being drunk while on duty. His West Point classmate Rufus Ingalls described what happened then: "Grant's friends at the time urged him to stand trial, and were confident of his acquittal; but, actuated by a noble spirit, he said that he would not for the world have his wife know that he had been tried on such a charge. He therefore resigned his commission and returned to civil life."

As soon as Grant was back with his beloved Julia and their sons, his drinking ceased, despite the struggles to make a living that he experienced in the years before the Confederates fired on Fort Sumter.

―――

While Ulysses S. Grant sat bored in his father's harness and leather goods shop in Galena at the age of thirty-eight, as the collision between North and South grew imminent, 750 miles to the south of him another West Pointer who had left the army as a captain, William Tecumseh Sherman, was reaching a dead end in one more of the careers he had tried since resigning from the service a year before Grant did.

The forty-year-old Sherman was failing dramatically, and this brilliant, nervous, ambitious man felt the frustration and insecurity that marked many phases of his life. When he was nine and living in Lancaster, Ohio, his father, a respected judge of the state's Supreme Court who was honorably paying off a large debt instead of declaring bankruptcy, suddenly died, leaving his wife and eleven children nearly penniless. Relatives and family friends offered to have the older children live with them, and off they went. He was taken into the handsome big house, a hundred yards away, of Thomas and Maria Ewing, a prominent lawyer and his wife who were raising four children and two nieces and a nephew, all of whom had always played with the young Shermans. When Ewing, a self-made man who stood over six feet tall and weighed 260 pounds, walked in with his new foster son, his pretty brown-haired daughter Eleanor, aged five and always called Ellen, somehow understood that she and this red-haired boy from next door were now going to be raised as brother and sister. "I peeped at him with great interest," she said. Twenty-one years later they would become husband and wife.

Living in his new home, the boy found that Thomas and Maria Ewing treated him as if he were one of their own children, and yet he always had the feeling of being different. His own mother and youngest brothers and sisters remained in his old house just down the hill, and he frequently ate there. Thomas Ewing had no thought of having the youngster change his name from Sherman to Ewing, but his wife, Maria, was a staunch Catholic and insisted that this new member of her family be baptized. Near the beginning of the christening ceremony the priest asked the nervous boy's first name, and learned that it was Tecumseh, Tecumseh Sherman—a name given him by his late father, who admired the great Indian chief—and nothing else. The priest pointed out the need to add a Christian saint's name, stated that the day was the Feast of Saint William, and baptized him as William Tecumseh Sherman.

When young Sherman was taken to Mass, the service meant little to him, but he felt other influences strongly. Two years after he entered the family, the Ohio legislature named his foster father to the United States Senate. Thomas and Maria Ewing believed that their children should work hard at school, they expected success in life for themselves and their entire family, and they tried always to be well-informed (when Senator Ewing was in Washington, his letters home, read to all the children, described dinners at the White House with President Andrew Jackson and conversations with Vice President John C. Calhoun).

Besides these influences, life had instilled some fears in young William Tecumseh Sherman. He had a horror of debt—as he saw it, if his father had not died owing so much money, he and his brothers and sisters would all still be living together with their mother. This feeling about debt extended to a dislike of being dependent on others. He also knew that the family from which he sprang had a history of mental disorders: his maternal grandmother and uncle both spent time in what were then called asylums. This produced a conflict: he needed friendship and love but felt that his world might betray him—fathers died, debts were presented, people became sick in all sorts of ways—and that he could rely only on himself. He yearned for the serenity of his early childhood, but those years would not return. Occasionally rebellious, he was untidy, and his mind leapt from one subject to another, but, in a near guarantee of future frustration, he wanted the world to be a predictable, well-behaved place.

As Cump—a nickname derived from Tecumseh—grew into a gangly, beak-nosed, animated youth, a teenager with a high bulging forehead, a pitted complexion, and a shock of coarse red hair, Thomas Ewing began to think about his foster son's future. A United States senator has the power to make appointments to West Point: at the age of seventeen, Sherman entered the Military Academy, one of 119 plebes. He excelled in his studies, graduating sixth among the forty-two who completed the four years. (He would have been fourth but for a heavy total of demerits for minor offenses that ranged from holding parties in his room after lights-out to chatting in ranks while on parade.)

In Sherman's last year at the academy, one of the entering plebes was a Cadet Grant, also from Ohio, who was to have his own problems about his name: his real name was Hiram Ulysses Grant, but the congressman who appointed him mistakenly sent his name in as Ulysses S. Grant, and Grant was told that could not be changed. Whenever anyone thereafter asked what the "S" in his name stood for, Grant answered, "nothing," but Sherman recalled how Grant came by the nickname Sam.

> I remember seeing his name on the bulletin board, where the names of all the newcomers were posted. I ran my eyes down the columns, and saw there "U.S. Grant." A lot of us began to make up names to fit the initials. One said "United States Grant." Another "Uncle Sam Grant." A third said "Sam Grant." That name stuck to him.

Graduating from West Point in 1840, Sherman was sent to Florida and joined the campaign against the Seminole Indians; despite serving ably and conscientiously, he experienced none of the hit-and-run fighting that defined this guerrilla war. Subsequent postings took him to a fort near Mobile, Alabama, and to Fort Moultrie, at Charleston, South Carolina. His time with the hospitable and charming citizens of Mobile and Charleston, combined with a brief visit to New Orleans, left him with a great affection for the South. Short assignments to places such as Marietta, Georgia, gave his retentive mind the opportunity to study territory that he would one day revisit under circumstances he would then have considered unthinkable.

In 1846, the Mexican War came; 523 graduates of West Point fought in those battles, many distinguishing themselves in ways that influenced future assignments and promotions, but despite his efforts to get into this second war in six years, Sherman was ordered to duty in California. He served there first as a supply officer and later in various assignments as aide to commanding officers, combining this with responsibilities as an adjutant in charge of paperwork. Writing to Ellen Ewing, to whom he was now engaged, he expressed his reaction to the impressive American victories in Mexico: "These brilliant scenes nearly kill us who are far off, and deprived of such precious pieces of military glory." In a letter he wrote her in 1848 after the war's end, Sherman added, "I have felt tempted to send my resignation to Washington and I really feel ashamed to wear epaulettes after having passed through a war without smelling gunpowder, but God knows I couldn't help it so I'll let things pass."

In the meantime, a major event occurred in this California that Sherman considered such a backwater, and he was among the first to learn of it. His commanding officer at the army post in Monterey called him into his office, pointed at some glistening stones brought in by two messengers from a Swiss-born California landowner named Sutter, and asked, "What is that?"

Three years after graduating from West Point, Sherman had seen gold in north Georgia; remembering what he could from his mineralogy course at the academy, he tested one stone and found it so malleable that he could hammer it flat. They were looking at large gold nuggets. The California Gold Rush began. Men of every description left their jobs, heading for the western slopes of the Sierra Nevada to hack through rocks and pan streams in hope of making their fortunes: sailors abandoned their ships; farmers threw aside their plows.

Sherman and his commander, accompanied by four soldiers and the commander's black servant, set off for the American River near Sacramento to find out for the United States government just what was happening. On Weber's Creek, a branch of the American River, they were shown a small area where two men had mined gold worth seventeen thousand dollars in a week, at a time when even with rising prices a servant could be hired for ten dollars a day. A little nearby ravine had yielded twelve thousand dollars.

When Sherman returned from the freshly discovered goldfields where thousands of men lived in tent cities as they took more gold from the ground every day, he found that soldiers were deserting in the "gold fever" frenzy. (Sherman's own clerk at headquarters soon vanished.) All this produced yet another of the frustrations that plagued his life: some men were making fortunes in hills just a few days' ride away, while as an officer he felt duty bound to keep order with a shrinking number of troops, resisting the impulse to ride out and grab his share of the gold being drawn from the earth. As tens of thousands of men poured into California, bringing with them a demand that outstripped the supply of everything from shovels to food, whiskey, and women, prices soared. Sherman, living on an officer's fixed pay, feared that the inflated prices for everything would make him, ironically, poor amid this explosion of riches.

A few chances to make money did come his way. He and two other officers invested in a store in the gold-mining area of Coloma, and he took a two-month leave during which he was paid as a surveyor in San Francisco and also profitably bought and sold some land in Sacramento. Nonetheless, he always remained aware of the difference between his undistinguished life and that of the Ewings, to whom he felt so indebted. Senator Ewing was now secretary of the interior in the cabinet of President Zachary Taylor, while Sherman was an obscure lieutenant, having had "peculiarly bad luck during the past four years serving in a distant land," as he put it, missing out on the victorious war that propelled General Zachary Taylor straight from the battlefield to the White House. He was well aware that while he was marking time, his younger brother John Sherman, back in Ohio, had become a successful lawyer and was making significant amounts of money in real estate and the lumber business while planning to enter politics.

Letters from Ellen, often taking six months to reach him, left him worried about her various ailments, which included frequent headaches, boils, and gynecological problems. She made it clear that she looked forward to marrying him but did not look forward to being an army wife. At twenty-six, Ellen knew her mind: she wanted to stay near her parents, with whom she now lived in Washington, and she wanted her husband to become a financially successful civilian. A Catholic as ardent as her mother, she could not understand why her fiancé, who had after all been baptized in her faith, virtually never went to Mass. In answer to her troubled inquiries about his religious feelings, he had earlier written her that he believed "firmly in the main doctrines of the Christian Religion, the purity of its morals, the almost absolute necessity for its existence and practice among all well regulated communities to assure peace and good will amongst all." But, he had added, "I cannot, with due reflection, attribute to minor points of doctrine or form the importance usually attached to them." It was a cerebral, unsatisfying response to a young woman who believed in the Real Presence of the body and blood of Jesus Christ in the Eucharist at Mass.

As if confronting these thoughts and problems were not enough for Sherman, who now began to suffer from asthma, his older sister Elizabeth, living in Philadelphia, was married to an alcoholic. When Sherman sent her fifteen hundred dollars of the money he had made in California, her husband promptly lost it in the latest of his schemes for becoming rich, and Sherman had reason to wonder if their children had enough to eat.

In 1850, Sherman's time in California was up. Happy to be reunited with Ellen in the East, he nonetheless had no illusions as to why certain marks of favor were shown him. Asked to dine with the army's commander, the legendary General Winfield Scott, Sherman realized that the great leader was less interested in him as a young officer than as the man soon to marry the daughter of Thomas Ewing, whose support Scott needed for the presidential bid he intended to make in two years. (During dinner, Scott, who had been a general since the War of 1812, told Sherman, eleven years before it occurred, that the nation was heading toward "a terrible Civil War.") Scott was only the first of a galaxy of prominent Americans he soon saw face-to-face. At his wedding to Ellen in Washington on May 1, with Sherman wearing a full dress uniform that included a saber and boots with spurs, the guests at the brilliant social event included President Taylor and his cabinet; Senator Henry Clay, who gave Ellen the silver basket in which she carried her wedding bouquet; Representative Daniel Webster; and the justices of the Supreme Court. (There is no record of Sherman's mother being at the wedding; he and Ellen did, however, go to Lancaster as part of their long honeymoon and spent a month there. During that period, Sherman signed an agreement with his brother John to support their mother. She also later went to live with a married daughter.)

The first year of his marriage to Ellen set a pattern of almost constant stress for Sherman. Even during their long honeymoon, Ellen's possessive parents wrote her letters from Washington saying how much they missed her and urged the couple to live near them in Washington when they returned. Even though Ellen referred to him as her "protector," she and her parents repeatedly asked Sherman to resign from the army, adding what they felt was an inducement: he could manage the saltworks they owned near Lancaster, Ohio, where the Ewings kept the large house they lived in when not in Washington. That proposal would have made him his father-in-law's employee, and the saltworks became for Sherman a symbol of the Ewings' desire to keep him in a state of friendly, comfortable captivity.

Sherman resisted. In the autumn, the army ordered him to St. Louis and promoted him to captain with extensive responsibilities as a supply officer, but Ellen, now pregnant with their first child, refused to go. When their baby, a girl, was born in January of 1851, Sherman was on duty in St. Louis while his wife and child were in Washington. Within two months he succeeded in getting them to St. Louis, where Sherman found time apart from his military duties to manage lands near the city that had been acquired by his powerful father-in-law, but the family tensions never ceased. To maintain a household in any style, Sherman had to borrow money from both his father-in-law and his brother John. In September of 1852, the army ordered him to his next post as a supply officer in New Orleans, where he found the city and its people as charming as they had been during his brief visit earlier in his military service, but Ellen, pregnant again, did not accompany him, remaining back in Lancaster for the birth of their second child, another daughter. It was during this pregnancy, with Sherman first in St. Louis and then in New Orleans, and Ellen clearly reluctant to leave her family in Lancaster, that Sherman vented his frustration with his in-laws in a harsh, nearly insulting letter to his brother-in-law Thomas Ewing Jr., who was a lawyer. Referring to the entire situation, he said, "This is too bad and is only due to the immense love you all bear Lancaster. I have good reason to be jealous of a place that virtually robs me of my family and I cannot help feeling sometimes a degree of dislike for the very name of Lancaster . . . As to her [Ellen] being home next summer when you get there I doubt it exceedingly—I think she has been at Lancaster too much since our marriage, and it is time for her to be weaned."

Soon after Sherman succeeded in prevailing upon Ellen to bring the two girls and join him in New Orleans, his old friend Henry Turner, a banker in St. Louis who knew of Sherman's military service in California and his recent commissary experience and management of Thomas Ewing's lands, offered him the opportunity to manage the branch that his St. Louis bank, Lucas and Turner, would soon open in San Francisco.

Always fretful, Sherman was torn by the possible risks and benefits of this unexpected opportunity. He understood that, initially, Ellen would refuse to come and bring their two daughters to California: sailing around Cape Horn could be hazardous, and, as Ulysses S. Grant's experience had demonstrated, the other route, which involved crossing the Isthmus of Panama by land, exposed travelers to deadly diseases. Even though Sherman could arrange a six-month leave from the army to try this venture, there was no guarantee of business success in that far-off city that had seen both a gold boom and widespread bankruptcies.

Sherman decided to gamble. As things stood, he and Ellen were constantly at odds about where they should live and whether he should continue his low-paying army career. Her family dominated both their lives. Perhaps he could make a significant success as a banker in California, become his own man, financially independent of the Ewings, and bring Ellen and the children to live there in harmony, far from her parents. If he made a lot of money and Ellen refused to join him, he could at least return east able to support her in style. If he could look ahead, while in California, and see any of that happening, he would willingly resign from the army, but if things did not go well there, he could continue as an officer.

Receiving a six-month leave from the army, Sherman took the job and headed for San Francisco, while Ellen and their two daughters returned to Ohio to live with her parents, who were as thrilled to have them under their roof as she was to be there. The night before Sherman's ship was to land in San Francisco in April of 1853, the vessel ran aground on a rock. Sherman transferred to a small lumber schooner, which capsized in San Francisco Bay. Coming ashore "covered with sand, and dripping with water," as he described it, he found a city that had in five years grown from its pre–Gold Rush population of nine hundred to a city of fifty thousand, with some millionaires building mansions while its harbor filled with ships whose sailors drank and whored in the waterfront area known as the Barbary Coast. Within days, Sherman discovered that California had no banking laws: anyone could open an office and start lending money on any terms.

Having sized up the situation in San Francisco as best he could, in July of 1853 Sherman returned east and committed himself to serve as the head of the San Francisco branch of Lucas, Turner & Co. He resigned from the army at the end of his leave in September and, accompanied by Ellen and their year-old younger daughter Lizzie, set out on a five-week trip from New York to California, which involved taking a ship to Nicaragua, crossing to the Pacific by land, and boarding another ship for San Francisco. (Unlike Grant's regiment, the Sherman party made the land passage without complications.)

For four years, Sherman served as president of Lucas and Turner's San Francisco branch, battling both the wild swings of California's economy and the attitudes of Ellen and her possessive parents. Ellen had left their two-year-old daughter Minnie at home with her parents: her diary entries for New Year's Day 1854 and the following day speak of her having "a cry about Minnie" not being with her. In the spring of 1855, Ellen went home for seven months, leaving with Sherman their daughter Lizzie and an eighteen-month-old son born in San Francisco, Willy, in whom Sherman immediately took a greater interest than he did in his daughters. When Ellen returned from Ohio, she again left their older daughter Minnie with her parents. Despite living in a handsome house and having three and sometimes four servants, Ellen detested the bustling city on the Pacific coast; writing home to her mother, she said of a cabin that belonged to her family in Ohio, "I would rather live [there] than live here in any kind of style." Sherman stated the other side of the matter. In a letter to his father-in-law, he tried to explain his craving to become a success on his own: "I would rather be at the head of the bank in San Francisco, a position I obtained by my own efforts, than occupy any place open to me in Ohio."

As for their domestic life, Ellen was constantly unwell, suffering from the headaches and boils that had plagued her for many years, in addition to having colds and some affliction for which she took tincture of opium. Sherman had relentless attacks of asthma, which he at different times thought might be caused by San Francisco's sea air, or moisture in the walls of their house, or "carbonic acid" being released by nearby trees at night. In the summer of 1854, he wrote his friend and employer Henry Turner, "For the past seven months I have been compelled to sit up, more or less each night, breathing the smoke of nitre paper"—a practice that bothered Ellen—and went on to say that he knew that "the climate will sooner or later kill me dead as a herring." Through all this, they shared the same bed: a diary entry of Ellen's in early May of 1854 recorded that "Cump rubbed me with whiskey." Along with entertaining their friends, mostly old army comrades and their wives now stationed in California, there were quiet evenings at home: "Cump & I sat upstairs in the evening, Cump reading and nodding and I sewing." At times it may have been livelier than that. Years later, when Sherman wrote Ellen about some good news, he said that on hearing of it, "I have bet you will get tight on the occasion, à la fashion of Green Street California."

As had been true in Ohio, Washington, and St. Louis, there was always Ellen's involvement with her Catholic faith. Soon after coming to San Francisco, she began making calls on members of the clergy. Her diary entry for February 11, 1854, said, "sent jellycake to Bishop," and on March 28, "Archbishop called." Ellen never gave up on her efforts to bring her husband into the church in which he had been baptized. A diary entry in March says, "Prayed for the conversion," and leaves it at that. Sherman's children became used to attending Sunday morning Mass while their father went horseback riding.

Ironically, Sherman was finding that as a businessman in civilian life he was saving no money at all. Keeping up the kind of domestic establishment that was expected of the head of a San Francisco bank, and wanting to provide Ellen with everything that might make her enjoy living in California, he found himself writing his friend and employer Turner that Ellen was extravagant, but she must not learn that he was no more able to save money now than when he had been in the army. Although Ellen never liked San Francisco, she was with her husband the great majority of his time there, gave birth to two sons while there, ran the kind of household expected of a family in their position, and was respected by all who knew her. There were no outward signs of affection between Ellen and her husband, and no lessening of their differences about where to live and how to worship, but whenever he came under any kind of criticism, she was solidly by his side and on his side. Cump and Ellen Sherman might seem an unlikely pair to be married to each other, but it was difficult to imagine either of them being married to anyone else.

Sherman occasionally had reason to wonder about the wisdom of his resignation from the army. One unforeseen event followed another: the failure of the home office of another bank back in St. Louis started a massive run on all the banks in San Francisco. By noon of the day the news swept Montgomery Street, Sherman's bank had honored withdrawals totaling $337,000, but his adroit management of the crisis enabled him to close his next day's business with a balance of $117,000, at a moment when seven of the nineteen banks in San Francisco failed. Then a prominent lumberman, financier, and leading citizen known as "Honest Harry" Meiggs suddenly left for Chile, leaving behind debts secured with forged paper totaling close to a million dollars—a huge fraud in an era when Sherman's bank had been able to open in San Francisco with assets of a quarter of a million. Sherman wrote that he had seen "no symptoms of dishonesty" in Meiggs, but he had kept his relatively modest loans to Meiggs under close review, and his bank was the one least hurt in the scandal.

The problems continued. Now deep in the unpredictable commercial life of the growing city, Sherman maneuvered his bank through situations ranging from a local panic caused by the loss of an inbound ship carrying half a million dollars, to credible reports that California might default on its state bonds. While dealing with these crises, any one of which might ruin all that he tried to accomplish, Sherman was aware of the difference between his tenuous situation and the growing influence of his brother John, who in the fall of 1854 had been elected to Congress as a member of the newly formed Republican Party and was steadily rising in the ranks of the House leadership.

In 1856, at a time of relative calm in the family—Ellen and their three children were with him in San Francisco, and she was pregnant with their fourth child—a situation occurred that tested Sherman's judgment and character, leaving him feeling defensive and troubled by the result. As an increasingly important citizen and a former army officer, he had reluctantly accepted the nominal role of commander of a division of the state militia, an organization that existed almost entirely on paper. When the editor of one San Francisco newspaper shot and killed another, a Vigilance Committee, sometimes known as the Vigilantes, sprang up and, among other acts, hanged the killer. The committee, soon numbering more than five thousand men, including both riffraff and prominent citizens, became the de facto force of law and order.

Sherman's duty, as well as his craving for order, required him to restore the authority of the state and city's elected officials, but Sherman had an equally powerful need to continue the success he was finally having as a banker. He did not want to antagonize the many prominent businessmen who believed that the only way to have a peaceful San Francisco was to enforce the law themselves. Although he worked to enlist militiamen and tried to maneuver behind the scenes for conciliation between the governor and the Vigilance Committee, he avoided an armed confrontation and soon resigned his militia commission. Sherman himself knew all too well that even his limited anti-Vigilance position made him unpopular with many in the business community whose goodwill his bank needed; his resignation stemmed from a combination of expediency and angry frustration that society would not conform to his vision of a wisely self-regulated world.

To this point in his time as a banker, the San Francisco newspapers had favorably mentioned Sherman, and Ellen had started a scrapbook of these clippings. Now the pro-Vigilance press attacked him for not supporting their position, while the governor of California publicly deplored Sherman's resigning his commission at a time of crisis. If more were needed to upset him, when Sherman was named foreman of a grand jury that indicted San Francisco's *Daily Evening Bulletin* for libeling the Sisters of Charity in stories criticizing the way they ran the County Hospital, the *Bulletin* struck at him by saying that he was motivated by the fact that he was a Catholic. All of this might have rolled off the back of a seasoned politician, but it wounded Sherman, giving him a suspicious dislike of journalists that would work against him at a future day.

Sherman's San Francisco experiences were taking their toll. His nervous behavior was noted by a man who saw him walking various employees out of the bank on their way to meetings and transactions at other offices.

> In giving his instructions, he will take a person by the shoulder and push him off as he talks, following him to the door all the time talking . . . His quick, restless manner almost invariably results in the confusion of the person whom he is thus instructing, but Sherman himself never gets confused. At the same time he never gets composed.

At times his state of mind was considerably worse than "he never gets composed." Writing to Turner in St. Louis in early 1856, before the Vigilante crisis and the grand jury matter, Sherman said that he had slept for only three hours during the past twenty-four. In words indicating a fear of unspecified but well-publicized failure, he urged Turner to replace him in San Francisco with someone else. In a subsequent letter he apologized for having been so dramatic about his "depression," saying that it was due to "the effects of a disease which I cannot control," presumably asthma, and bad business conditions. Ellen, later writing of some moment during their time together in California, made this reference: "Knowing insanity to be in the family and having seen Cump in [sic] the verge of it once in California . . ." She never expanded on that, but, whatever Sherman and Ellen were experiencing, they were experiencing it together.

In early 1857, his bank's home office in St. Louis studied California's fluctuating economy and recent explosive history, and decided to close its San Francisco branch. It did, however, plan to open a branch in New York City, and offered Sherman the opportunity to be its manager. No sooner did he travel to New York and launch this branch than the Panic of 1857 hit Wall Street in August, immediately staggering the nation's economy. In October, word came from St. Louis to close the office in New York. Bitter at this end of his ambitions to become a prosperous banker who could be independent of his in-laws, Sherman wrote this to Ellen, who deserved kinder words: "No doubt you are glad to have attained your wish to see me out of the army and out of employment."

Back in St. Louis in late 1857 for discussions preparatory to making a final trip to San Francisco to untangle and salvage the bank's assets there, Sherman was walking down the street when he encountered Ulysses S. Grant, who had just moved to the city after his failure as a farmer. The two men had never served together in the army but recognized each other from their days at West Point.

If ever there was a commonplace meeting that nonetheless foreshadowed great events, this was it. Anyone watching the brief conversation between the shorter, brown-haired Grant in his rumpled clothes, who was then thirty-five, and the tall, red-headed, constantly gesturing thirty-seven-year-old Sherman could never have dreamt what lay ahead for them. Within five years the two would be winning immense military victories that preserved the American nation as one country; eleven years after their brief chat, the shabby shorter man would be elected president of the United States. As for what they discussed that day on the street, all Sherman could recall of their talk was that he walked on feeling "that

West Point and the regular army were not good schools for farmers [and] bankers."

On what proved to be Sherman's final return from San Francisco, Ellen and her parents once again urged him to take on the job of running the family saltworks near Lancaster. He still refused and in effect fled to Leavenworth, Kansas, where Thomas Ewing had set up his sons Thomas Jr. and Hugh Boyle Ewing in a combination of law firm and real estate management business. It was hardly a meaningful act of defiance: Sherman dabbled in a few legal matters, but his primary job there was to manage large tracts of farmland owned by his father-in-law. That entire venture failed; during the winter in Kansas, laboring in the wind and snow as he and some hired hands built storage barns to house corn, he wrote to Ellen, who was pregnant with their fifth child, in despairing terms, using the language of cockfighting, where the birds fought to the death: "I look upon myself as a dead cock in the pit, not worthy of further notice."

When Sherman came back from his galling struggles in Kansas, through Thomas Ewing's influence he received an offer to be the manager of an American bank in London. Burnt by his experience with banking, and still wanting to accomplish something on his own in a position that did not come to him because of his father-in-law, Sherman now tried to get back into the United States Army. This proved fruitless: the small peacetime army could not even retain all the junior officers who had stayed in the service since graduating from West Point, and had no way of bringing back in those who had resigned. Out of these efforts to reenter the Regular Army, however, he learned that Louisiana had created a new school to be known as the Louisiana State Seminary of Learning and Military Academy, a name that Sherman considered to be awkward and pretentious. Its Board of Supervisors was accepting applications for the job of leading this institution with the title of superintendent.

Sherman applied for the position, was accepted, and on November 12, 1859, two months after the birth of his and Ellen's fifth child, a daughter, arrived by himself at the school in Alexandria, Louisiana. At that moment the newspapers were filled with stories about the recent seizure of the federal arsenal at Harpers Ferry led by the fanatical abolitionist John Brown—an abortive raid put down by a company of United States Marines under the overall command of Lieutenant Colonel

Robert E. Lee of the United States Army, assisted by an army lieutenant nicknamed "Jeb" Stuart. The bloodshed at Harpers Ferry was small—ten of John Brown's nineteen followers were killed, including two of his sons, along with five townspeople shot by his men—but its portents were immense. Brown's purpose had been to seize the weapons at the federal arsenal and distribute them to slaves to use against their owners in an uprising. This threat of a slave revolt was an old nightmare in the South, and Southerners were shocked not only by the raid itself but also by the way many Northern abolitionists hailed Brown, who was tried and hanged for his act, as a martyr in the cause of freedom. Increasing numbers of white Southerners began to feel that the only way to preserve slavery would be to form a separate nation. This would involve seceding from the Union. Millions of Americans outside the South were not particularly incensed about slavery but were prepared to fight, if necessary, to preserve that Union.

Considering that Louisiana's military academy had hired a Northerner at a time when many in the North and South were taking irreconcilable political positions, things at the school went surprisingly well for nearly a year. As a result of Sherman's past postings in the South, he liked and admired its people and felt comfortable among them. As for the great issues agitating so many Americans, Sherman regretted that slavery existed but did not want to see war waged to abolish it, and he was content to live among white slaveholders; as for their black slaves, he considered them to be inferior beings and sympathized with Southern fears of a slave uprising. Secession from the Union, on the other hand, offended Sherman's need for the world to be a logical place. He wrote to Ellen, "I have heard men of good sense say that the union of the states any longer was impossible, and that the South was preparing for a change. If such a change be contemplated and overt acts be attempted of course I will not go with the South." In a later letter he continued to express his anxiety: "All here talk as if a dissolution of the Union were not only a possibility but a probability of easy execution. If attempted we will have Civil War of the most horrible kind." As the crisis heightened, with statements of some bellicose Southerners proclaiming that the North had no stomach for a war, and that if it came, one Southern soldier would prove to be equal to two or more Northern men, Sherman argued this to David French Boyd, the seminary's professor of ancient languages and a Virginian who liked and admired Sherman:

You mistake, too, the people of the North. They are a peaceable people, but an earnest people and will fight too, and they are not going to let this country be destroyed without a mighty effort to save it . . . The North can make a steam-engine, locomotive or railway car; hardly a yard of cloth or a pair of shoes can you make. You are rushing into war with one of the most powerful, ingeniously mechanical and determined people on earth—right at your doors. You will fail.

Seven hundred and fifty miles to the north, Ulysses S. Grant had already written a friend, "It is hard to realize that a State or States should commit so suicidal an act as to secede from the Union, though from all the reports, I have no doubt that at least five of them will do it." What both Sherman and Grant failed to comprehend was the degree of military skill that many of their fellow West Pointers would bring to the Southern cause, as well as the historical heritage that led so many Southerners to see themselves as both right and invincible. By 1860, Southern presidents had led the nation during sixty of its eighty-four years. Of the twenty-nine men who had served on the Supreme Court, eighteen were from the South. More than twice as many presidents pro tem of the Senate, speakers of the House, and attorneys general had been from the South as from the North. In the army and navy, the great majority of the higher ranks had invariably been filled by Southern men and still were. The South believed that its men would prevail, because they always had.

As the national crisis grew, at the seminary in Louisiana, Sherman, the Board of Supervisors, the faculty, and the new cadets all acted as if the only business at hand was that of teaching, studying, and participating in military drills preparing them to engage an unspecified enemy. (At this point, the academy's muskets were being supplied by the federal government.) Sherman, who was first addressed as Major and later as Colonel, worked hard and effectively. The board and faculty liked him and appreciated his efforts, and the cadets admired and came to be fond of him. His desire to have a success at last kept him from recognizing just how fragile his own position was, no matter how accurately he foresaw the growing crisis and how efficiently he ran the seminary, and reality descended upon him at the end of 1860. Abraham Lincoln and the Republican Party had won the presidential election; at just the time Ulysses S. Grant was writing a friend about his job in the family leather goods store in Galena, the nation learned that South Carolina had seceded from the

Union. Sherman's cadets remained quiet, while Sherman, who burst into tears on hearing of South Carolina's action, wrote a Southern friend, "You are driving me and hundreds of others out of the South, who have cast [our] fortunes here, love your people and want to stay."

In January of 1861, Louisiana state militia units seized the United States Arsenal at Baton Rouge and sent the captured weapons to be stored at the seminary. Sherman resigned. Even in this hour of hot feeling and impending bloodshed, the board passed two resolutions praising him and thanking him for his services. Men from the governor of Louisiana on down wrote him that they wished he would continue as superintendent. Overtures were made to Sherman, suggesting that high rank awaited him if a separate Southern army came into being, but everyone soon understood and respected his need to go. When Sherman said good-bye to his assembled cadets, all of them clearly sad to see him leave, his emotions overcame him: trying to speak, all he could manage was to point to his heart, say, "You are all in here," and stride away. (When he later encountered some of these young men as prisoners captured by Union troops he commanded, he did everything in his power to help them, including giving them some of his own clothes; he would write his daughter Minnie that she must remember that he was fighting those "whom I remember as good, kind friends.") Heading north to Ellen and their four children, even now Sherman hoped for peace.

2

GRANT AWAKENS

On April 12, 1861, artillery belonging to the seceded state of South Carolina began firing on Fort Sumter, the United States Army post in Charleston Harbor, and the fort surrendered two days later. The forces of the United States of America were engaged in combat with those of the Confederate States of America.

Abraham Lincoln, who had taken the oath as president of the United States five weeks before, still hoped to avert a large-scale war, but he issued a call for seventy-five thousand volunteers from the North, to augment the small Regular Army, then numbering sixteen thousand men, in which Grant and Sherman had once served. The Confederacy, its central government coming into existence overnight, brought its army into being from a collection of militia organizations and companies of volunteers.

In Washington, the federal government kept its few Regular Army regiments intact and authorized the creation of additional Regular units, but many experienced officers of the peacetime army were quickly moved into posts commanding regiments filled with the new volunteers. In the Union Army as a whole, an important distinction existed between those holding Regular Army commissions—men who had graduated from West Point, or who had in a very few cases been given Regular commissions as they were brought in from civilian life—and those officers holding Volunteer commissions, which, while carrying real responsibilities, were appointments frequently made as a political favor to men with little or no military experience. With the wartime expansion, a West Pointer who had served for years in the Regular Army might find himself advanced several ranks, to make use of his ability and knowledge: a man would, for example, continue to hold his Regular commission as a captain, a rank in which he previously commanded no more than a hundred men, but soon be given the rank of colonel of Volunteers and become the commander of a regiment of a thousand. As for marching into battle, it was all the same Union Army.

It was a time of fateful, painful decisions: the South gained the services of Robert E. Lee, who declined an offer by an intermediary acting for President Lincoln that he should, in Lee's words, "take command of the army that was to be brought into the field." Three months before Fort Sumter was fired upon, Lee had written about the agonizing issue of conflicting loyalties to his son Lieutenant Custis Lee, who had graduated first in the West Point class of 1854: "I can anticipate no greater calamity for the country than a dissolution of the Union." But when his beloved native Virginia seceded, he resigned from the army he had entered as a West Point cadet thirty-five years before, stating, "I could take no part in an invasion of the Southern States." Lee urged his son to make his own decision in the matter, but Custis also resigned, to fight for the Confederacy. Of the 1,108 officers serving in the United States Army, a third chose to join the forces of the South; in the navy, a quarter of the officers resigned, to reappear in the new Confederate States Navy.

In the eyes of professional military men, another great loss to the Union was the decision to go with the South made by the exceptionally able Joseph E. Johnston, who had been Lee's West Point classmate. The Confederacy elected as its president Jefferson Davis, a Southerner who had graduated from West Point, had led with distinction a regiment of Mississippi volunteers in the Mexican War, and had gone on as a civilian to serve as a United States senator and subsequently to become secretary of war in the cabinet of President Franklin Pierce, dealing with reams of paperwork that included accepting the resignations of Captains Sherman and Grant. In early 1861, he was again a United States senator from Mississippi. On the day before he made his farewell address in the Senate and headed south, Davis said, "Civil war has only horror for me, but whatever circumstances demand shall be met as a duty."

In Galena, Illinois, President Lincoln's call for volunteers produced a mass meeting; Ulysses S. Grant, the only man in town who had served as an officer in the Regular Army, was pressed into duty as chairman. The citizens voted to form a company of foot soldiers to be known as the Jo Daviess Guards, named for Jo Daviess County, of which Galena was the county seat. Asked if he would take command of what soon became a hundred volunteers, Grant declined, saying that he intended to offer his services at a higher level, but he threw himself into the business of organizing the town's company and readying these recruits to proceed to a camp outside the state capital of Springfield for training. "I never went into our leather store after that meeting," Grant said, "to put up a package or do other business."

Suddenly this quiet man was everywhere, helping the patriotically minded ladies of Galena order the right kind of cloth for uniforms from a dry-goods merchant appropriately named Felt, and telling the tailors at Corwith Brothers what the dark blue uniforms should look like. He showed the company's newly elected captain how to drill the men: the entire state of Illinois had only 905 muskets and rifles on hand, 300 of which needed repairs, so the Jo Daviess Guards had their first instruction in the manual of arms using wooden laths instead of real weapons.

By the end of the first week, Grant had a tentative plan for himself: Galena's congressman Elihu Washburne, who had given a fiery patriotic speech at the meeting that voted the Jo Daviess Guards into being, told Grant that he should go with the new company when they went to Springfield. At the state capital, Washburne told Grant, he would use his influence with the governor to find him a suitable position in the state's effort to mobilize. Writing to his father in Kentucky, Grant urged him to come north from that border state, which might explode in violence at any time, and added that his own duty was clear: "Having been educated for such an emergency, at the expense of the Government," he must offer his services in the conflict that had begun. At the moment he thought he might be gone for as long as three months. Of his wife's reaction to both the national crisis and his intention to serve, he told his father that "Julia takes a very sensible view of the present difficulties. She would be sorry to have me go, but thinks the circumstances may warrant it and will not through [throw] a single obsticle [sic] in the way."

That was not the entire picture of Julia's feelings. A woman from a slaveholding family, married to a man who might soon be fighting against the South, she hoped that her home state of Missouri could be kept in the Union, and she had followed closely the events leading to the attack on Fort Sumter. Julia later wrote:

Oh! how intensely interesting the papers were that winter! My dear husband Ulys read aloud to me every speech for and against secession. I was very much disturbed in my political sentiments, feeling that the states had a right to go out of the Union if they wished to, and yet thought it the duty of the national government to prevent a dismemberment of the Union, even if coercion should be necessary. Ulys was much amused by my enthusiasm and said I was a little inconsistent when I talked of states' rights, but that I was all right on the duties of national government.

The news of Confederate shells landing on a United States Army post at Fort Sumter evidently resolved the question of Julia's loyalties. "I remember now with astonishment the feeling that took possession of me in the spring of '61. When reading patriotic speeches, my blood seemed to course more rapidly through my veins." She added, "Galena was throbbing with patriotism."

Two days before his thirty-ninth birthday, Grant said good-bye to Julia and their four children and headed downtown, wearing a tired old civilian suit, a slouch hat, and the faded army overcoat he had worn peddling firewood in St. Louis, and carrying an old bag that had little in it. The Jo Daviess Guards were being sent off in a large and enthusiastic parade through town and across the bridge to the railroad depot, where the recruits would board the train for Springfield to join the many volunteer companies converging there. Grant watched from a sidewalk as different organizations—the Masonic Assembly, the city's fire companies with their horse-drawn engines, the Odd Fellows, the mayor and various civic groups, all interspersed with brass bands—paraded down the street, followed by the hundred newly uniformed recruits he had equipped, many of them waving high-heartedly to the cheering crowds. As the last of the Jo Daviess Guards passed, the brother of the company's captain watched Grant standing there on the sidewalk. A man to whom he later spoke of the moment remembered him describing how Grant "fell in behind the column and quietly, with head pensively drooping, marched in their wake across the bridge, and entered the train for Springfield."

When Grant arrived with the Jo Daviess Guards at Springfield sixteen days after Fort Sumter was fired on, he found a military nightmare. He knew that the Volunteer companies, units of a hundred recruits apiece, had elected their officers, who might or might not lead them well, but now Grant found many men at the state capital, some with no military experience, seeking political appointments to be commanders of the ten regiments whose formation the Illinois legislature had authorized. This meant that, although even colonels were nominally elected by ballot, "candidates" named by the governor of Illinois would lead regiments composed of ten Volunteer companies, each regiment having a thousand men. With the Regular Army still trying to keep many of its officers with their prewar Regular regiments, the new Volunteer regiments desperately needed qualified commanders, wherever they might come from, but Grant was appalled by the inadequacy of the applicants he saw. To his father he wrote, "I might have got the Colonelcy of a Regiment possibly, but I was perfectly sickened at the political wire-pulling for all these commissions, and would not engage in it."

Grant took a civilian job that Congressman Washburne found for him in the office of the state adjutant general, where he efficiently processed paperwork for the mobilization of the Illinois regiments, using forms that were in some cases the same ones he had often filled out during his army service. He would soon write a letter to Washington, trying to get back into the Regular Army, and it was known that, although he would not enter the political dogfight, he wanted command of one of the Illinois regiments. The elected captain of the Jo Daviess Guards, who saw him working "at a little square table, of which one leg was gone and which had been shoved into a corner to keep it upright," said that Grant, wearing his "one suit that he had worn all winter, his short pipe, his grizzled beard and his old slouch hat did not . . . look a very promising candidate for the colonelcy."

When, frustrated and discouraged, Grant finished his various duties involving mobilization and was once again unemployed, he took a train for Cincinnati and appeared at the office of Major General George B. McClellan, who had been three years behind him at West Point and whom he had known during the Mexican War. In 1855, McClellan had been one of three United States Army officers sent to Europe to observe the war being fought on the Crimean Peninsula between Russia on the one hand, and the armies of Britain, France, and the Ottoman Turks on the other. Then a captain, McClellan had been present during the siege of Sebastopol, and during his year abroad had the opportunity to study other European armies as well; his modification of the Hungarian saddle used by the Prussian army became known as the McClellan Saddle that the army adopted in 1859 and would use for generations. In 1857, he had resigned his captain's commission to enter a business career that brought him to prominence: on the eve of war, he had the largest salary of any railroad executive in the United States.

Here was an example of a combination of solid military experience and superb political connections: "Little Mac," who had left the army with the same rank of captain as had Grant and Sherman, was appointed by the governor of Ohio to organize and lead that state's regiments, holding the rank of major general. Able, painstaking, and vain, his efficiency, coupled with a flair for dramatic appearances and confident statements, was swiftly gaining him wide recognition not only in Ohio but also in Washington, where President Lincoln was one of his acquaintances. Although Grant had wanted a regiment of his own, he too held McClellan in high regard and was now ready to serve under him. "I thought he was the man to pilot us through," Grant recalled, "and I wanted to be on his staff."

At McClellan's headquarters, Grant was greeted by Major Seth Williams, a West Pointer from the class ahead of his who had been Robert E. Lee's adjutant when Lee served as superintendent of West Point from 1852 to 1855. Word was sent in to McClellan's office that Grant would like to see him. After waiting for two hours, Grant left, telling someone that he would come by the next day. The following day, Grant reappeared, was asked to wait, and once again left after two hours, later writing that "McClellan never acknowledged my call."

In his determination to enter the war in some capacity, there was something that Grant may have forgotten, or blocked from his mind. Eight years before, while serving on the West Coast as quartermaster of the Fourth Infantry Regiment, one of Grant's responsibilities had been to equip a survey party led by McClellan, then a captain in the army's elite Corps of Engineers, who was setting out to map the Cascade Range in the Oregon Territory. According to another officer, during the time Grant supervised the issuing of supplies and assignment of horses required by McClellan's detachment, he "got on one of his little sprees, which annoyed and offended McClellan exceedingly, and in my opinion he never quite forgave Grant for it, notwithstanding the necessary transportation was soon in readiness."

This rebuff plunged Grant into gloom. "I've tried to reenter service in vain," he told a friend in Ohio. Recalling his varied experiences as a regimental supply officer in both Mexico and California, he added, "Perhaps I could serve the army by providing good bread for them. You remember my success at bread-baking in Mexico?"

When Grant returned to Springfield after his humiliating experience at McClellan's headquarters in Cincinnati, his fortunes changed. Among his various duties during the Illinois mobilization had been a temporary appointment as a "mustering officer and aide," during which he had the task of swearing several of the new regiments into the service of the state. Doing this, he had spent two days in Mattoon, Illinois, eighty miles west of Springfield, with the Twenty-first Illinois, whose officers liked him. One of them, who had spent two years as a West Point cadet, observed, "[We] saw that he knew his business, for everything he did was done without hesitation. He was a little bit stooped at the time, and wore a cheap suit of clothes and a soft black hat. Anyone who looked beyond that recognized that he was a professional soldier."

At the time Grant spent two days with the Twenty-first, it was commanded by a man whose first appearance among them, striding into camp at the head of the Volunteer company from Decatur who made him their captain, had so impressed everyone that he was in effect elected colonel by acclamation, rather than becoming a "candidate" through political influence at Springfield. Tall, erect, shooting piercing glances in every direction, newly elected Colonel Simon S. Goode wore high boots and a broad-brimmed hat, and for weapons carried a big bowie knife and no fewer than six small pepper-box revolvers. It soon became apparent that he was a drunk, given to moving around the camp at night in a long cloak while he quoted Napoleon and told bemused sentries, "I never sleep." After Grant left at the end of his two-day visit, under Goode's unsteady hand the regiment rapidly deteriorated: the new recruits rioted, protesting the lack of proper food, and when the guardhouse became infested by vermin, they burnt it. Men dug tunnels under the fence at night to carouse through the streets of Mattoon and roam the countryside, stealing food: an old sergeant commented that "there wasn't a chicken within four miles of us." At one point, Colonel Goode went to a tavern with the men who had been assigned to guard duty that night and were supposed to be at their sentry posts. Scores of men of the Twenty-first began to desert.

In response to vociferous complaints from the authorities and citizens of Mattoon, the governor's office ordered the Twenty-first Illinois to be brought to Springfield by train; on the way, they created disturbances in the coaches carrying them. Once in camp at Springfield, Colonel Goode tried unsuccessfully to restore discipline by surrounding their regimental area at the state fairgrounds with a guard detachment of eighty men who wielded clubs in an effort to keep them from breaking out of what they regarded as a prison.

Desperate to improve the situation, two lieutenants of the Twenty-first went to call on the Illinois secretary of state. They were ushered in to see Governor Richard Yates, who was aware that all the Illinois volunteers, currently in the status of state militia, believed that they would soon have the option of going home or being sworn into federal service for a three-year term of duty as was intended. Yates took the complaints about the worst-behaving Illinois regiment at face value and convened a meeting of the regiment's officers. They told him they wanted a new colonel, "preferably Captain Grant," and Grant was offered the colonelcy. He accepted.

Ulysses S. Grant, who at this moment had neither a uniform nor a horse, went out to his new command by riding on the horse-drawn trolley to the state fairgrounds. A man who saw him walk into the encampment said that the new colonel "was dressed very clumsily, in citizen's clothes—an old coat, worn out at the elbows, and a badly dinged plug hat."

As Grant headed toward headquarters and the word spread that this was the new commanding officer, the recruits began to jeer, shouting, "What a colonel! Damn such a colonel!" One private asked another, "What do they mean by sending a little man like that down to command this regiment? He can't pound sand in a dry hole." According to an observer, "Rustic jokes were passed upon him, and one young fellow made insulting gestures behind his back. Another daredevil slipped up behind him, and flipped his hat from his head. Grant turned and said, 'Young man, that's not very polite,' and walked on."

Grant took command on June 17, 1861. He had eleven days in which to turn this insubordinate mob into a unit that would, man by man, choose either to go home or to sign up for three years of dangerous service. After Grant's first night in camp, there were twenty men under arrest for leaving the post without permission, some facing additional charges of being drunk and disorderly. In addition to those arrested was a notorious troublemaker known as "Mexico," who appeared drunk in front of Grant's tent, defying anyone to touch him. When Grant had him tied to a post, Mexico shouted at him, "For every minute I stand here I'll have an ounce of your blood!" Grant turned to a sergeant, said, "Put a gag in that man's mouth," and went about his duties. When Grant decided that Mexico had stood there long enough in the June weather, tied to a post with a gag in his mouth in the middle of camp where everyone could see him, he took off the gag and the ropes himself, and stood back waiting to see what Mexico would do next. The man saluted and silently walked away. A sentry greeted Grant by saying, "Howdy, Colonel?" while standing with his musket at his side. Grant asked the man to hand him his musket, which Grant then snapped up to the saluting position of present arms. Handing it back, he said, "That is the way to say 'how do you do' to your Colonel." When the different companies all held morning roll call an hour late, with the men getting up whenever they pleased, they found no breakfast waiting for them.

Within forty-eight hours Grant had set up a simple daily schedule, understood by all: the men would drill in small groups as squads from six to seven in the morning and as companies from ten to eleven, and again as companies from five to six in the afternoon. Other than these times, the men could go into Springfield during the daylight hours, as they wished. Grant's words regarding their conduct were set forth in his Orders No. 8: "All men when out of Camp should reflect that they are gentlemen—in camp soldiers; and the Commanding Officer hopes that all of his command, will sustain these two characters with fidelity."

The men began to feel that Grant considered them responsible individuals. The regiment's chaplain spoke of Grant's "unostentatious vigor and vigilance," saying that he "would correct every infraction on the spot," and do it in a "cool and unruffled manner." Each day, there were fewer disciplinary cases. Some soldiers who thought that they were still back in Mattoon, with a colonel who would go out drinking with them when they slipped away from guard duty, left their sentry posts and found themselves under arrest, with Grant's Orders No. 14 stating that they could be fined ten dollars apiece and face "corporal punishment such as confinement for thirty days with ball and chain at hard labor." Bearing in mind the recent history of this regiment and its deficient commander, Grant let the offenders off lightly but reminded them that if they left a sentry post in the face of the enemy, "the punishment of this is death."

As Grant took his regiment out on a short route march, someone told him that many of the men's canteens were "loaded," filled not with water but whiskey. He halted the column, ordered everyone to pour out the contents of his canteen, and resumed the march. A lieutenant wrote his wife that the guardhouse was packed with miscreants for the first few "nights and days but yesterday there was but two or three in and to day none." The colonels of the other new regiments began coming around to see what Colonel Grant was doing with the Twenty-first.

Grant made a swift trip home to Galena and returned wearing a new uniform, riding a newly bought horse named Rondy, and accompanied by his oldest son, eleven-year-old Fred, who Julia felt should see what his father was doing. Julia had always believed that her Ulys would eventually do splendid things, and she wanted their son to see him commanding his regiment. Likening her husband to Philip of Macedon and their son to Alexander the Great, going off to conquer in ancient campaigns, she wrote Grant, "Alexander was not older when he accompanied Philip. Do keep him with you." For his part, Grant was thinking not of Julia's romanticized view of history but of his daily work with one steadily improving regiment of recruits in Illinois. In a letter to Julia that he signed, "Your Dodo," Grant said, "The men I believe are pleased with the change that has taken place in their commander," and added that the greatest change was "the order in camp."

On June 28, 1861, after patriotic speeches by two Democratic congressmen, the soldiers of the Twenty-first Illinois had their opportunity to go home or to sign up to be in the Union Army for three years. As Grant put it, "They entered the United States service almost to a man."

Five days after this, Grant started moving his regiment west toward Quincy, Illinois, on the Mississippi River, to aid the Union forces in Missouri who were attempting to prevent that state from joining the Confederacy. The movement was taking him into the region where his and Sherman's military destiny lay. Other than Lincoln, Grant and Sherman would have more to do with winning the war that preserved the Union than anyone else, yet at this moment Grant commanded fewer than a thousand men in an army that he would command when it would number more than a million, and he had an unrealistic view of what lay ahead. Two months before, writing to his undemonstrative father, a man he yearned to impress, Grant offered these views, in words that were frequently misspelled:

> My own opinion is that this War will be but of short duration. The Administration has acted moste [sic] prudently and sagaciously so far in not bringing on a conflict before it had its forces fully martialed [sic]. When they do strike[,] our thoroughly loyal states will be fully protected and a few decisive victories in some of the southern ports will send the secession army howling and the leaders in the rebellion will flee the country. All the states will then be loyal for a generation to come, negroes will depreciate so rapidly in value that no body will want to own them and their masters will be the loudest in their declamations against the institution [slavery] in a political and economic view. The nigger will never disturb this country again. The worst that is to be apprehended from him is now; he may revolt and cause more destruction than any Northern man, except it be the ultra abolitionist, wants to see. A Northern army may be required in the next ninety days to go south to suppress a negro insurrection.

Sixty days had elapsed since Grant wrote that letter to his father, who cared far more about the injustices of slavery than he did. No "negro insurrection" had occurred, but in his indifference to the condition of blacks, an attitude similar to Sherman's, Grant echoed a widespread Northern point of view: what was bringing volunteers forward, what had compelled Sherman to leave Louisiana, was not a desire to eradicate slavery but the conviction that secession was treason and that the Union must be preserved as one nation by force of arms if necessary. Grant's and Sherman's views on several issues would change—indeed, Lincoln himself was not yet the Lincoln of the Emancipation Proclamation—but Grant's mind was now devoted entirely to daily military matters and decisions. When state officials started making arrangements to move his Twenty-first Illinois by rail to Quincy, 116 miles from Springfield, Grant startled Governor Yates by saying that his men would go there on foot. "This is an infantry regiment," he said. "The men are going to do a lot of marching before the war is over and I prefer to train them in friendly country, not the enemy's."

And so Grant marched his men to war, riding one horse while his eleven-year-old son Fred rode beside him on Rondy, the horse he had bought for himself. In a letter to Julia written several days into the regiment's movement west through peaceful farmland, Grant said, "Fred enjoys it hugely . . . The Soldiers and officers call him Colonel and he seems to be quite a favorite." He closed his letter: "Kisses to you. Ulys."

3

SHERMAN GOES IN

In the weeks before the attack on Fort Sumter, while Grant sat bored in the family leather goods store at Galena, Sherman returned to his family in Ohio after resigning from his leadership of Louisiana's military academy. He found two letters waiting for him. The first offered him the presidency of the Fifth Street Railroad in St. Louis, a company that ran the city's horse-drawn trolleys. The second came from his ever-devoted brother, United States Senator John Sherman, urging him to come to Washington as soon as he could. John's newly elected Republicans had many choice appointive posts to fill, and in Washington Sherman could also investigate the possibility of reentering the army on what might be the eve of war.

Sherman arrived in Washington in a frustrated state. On his recent trip north from Louisiana, passing through Southern areas that were seceding from the Union, he had seen excited preparations for war; in the North, he saw calm street scenes and a population that seemed unaware of the crisis. Sherman did not want war, nor did he wish to rush into a position of military leadership, but he felt that those who believed in the Union should make an accurate assessment of the situation.

On March 8, 1861, two days after Abraham Lincoln was sworn into office and thirty-five days before the Confederates fired on Fort Sumter, Sherman's brother John steered him into the White House, and he found himself shaking hands with the new president. Introducing him to the lanky, sallow-skinned Lincoln, John Sherman used the rank given him at the southern military school: "Mister President, this is my brother, Colonel Sherman, who is just up from Louisiana. He may give you some information you want."

To this the affable Lincoln replied, "Ah! How are they getting along down there?"

"They think they are getting along swimmingly," Sherman answered, eager to convey his sense of urgency to the nation's new commander in chief. "They are preparing for war."

"Oh, well!" Lincoln spoke cheerfully. "I guess we'll manage to keep house."

A few moments later, possibly in response to some remark that Sherman be considered for reinstatement in the army, Lincoln remarked dismissively that he would not be needing "military men" and indicated that the secession crisis could be solved peacefully.

Emerging from the White House, Sherman turned angrily to his brother. "I was sadly disappointed and I broke out on John damning the politicians generally . . . adding that the country was sleeping on a volcano that might burst any minute, but that I was going to St. Louis to take care of my family and would have no more of it. John begged me to be patient, but I said I would not, that I had no time to wait; that I was off for St. Louis; and off I went."

And so it was that, when Fort Sumter was attacked, with Lincoln's views changing from "I guess we'll manage to keep house" to what he said the day of Fort Sumter's surrender—"I shall, to the extent of my ability, repel force by force"—Sherman was in an office in St. Louis running the Fifth Street Railroad's trolley service, being paid well and swiftly improving its efficiency and profits. While Ulysses S. Grant trained Galena's company of Jo Daviess Guards, accompanied them to Springfield, Illinois, and began looking for ways of entering military service, Sherman remained in his civilian job, his fourth job in four years. In a letter to his brother John, who with his Ewing in-laws was using every kind of influence on his behalf in Washington, he expressed his desire to make enough money "so as to be independent of any body so I can not be kicked around as heretofore." He added, with the air of being above the fray, "If the country needs my services, it can ask for them."

Ellen Sherman, pregnant with their sixth child, began to realize that, no matter what her husband said, he wanted to be back in uniform. Writing to John Sherman, who was becoming impatient with his brother's sensitivity and entire stance in what was now a time of war, she said, "I am convinced that he will never be satisfied out of the army & I know that you can obtain for him a high position in it."

As a result of the family's lobbying on his behalf in Washington, Sherman was offered an important War Department civilian position, which he turned down, but on May 14, 1861, his brother telegraphed him that he had been appointed colonel of one of the newly authorized Regular Army regiments, which was yet to be organized. "Of course I could no longer defer action," Sherman said of this moment. Arriving in Washington in early June, he was not sent to the still-forming regiment to which he had been assigned but was immediately utilized in the inspection of the capital's defenses. This task involved reporting daily to the army's infirm seventy-five-year-old commander, General Winfield Scott, who was well aware of Sherman's political connections and formed a good opinion of him as one of the badly needed officers who had Regular Army experience.

Sherman was soon given a brigade to command, of units all encamped in the Washington area: the Thirteenth, Sixty-ninth, and Seventy-ninth New York Volunteer Infantry, and the Second Wisconsin, along with a battery of the Regular Army's Third Artillery. His units totaled thirty-four hundred men, a force roughly four times the size of the Twenty-first Illinois that Grant was leading toward Missouri.

Sherman was later to speak as if these regiments had been good units at the time he took command of them, but when he walked in to take over his brigade on June 30, he encountered the same kind of disdainful reactions that Grant experienced a few weeks earlier when the enlisted men of the Twenty-first Illinois had their first look at him. One soldier remembered a "tall gaunt form in a thread bare blue coat, the sleeves so short as to reveal a bony wrist, the trousers at least four inches shorter than the usual length." Others recalled that the troubled, prematurely wrinkled face looking out from under a most unmilitary "broad brimmed straw hat"—quite sensible to wear in a Washington summer—had "hollow cheeks," "a bushy untrimmed beard," and "a pair of piercing eyes."

The new commander was no more impressed by his men than they were with him, referring to them in his letters as "rabble"; at one point Sherman wrote Ellen that he commanded "volunteers called by courtesy Soldiers, but they are all we have got." Two weeks after taking command, he received orders from Brigadier General Daniel Tyler to begin moving his green troops from their base on the south bank of the Potomac, marching slowly west into Virginia. The Union Army was about to take the offensive, and Sherman faced the possibility that he might soon be killed. On July 16, about to put his columns in motion, he wrote to Ellen, who had given birth to a baby daughter a few days before, and told her, "Whatever fate befals [*sic*] me, I know you appreciate what good qualities I possess—and will make charitable allowances for defects, and that under you the children will grow up on the safe side." Speaking of his two sons, his favorite child Willy, age seven, and four-year-old Tommy, he said, "Tell Willy I have another war sword, which he can add to his present armory . . . when I come home again . . . though truly I do not choose for him or Tommy the military profession. It is too full of blind chances to be worthy of a first rank among callings." Sherman closed this letter to Ellen with, "Goodbye—and believe me always most affectionately yours." He signed it, as he did all his letters including those to his ten-year-old daughter Minnie, "W. T. Sherman." Then he led his brigade to a succession of night encampments in the field. "The march," Sherman recalled, "demonstrated little but the general laxity of discipline."

The first big movements of the war were now under way. After Fort Sumter fell, the United States Navy had dispatched its ships in an effort to blockade Southern ports, but in the land war, both sides had initially devoted most of their efforts to organizing their armies rather than to attacking each other. (It was during this period that Sherman wrote, "As soon as real war begins, new men, heretofore unheard of, will emerge from obscurity, equal to any occasion.") Then, after some movements in Missouri that brought most of that state under Union control, McClellan won some minor engagements in western Virginia, feeding the Northern hope for a quick series of bigger victories. The Washington newspapers were clamoring for a large and decisive battle. The slogan "On to Richmond!" expressed a widespread belief that Union forces could thrust aside any opposition and sweep on to capture the Confederate capital, 105 miles to the south. (Horace Greeley's *New York Tribune* was both explicit and demanding, with a headline that read "FORWARD TO RICHMOND!" and a subheadline that said of the Confederate Congress scheduled to meet there on July 20, "*By That Date the Place Must Be Held by the National Army.*")

At the time Sherman's brigade, part of a force of thirty-five thousand, made the first of two overnight bivouacs at Centreville, Virginia, seventeen miles west of Washington, they knew that large Confederate units were in that area. But there was something the Union commanders did not know: several days before, a beautiful dark-haired Southern intelligence courier named Bettie Duvall, wearing a smart riding habit and with her long black hair swept up under her hat, had come out of Washington and ridden into Confederate headquarters at Fairfax Court House, a few miles from Centreville. Taking off her hat and loosening her tresses, she pulled out something hidden in her hair: a tiny package wrapped in black silk, containing the Union Army's plan for its advance into Virginia and the approximate time of the movement. This, and a later more specific message brought by another courier, swiftly reached the headquarters of Confederate general P.G.T. Beauregard southwest of Centreville at Manassas Junction, an important railroad hub located west of a wide, slow-moving stream called Bull Run. Alerted, Beauregard, who had twenty-two thousand men, telegraphed Jefferson Davis in Richmond, and the Confederate president immediately sent him by railroad a reinforcement of twelve thousand troops led by Joseph E. Johnston, the general whose decision to go with the South so many knowledgeable Union officers regretted. Not only were the Confederates in the area now equal in numbers, but thirteen of the fifteen senior Southern commanders were West Pointers of exceptional ability, including Ulysses S. Grant's best man and cousin by marriage, James Longstreet, and the gifted cavalry officer Jeb Stuart. Taking the role of principal commander and knowing what to expect, Beauregard deployed his forces on favorable higher ground and awaited the clash.

At dawn on Sunday, July 21, Union forces under General Irvin McDowell, who had no idea the Confederates were there in such strength, began to attack, and the federal troops poured across Bull Run at Sudley Ford. During the morning, both sides committed more regiments to the battle, and by noon the Confederates had set up what proved to be their final line of defense, on a wooded ridge.

It was at this point that Sherman's brigade crossed Bull Run, some of his men fording the stream while others marched over a stone bridge, and entered the thickly wooded area in which the battle was already raging. At the age of forty-one, twenty-one years after graduating from West Point, Sherman said that "for the first time I saw the Carnage of battle—

men lying in every conceivable shape, and mangled in a horrible way," as well as horses "with blood streaming from their nostrils, lying on the ground hitched to [artillery] guns, gnawing their sides in death." Bullets grazed his shoulder and knee, and his horse was shot through the fore-leg, but Sherman kept moving his men forward.

Three hours into his brigade's part in the battle, with Sherman about to make the mistake of sending his four regiments up Henry Hill one af-ter another, rather than making one massive attack with his entire force, he encountered a Confederate brigade that had come into position op-posite them on higher ground and was mowing down his men. Speaking of his troops, Sherman said, "Up to that time all had kept their places, and seemed perfectly cool . . . but the short exposure to an intense fire of small arms, at close range, had killed many, wounded more, and had produced disorder in all of the battalions that had attempted to en-counter it." His men were now facing the regiments commanded by Thomas Jonathan Jackson, whose performance that day in defense of Henry Hill prompted a general from South Carolina to cry admiringly, "There stands Jackson like a stone wall!"—a tribute that gave the com-mander and his brigade the name by which they were thereafter known.

Sherman described how he became aware of an even bigger problem than the effect being produced by Jackson's brigade. Referring to the past hours, he said, "After I had put in each of my regiments, and had them driven back . . . I had no idea that we were beaten, but reformed the reg-iments in line in their proper order, and only wanted a little rest, when I found that my brigade was almost alone . . . I then realized that the whole army was 'in retreat,' and that my men were individually making back for the stone bridge."

What Sherman was seeing as a commanding officer was mirrored by the experience of one of his soldiers. Private Alexander Campbell be-longed to the Seventy-ninth New York, a regiment known as the High-landers because its ranks were filled with men born in Scotland, or of Scottish ancestry; this unit had a bagpipe band, and early in the war some of its soldiers wore kilts. Campbell, whose brother James served in the Confederate Army's First South Carolina Battalion, wrote his wife, Jane, of this same moment of the Union collapse: "We could see our army retreating and the men cutting their horses Loose from the wagons and mo[u]nting there [sic] backs and galloping off as fast as they could . . . Then we came across a field running across as fast as we could . . .

[Later] we came into centervall [Centreville] and the regiments that was at the fight tried to get themselves together but it was impossable [*sic*]." After trying vainly to find some of his comrades, "I gave them up for Lost then started with a small party for arlington heights"—all the way back to the Potomac River and Washington.

Sherman's official report gave more glimpses of what became a chaotic nightlong flight back to Washington, with the carriages of civilian spectators who had come out to see the battle mixed in with horse-drawn ambulances and carts filled with groaning wounded men. The mercurial Sherman, always craving a sense of organization, conveyed his own bewilderment and inability to change the course of events.

> There was no positive order to retreat, although for an hour it had been go-ing on by the operation of the men themselves. The ranks were thin and ir-regular, and we found a stream of people strung . . . across Bull Run, and far toward Centreville . . . About nine o'clock that night [at Centreville] I re-ceived . . . the order to continue the retreat to the Potomac. This retreat was by night, and disorderly in the extreme. The men of the different regiments mingled together . . . reached [the Potomac River, at a point opposite Wash-ington] at noon the next day, and found a miscellaneous crowd crossing over the aqueduct and ferries.

Despite energetic efforts by Sherman and other commanders to keep the men on the battlefield, and then to reorganize them at various places during the nightlong retreat, the first major battle of the Civil War had ended in a rout. At times during the retreat, Sherman himself was sepa-rated from his command, and there was a question about his own behav-ior: the men of the Seventy-ninth later successfully petitioned to be removed from his command because of an alleged incident in which, on the rainy day after the defeat, he had some of the Highlanders ejected from a barn so some horses could be sheltered there. In the wake of this defeat and his own baptism by fire, Sherman, the lover of orderly proce-dure, poured out his descriptions and feelings in three letters to Ellen, telling her of the "Shameless flight of the armed mob," and said that he had in the past "seen the confusion of crowds of men at fires and Ship-wrecks, but nothing like this. It was as disgraceful as words can portray."

There was confusion indeed. Leaderless soldiers wandered the streets of Washington, some begging for food, while many saloons were packed with officers getting drunk instead of trying to find and care for their demoralized men. Other soldiers boarded trains north and were never seen again. Sherman later summed up Bull Run in these terms: "Though the North was overwhelmed with mortification and shame, the South had not much really to boast of, for in the three or four hours of fighting their organization was so broken up that they did not and could not follow our army, when it was known to be in a state of disgraceful and causeless flight."

Back in camp after they finished pouring into the Washington area, some of Sherman's troops "were so mutinous, at one time, that I had ordered the [Regular Army artillery] battery to unlimber, threatening, if they dared to leave camp without orders, I would fire upon them." His brigade was soon visited by President Lincoln, accompanied by Secretary of War Simon Cameron, whose brother James, the colonel commanding the Seventy-ninth New York, had been killed at Bull Run. During the visit, Lincoln stood in his carriage, surrounded by hundreds of soldiers, and spoke to them in a way that Sherman described as "one of the . . . best, most feeling addresses I ever listened to, referring to our late disaster at Bull Run, the high duties that still devolved on us, and the brighter days yet to come." The talk steadied and encouraged the shaken young soldiers; as for his own feelings and those of his fellow commanders during these days, Sherman said, "We were all trembling lest we should be held personally accountable for the defeat."

Still in this troubled mood, Sherman was speaking one evening with several other worried colonels in a large, high-ceilinged room at Arlington House, the Custis family mansion that until a few weeks before had been the home of Robert E. Lee and his family. (Arlington House and its farmland, sitting directly across the Potomac from Washington in Virginia and clearly visible from the Capitol, was one of the first places in Virginia to be occupied by federal troops. Some of the dead from Bull Run were being buried on the extensive Arlington lands, which later became the Arlington National Cemetery.) As the group of colonels talked in this room that was now serving as an adjutant general's office, important news arrived in a way that Sherman later described.

Some young officer came in with a list of the new brigadiers just announced at the War Department, which embraced the names of Heintzelman, Keyes, Franklin, Andrew Porter, W. T. Sherman, and others, who had been colonels in the battle, and all of whom had shared the common stampede. Of course, we discredited the truth of the list, and Heintzelman broke out with, "By —— ——, it's all a lie! Every mother's son of you will be cashiered." We all felt he was right, but, nevertheless, it was true, and we were all announced in orders as brigadier generals of Volunteers.

Another name on this list was that of an officer serving far to the west who had nothing to do with the Bull Run disaster: Ulysses S. Grant. Ten days after receiving his own promotion, Sherman opened a note from a more senior brigadier general, Robert Anderson, commander of the Union force that had, after surrendering Fort Sumter, been allowed to come north. Early in his army career, Sherman had served under then-Captain Anderson, who found him impressive: when they met now in Washington, Anderson told Sherman that, as Sherman recounted it, he had been "offered the command of the Department of the Cumberland, to embrace Kentucky, Tennessee, etc., and that he wanted . . . me as his right hand." This led quickly to a meeting between the two officers and President Lincoln, during which Sherman made a prophetic request. "In this interview with Mr. Lincoln, I explained to him my extreme desire to serve in a subordinate capacity, and in no event to be left in a superior command. He promised this with promptness, making the jocular remark that his chief trouble was to find places for the too many generals who wanted to be at the head of affairs, to command armies, etc."

Sherman was indeed on his way to Kentucky, but his request to remain as a second in command betrayed a lack of self-confidence. Perhaps under there was still the boy whose father had died when he was nine, the boy sent to a prominent family by whom he had always felt overshadowed, or it could be that the disaster at Bull Run made him wish that there would always be someone higher in command to take the blame if things went wrong.

Arriving in Louisville, Kentucky, Sherman threw himself into the work of assisting Anderson in trying to assemble a new army in a new theater of war. Here he now found more confusion, in a border state of great strategic importance with a population whose loyalties were mixed. There was little fighting, but Sherman struggled with shortages of trained personnel, weapons, and supplies, an ill-organized command structure, no clear picture of when and where additional troops would arrive, and a volatile political situation.

Sherman began to exaggerate things; there were indeed Confederate spies and sympathizers about, but he saw the Kentuckians as "nearly all unfriendly." Military intelligence was poor on both sides: on the same day that Sherman reported that he had four thousand men to oppose a force led by Simon Bolivar Buckner that he estimated to number fifteen thousand, Buckner was telling his superiors that the six thousand men he actually had could easily be defeated by the thirteen thousand he felt certain were with Sherman. On October 5, Sherman wrote his brother John, "I'm afraid you are too late to save Kentucky. The young active element is all secession, the older stay at homes are for Union & Peace. But they will not take part." To Ellen he wrote the following day, "I don't think I ever felt so much desire to hide myself in some obscure place, to pass the time allotted to us on earth, but I know full well that we cannot if we would avoid the storm that threatens us, and perforce must drift on to the end. What that will be God only knows."

He soon found his deepest fear realized: Anderson, in poor health since his ordeal at Fort Sumter and overwhelmed by the night-and-day task of trying to organize this new Department of the Cumberland, resigned from his command on October 8 and went home. Sherman wrote that Anderson "said he could not stand the mental torture of his command any longer, and that he must go away or it would kill him . . . I had no alternative but to assume command, though much against the grain, and in direct violation of Mr. Lincoln's promise to me."

Already under the strain of constant work, Sherman became increasingly apprehensive about the balance between the unreliable forces at his disposal and the unknown numbers of Confederates opposing him. As he assessed the placement of federal forces along the Union's east-to-west front from the Atlantic to the Mississippi River, a distance of 825 miles, he found that of the 175,000 men defending this line, he had a total of 14,000 to cover approximately a third of that area. He believed himself to be facing 55,000 Confederates.

Going completely out of the chain of command, Sherman, who had earlier written Ellen that he intended "to meddle as little as possible with my superiors, and to give my opinion only when asked for," sent a telegram to President Lincoln. It said in part, "My own belief is that Confederates will make a more desperate effort [to] join Kentucky [to them] than they have for Missouri. Force now here or expected is entirely inadequate[.] The Kentuckians instead of assisting, call from every quarter for protection against local secessionists." It closed with the one-word imperative, extraordinary to be coming from a brigadier general to the commander in chief: "Answer." This produced a response, also out of any normal chain of command, from Secretary of the Treasury Salmon P. Chase, one of Senator John Sherman's political allies. Chase told him that keeping Kentucky in the Union was indeed crucial, but that Lincoln thought Sherman already had enough troops. Sherman replied, "I am sorry if I offended the President, but it would be better if all saw things as they are, rather than as we would they were."

As his anxiety mounted, Sherman sent Ellen letters so pessimistic that she wrote back, "Do write me a cheerful letter that I may have it to refer to when the gloomy ones come." To this, Sherman answered, "How any body can be cheerful now I cant tell . . . Give my love to all at home and tell Willy that I am very anxious to leave him a name of which he will not be ashamed if the tools are furnished me for the task to which I am assigned."

On the same October day that Sherman wrote Ellen he feared he might leave his son Willy a shameful legacy, Brigadier General Ulysses S. Grant, now commanding the military district headquartered at Cairo, Illinois, a grimy, bustling port located where the Ohio River entered the Mississippi, issued orders referring to "our Gun Boat Fleet." These were flat-bottomed paddle-wheeler riverboats, each with two tall side-by-side funnels, that had cannon poking out of their slanted dark armor super-structure; the men called them "mud turtles." Officers and sailors of the United States Navy manned these ships. The other vessels now at Grant's disposal, to carry troops and supplies, were the colorful riverboat steam-ers of the type immortalized by Mark Twain, each also with two funnels, some propelled by one large paddle wheel at the stern, and others with a paddle wheel on each side.

In contrast with Sherman's recent war experience, which began with the stunning rout at Bull Run and was continuing with what he saw as an impending disaster in Kentucky, Grant was having a varied and produc-tive apprenticeship in command. He had sent his son Fred home to Julia after a week of his comradeship as he led his Twenty-first Illinois west. Rather than being ordered into battle, Grant found himself peacefully

encamped with his regiment at different points in northern Missouri, en-
suring that the population did not take up arms against the Union. This
gave Colonel Grant time to train his regiment. To improve his men's already good morale, he organized a group of mail wagons to serve his
command alone, which increased the speed of communications between his soldiers and their families. At this point, Grant still believed
that the war would end within nine months, and that his time as a
colonel of Volunteers would be only an episode in his life. In answer to a
letter from his father, who asked Grant if he would not be wise to consider staying in the army as a career, he answered, "You ask if I should
not like to go into the regular army. I should not."

During this quiet period, James Crane, chaplain of the Twenty-first,
was sitting in a tent reading a newspaper when he came across Grant's
name in a list of newly promoted brigadier generals of Volunteers. Grant
said that he had no idea this was coming and commented that it must be
"some of Washburne's work." So many military matters were intertwined
with politics: Lincoln, who never forgot his original power base of
Illinois, had granted the Illinois congressional delegation the right to
appoint six brigadier generals of Volunteers—two more than he appor-
tioned to any other state. That gave Republican Representative Elihu
Washburne, Grant's congressman, the opportunity to urge his fellow Illinois representatives to include the man he had spotted as an obscure former Regular Army captain. This same list of promotions to brigadier
general of Volunteers, all backdated to May 17, contained the name of
William T. Sherman, with Sherman being senior to Grant because he
graduated from West Point three years before Grant did.

With this promotion, Grant was suddenly given important responsibilities. First he was sent to St. Louis to confer with Major General John
Frémont, a most interesting figure who had just been assigned to command the Department of the West—the critical and complicated theater
of war that had the Mississippi River at its heart.

Here was yet another case of a politically based military appointment.
Frémont (whom Sherman also went to see in his fruitless quest for reinforcements and the formation of a cohesive strategy, concluding that "I
could not discover that he was operating on any distinct plan") was a
man of considerable accomplishments and checkered background, but
he had no experience commanding large bodies of soldiers. Not a West
Pointer, at the age of twenty-five Frémont had been appointed as a sec-

ond lieutenant in the Army's Topographical Engineers. In that capacity, sometimes using Kit Carson as a scout, he made the first maps of the Oregon Trail, explored the Sierras, discovered Lake Tahoe, and later gave the name Golden Gate to the entrance to San Francisco's harbor. His prewar army career had ended in a court-martial for disobedience of orders; the military tribunal handed down a sentence that was remitted by President James K. Polk, but it was an affair that ultimately forced him to resign.

Known to the American public as "the Pathfinder," in civilian life Frémont had become one of California's first two senators and was the Republican Party's first candidate for president, being beaten in 1856 by the Democrats' Buchanan. Twenty years before the war began, he had married the attractive and ambitious daughter of Senator Thomas Hart Benton of Missouri; Frémont's national reputation, and his alliance with a family of influential Republican politicians, motivated Lincoln to entrust him with the complexities of the Department of the West.

In this first effort that Grant and Sherman made to establish themselves in their new western commands, Grant fared better with Frémont than Sherman did. This was partly because Frémont, who sent his wife to Washington to ask Lincoln for more troops for his own command, was intent on making a name for himself by thrusting down the Mississippi River rather than in diverting much of his strength to help Sherman defend neighboring Kentucky. After an initial Union defeat at Wilson's Creek in Missouri for which Grant had no responsibility, Frémont ordered Grant to organize an effective defense of Jefferson City.

When Grant did that and no Confederate attack materialized, Frémont then ordered him to go to the Union riverfront headquarters at Cairo, Illinois, and prepare to lead offensive actions from there. In picking Grant over a number of other brigadier generals, Frémont overruled those on his staff who reminded him of Grant's old reputation for drunkenness, and had only one fault to find: Grant was wearing a civilian suit, possibly the one he had worn most of the past year. Frémont told his chosen general to get into uniform. Chaplain Crane described how Grant obeyed that order: "He usually wore a plain blue [enlisted man's] blouse coat, and an ordinary black felt hat, and never had about him a single mark to distinguish his rank." (In fact, Grant sometimes wore pinned-on shoulder straps that had emblems of rank.)

The port where Grant had his headquarters in a hotel that the correspondent of the London *Times* found "almost untenable by reason of heat and flies" was on a strategically located south-pointing peninsula. To its west, the Mississippi moved downstream from St. Louis. To its east, the Ohio River ran down from Cincinnati, passing Kentucky's riverfront cities of Louisville and Paducah, and joined the Mississippi at Cairo. Back up the Ohio River near Paducah were the entrances to the Tennessee River and the Cumberland, both of which flowed north from hilly country to empty into the Ohio, resulting in a situation in which an advance into the South along those rivers had to be made by going upstream. (The war had turned Cairo into a rip-roaring Western town: in his General Orders No. 5, issued within a week of taking up his headquarters there, Grant deplored what he found: "It is with regret that the Genl Comdg sees and learns that the closest intimacy exists between many of the officers and soldiers of his command; that they visit together the lowest drinking and dancing saloons; quarrel, curse, drink and carouse generally on the lowest level of equality . . . Discipline cannot be maintained where the officers do not command respect and such conduct cannot insure it.")

Taking the military initiative under the authority given him by Frémont, Grant first quickly and bloodlessly seized Paducah, located thirty-two miles to the east of him at the point where the Tennessee River flows into the Ohio. Occupying the city on the morning of September 6, he issued a proclamation to its citizens in which he said, "I have nothing to do with opinions. I shall deal only with armed rebellion and its aiders and abbetors [*sic*] . . . The strong arm of the Government is here to protect its friends, and to punish only its enemies." (Grant's entrance into Kentucky ended the state's attempt at neutrality and brought war to the commonwealth.) Leaving a subordinate commander and two regiments to occupy Paducah, Grant was back in Cairo by late afternoon, ready to concentrate on the many matters involved in preparing to take the war into Confederate territory down the Mississippi.

This daily work of gathering forces and planning for an offensive brought Grant into several activities new to him. In his headquarters beside the Mississippi, he had daily contact with the officers of the United States Navy who commanded the "mud turtle" gunboats, and he also conferred with the captains of the paddle-wheeler riverboats that would be needed as transports and cargo vessels to support landings, crossings, and other movements along the shores of the Mississippi, Tennessee, and Cumberland rivers. Vessels carrying supplies for the Confederacy also plied these rivers, at some distance from Cairo, and on September 9 he reported the capture of three Confederate "Steamers . . . prizes just brought into this port by Gun Boat . . . The [civilian] officers and crew will be detained as prisoners until instructions are received from St. Louis what disposition to make of them."

In another aspect of his varied responsibilities, Grant became involved with intelligence activities: he acted on information received from a former Russian army officer who was a spy in the Confederate stronghold of Memphis, and read telegrams from Frémont that were sent from St. Louis in Hungarian and translated back into English at Grant's headquarters, on the assumption that even if the messages were intercepted the Confederates had no one who could read them. Twice in one week in September, Grant had to ask Frémont for money "required here to pay for secret services."

Despite his growing importance and the constant demands on his time, Grant thought frequently of home. The closing passage in one of his letters to Julia echoed Sherman's concern for what the world might think of his performance in the campaign to come.

> Remember me to all in Galena. Kiss the children for me and a hundred for yourself. You should be cheerful and try to encourage me. I have a task before me of no trifling moment and want all the encouragement possible. Remember that my success will depend a greatdeel [sic] upon myself and that the safety of the country, to some extent, and my reputation and that of our children greatly depends upon my acts.

Interestingly enough, the sequence of events that would soon draw Grant and Sherman together was caused in good part by ethical questions concerning General Frémont. When Sherman went to St. Louis in his vain effort to strengthen his shaky new Department of the Cumberland and bring it into concerted action with Frémont's Department of the West, he found the famous Californian surrounded by several San Francisco businessmen Sherman remembered from his days there as a banker. The man who ushered him in to see Frémont was a recently commissioned major Frémont had brought onto his staff: Isaiah C. Woods, who had been head of the San Francisco branch of the St. Louis bank whose failure had started the run on the other eighteen banks in the city, hurting all of them and causing six to collapse. Of Frémont's making Woods his commissary of subsistence, Sherman wrote Ellen that "Woods should not be appointed to an office of Trust, when money is to be handled." The next man he saw was another San Francisco banker, Joseph Palmer, a major contributor to Frémont's political campaigns whose fraudulent handling of state and federal funds entrusted to him caused his bank to close its doors for good. Another of Frémont's advisers was Abia A. Selover, an investor in mines and real estate owned by Frémont. At the hotel where Sherman stayed, he saw "old Baron Steinberger, a prince among our early California adventurers . . . His presence in St. Louis recalled the maxim, 'Where the vultures are, there is a carcass close by.' " Rounding out the picture was a Mormon from California named Beard, who had been awarded the contract for building a line of fortifications around the city. Sherman did not say that Frémont was involved in conflicts of interest but only that he "had drawn to St. Louis some of the most enterprising men in California."

News of what Sherman and others had seen and suspected reached Washington. Frémont had failed to achieve the military success Lincoln expected, and he had also issued an unauthorized political proclamation harmful to Lincoln's efforts to win over the border states of Delaware, Maryland, Kentucky, and Missouri. In it Frémont announced the confiscation of all "real and personal" property owned by Confederate sympathizers in Missouri, language that included the concept that slaves could be taken from their owners—something Lincoln intended to accomplish in time but not an idea he wished at the moment to force upon slaveholders who might remain neutral. Now came these reports of possible corrupt dealings involving public funds. Secretary of War Simon Cameron set out for St. Louis to investigate, accompanied by Lorenzo Thomas, adjutant general of the army.

On their return east, talking freely of the military disappointments, political ineptitude, and unresolved questions of disbursement that soon caused Frémont to be relieved of command, they met with Sherman in his rooms at the Galt House hotel in Louisville. Had it not been for Frémont's deficient performance and rumors of a version of the spoils system, the secretary of war would not have been within five hundred miles of Louisville, and Sherman would not have had the opportunity to see him face-to-face. (Cameron had lost his brother James, colonel of the Seventy-ninth New York, the Highlanders, when that officer was killed while under Sherman's command at Bull Run, and Cameron also had approved the Highlanders' request to be removed from Sherman's command after the alleged incident in which Sherman had some men of that regiment ejected from the shelter of a barn during the Bull Run retreat so some horses could be stabled there.)

By the time of this meeting, Sherman, not outnumbered but thinking he was, had become so worried about the situation in his large area of operations that he could be found pacing the corridors of the hotel at all hours, smoking eight to ten cigars a night, and waiting for dispatches at the telegraph office at three a.m. He drank too much; his hands sometimes shook. Sherman's experiences with the press in San Francisco had given him a permanent hostile mistrust of reporters, and he banned them from his headquarters. When journalists found the opportunity to ask him questions, he replied with a snarl, and on one occasion had a reporter jailed for disobeying his order that the man stay out of military camps. (When Sherman heard that the Confederates had shot two Northern reporters they considered to be spies, he expressed his pleasure, and said, "Now we'll have news from Hell by noon.")

As Cameron and General Thomas entered Sherman's rooms on October 17, they were accompanied by six or seven reporters, some from local papers and some from the East who were traveling with the secretary of war. After the hotel manager sent in what Sherman described "as a good lunch and something to drink," Cameron, who had arrived feeling sick, lay on Sherman's bed and said, "Now, General Sherman, tell us of your troubles." When Sherman remarked that he felt uneasy discussing military matters "with so many strangers present," Cameron answered expansively, "They are all friends, all members of my family, and you may speak your mind freely and without restraint."

Sherman stepped to the door, locked it, and started talking. He described the Union defenses in that part of Kentucky as being so weak that if Confederate general Albert Sidney Johnston chose to do so, "He could march to Louisville any day."

As Sherman described that moment, "Cameron exclaimed, 'You astonish me! Our informants, the Kentucky Senators and members of Congress, claim that they have in Kentucky plenty of men, and all they want are arms and money.'"

Sherman pressed on, describing the situation in the darkest terms. Holding up a large map of the United States,

I argued that, for the purpose of defense, we should have sixty thousand men at once, and for offense, should need two hundred thousand, before we were done. Mr. Cameron, who still lay on the bed, threw up his hands and exclaimed, "Great God! Where are they to come from?" I asserted that there were plenty of men at the North, ready and willing to come, if he would only accept their services . . . We discussed these matters fully, in the most friendly spirit, and I thought I had aroused Mr. Cameron to a realization of the great war that was before us, and was in fact upon us. I heard him tell General Thomas to make a note of our conversation, that he might attend to my requests on reaching Washington. We all spent the evening together agreeably in conversation.

That is the way it seemed to Sherman. Although Cameron sent telegrams from Louisville ordering that additional forces be sent to reinforce Sherman, on his way back to Washington Cameron told reporters at Harrisburg, Pennsylvania, that Sherman was "absolutely crazy." Thomas soon wrote a report of the meeting at the Galt House, something that was supposed to be a confidential War Department memorandum. On October 30, the *New York Tribune*, one of whose reporters had been at the Galt House meeting and had since been given an unauthorized look at Thomas's report, published an article that made no distinction between Sherman's estimate that he needed sixty thousand men for defense but that it would take two hundred thousand to mount and sustain a successful long-range offensive. The piece said only that when Cameron asked him how many men he had to have, Sherman "promptly replied 200,000."

Sherman saw the handwriting on the wall. Still hoping to salvage his disintegrating reputation, on November 1 he wrote Ellen that he was "riding a whirlwind unable to guide the Storm," and added, "God knows that I think of you and our dear Children all the time, and that I would that we might hide ourselves in some quiet corner of the world." He told her that "the idea of going down to History with a fame such as threatens me nearly makes me crazy, indeed I may be so now." Two days later he received a telegram sent from Washington by General George McClellan, who had just succeeded the aged and retiring Winfield Scott as general in chief of the United States Army, asking him to set forth the exact situation in Kentucky.

Sherman stuck to his assessment but saw the end coming for him as commander of the Department of the Cumberland. In a flurry of communications, he requested that McClellan relieve him of command, but his ordeal was not over. On November 5, McClellan sent him a letter saying that the highly regarded Don Carlos Buell would relieve him, but the letter took time to arrive, as would Buell. The following day, when Ulysses S. Grant was embarking three thousand soldiers aboard ships for his first battle of the war, an attack on the Confederate positions twenty miles south of Cairo at the riverfront town of Belmont, Missouri, McClellan sent Sherman a telegram asking for daily reports on all military affairs in Kentucky. While waiting for Sherman's replacement to arrive, the Union Army's commander was placing Sherman under close and mistrustful scrutiny. McClellan even quietly sent to Louisville Colonel Thomas M. Key of his staff, with instructions to observe Sherman's behavior; after some days, Key reported that Sherman was close to a nervous breakdown. In Washington, rumors about Sherman grew: at one point Assistant Secretary of War Thomas W. Scott was heard to remark, "Sherman's gone in the head, he's looney."

At the same time that Sherman's position was dramatically deteriorating, there was a shift in army commanders that would affect Sherman and also reflected the political battles Lincoln faced. In St. Louis, Frémont was relieved of command and would be sent to the soon-to-be-created Mountain Department, consisting mostly of what had been the Department of Western Virginia. (During the war, the Union Army frequently reorganized its geographical departments and renamed armies operating in various areas.) Frémont's transfer, in effect a demotion, was caused in part by his military ineptitude as well as the suspicions of corruption that Secretary of War Cameron had been sent out to investigate, but it was also an early example of the never-resolved tensions between Lincoln and the Radical faction of the Republicans in Congress. Both Lincoln and the Radicals shared the war aim of restoring the Union, but the Radical Republicans wanted the earliest possible end to slavery everywhere, while Lincoln continued to defer the emancipation issue in the interest of trying to keep the slaveholding border states of Delaware, Maryland, Kentucky, and Missouri in the Union. When Lincoln had rescinded Frémont's declaration that the slaves of Missouri were free, the Radicals were enraged, and now Lincoln added to their anger by this downgrading of one of their favorite generals.

The man taking Frémont's place and now commanding both Sherman and Grant was Henry Wager Halleck, a pop-eyed, gray-haired, portly forty-six-year-old, who before starting West Point earned a Phi Beta Kapa key at Union College. A scholar who continued his reading and writing while in the army, at the age of thirty-one he had published an important military textbook, *Elements of Military Art and Science*. He resigned from the army with the rank of captain, doing it in the same year that Grant did, and became a lawyer whose books on international law and land-title issues were widely praised. In San Francisco he served as a railway president and director of a quicksilver mine. The recently retired Winfield Scott had brought this capable administrator back into the army at the beginning of the war, giving him a Regular Army commission as a major general. What remained to be seen was how Halleck, a greatly ambitious man who was no stranger to intrigue and whose Regular Army commission made him one of the most senior Union officers, would handle Grant, Sherman, and the other generals in the Mississippi River theater of operations he now commanded.

On November 8, back in Lancaster, Ohio, Ellen Sherman opened a telegram from Louisville for her prominent father, who was in Washington. It was from Sherman's aide, Captain Frederick Prime, and said: "Send Mrs. Sherman and youngest boy down to relieve General Sherman's [*sic*] and myself from the pressure of business—no occasion for alarm." Ellen, always aware of the history of mental instability in her husband's family, immediately made the fourteen-hour trip by railroad to Louisville, taking along her older brother Philemon, a lawyer, and both boys, Willy and Tommy. When they arrived at three in the morning, Philemon said, they came upon Sherman "in a great, barnlike room with blazing lights, with a lounge at one end, on which he tried from time to time to catch snatches of sleep, and messengers rushing in at all hours bringing details of disaster or threat."

The next day Ellen wrote Sherman's brother John an agonized letter: "Knowing insanity to be in the family and having seen Cump in [*sic*] the verge of it once in California, I assure you I was tortured by fears, which have been *only in part relieved* since I got here . . . I have not been here long enough to judge well his state of mind. He wrote me that he felt almost crazy, and I find that he has had little or no sleep for some time." She added that he had been eating almost nothing, and that his officers, worried about him, had told her that "he thinks the whole country is gone irrevocably & ruin and desolation are at hand." The immensely loyal John Sherman immediately rushed to see his brother in Louisville. Within five days, Sherman's replacement, General Buell, arrived, and John and Ellen felt that they could return to Lancaster, leaving her somewhat calmer husband to continue his military duties.

After a few days spent showing Buell around, days during which Buell telegraphed McClellan that he saw no reason to expect a significant Confederate move on that front, Sherman received orders to report to Halleck in St. Louis. When they had served together in the army in California during the time of the Mexican War, a dispute had arisen between them about the proper placement of some coast artillery positions, and they had deliberately not spoken to each other for years, including the later period when they had both been businessmen in San Francisco. Halleck greeted Sherman in a friendly way, but intended to give him only a perfunctory task that would greatly reduce the pressures he had been feeling. Halleck sent him to inspect regiments placed in quiet areas to the west, but made the mistake of authorizing Sherman to take actual command of those regiments if he felt it necessary.

At Sedalia, Missouri, Sherman decided that General John Pope, the commander of the units he was inspecting, had his forces spread so wide that Confederate General Sterling Price, who was not in fact advancing, could fall upon and destroy them. That was a future possibility, but when Sherman assumed command and started ordering Pope's regimental commanders to consolidate their positions, Pope fired off a strong protest to Halleck, who sent his department's medical director out to judge Sherman's condition. The doctor reported that Sherman was in a state "of such nervousness that he was unfit for command." Halleck telegraphed Sherman to make no further movements of troops. At the same time, Ellen Sherman arrived in St. Louis, alone and terribly worried. She was now also concerned about the effect on her husband of additional newspaper reports criticizing him. Ellen went to consult Halleck, who had his adjutant send Sherman a message that his wife was at headquarters, and added that "General Halleck is satisfied, from reports of scouts received [in St. Louis] that no attack on Sedalia is intended. You will therefore return to this city, and report your observations on the condition of the troops you have examined."

When Sherman saw Ellen in St. Louis and learned that she intended to take him home for a rest, he at first resisted the idea, but when Halleck told him to take a twenty-day leave, Sherman recognized that a gently worded order had just been issued. As soon as Halleck saw Sherman off with the comforting observation to Ellen that a good workhorse needed an occasional rest in the barn, he expressed his real views in a message to McClellan in Washington. Halleck told the general in chief that officers in Sedalia had sent him word that Sherman was "completely 'stampeded,' and was 'stampeding' the entire army . . . I am satisfied that General S's physical and mental condition is so completely broken by labor and care as to render him for the present entirely unfit for duty. Perhaps a few weeks' rest will restore him . . . in his present condition it would be dangerous to give him a command here." To his wife, Halleck wrote that Sherman had without doubt "acted insane."

At the time Sherman returned home with Ellen to Lancaster, in a state of near collapse, the public was still hearing much of Grant's recent November 7 attack on Belmont, twenty miles down the Mississippi River from the headquarters at Cairo. Hailed in the Northern press as a victory, this attack was actually much less than that. Grant had no authority to bring on this battle, but this was not the Grant who sat indifferently in his family's leather goods store the year before. Weeks earlier, he had written Julia that "I would like to have the honor of commanding the Army that makes the advance down the river, but unless I am able to do it soon cannot expect it. There are too many Generals who rank me that have commands inferior to mine for me to retain it." In short, despite Frémont's confidence in him, some other general senior to Grant might appear, entitled to lead the big offensive. Even the soon-to-be-relieved Frémont, who had suffered the unexpected loss of Lexington, Missouri, 180 miles west of St. Louis on the Missouri River, now wanted Grant to delay any thrust down the Mississippi until he recaptured that strategic mercantile center. The moment Grant heard of Frémont's being relieved of command, he decided not to wait for any successor to Frémont to appear and, without orders to do so, launched his attack on the Confederate positions at Belmont.

At first things went smoothly. While a Union column made a large demonstration on the eastern side of the river, advancing against the Confederate general Leonidas K. Polk, Grant, after keeping a force of three thousand men hidden aboard ships overnight near Belmont on the western bank, poured his troops ashore and smashed through the Confederate ranks that quickly assembled to oppose him.

Grant, who had a horse shot from under him early in the bloody fighting, described his men's performance in the initial Union assault, which forced the Confederates to retreat through their camp and hide out of sight along the steep riverbank, demoralized and ready to surrender: "Veterans could not have behaved better than they did up to the moment of reaching the rebel camp . . . The moment the camp was reached our men laid down their arms and commenced rummaging the tents to pick up trophies. Some of the higher officers were little better than the privates. They galloped about from one cluster of men to another and at every halt delivered a short eulogy upon the Union cause and the achievements of the command." Even this account did not capture the festival atmosphere. Amid the looting of the enemy camp, the Stars and Stripes was raised on the enemy flagstaff, and at its base Union regimental bands played patriotic airs while soldiers cheered.

The Confederates, still out of sight along the riverbank and ready to surrender a short time before, now counterattacked and surrounded Grant's men. "The alarm 'surrounded' was given. The guns of the enemy and the report of being surrounded, brought the officers and men completely under control. At first some of the officers seemed to think that to be surrounded was to be placed in a helpless position, where there was nothing to do but surrender. But when I announced that we had cut our way in and could cut our way out just as well, it seemed a new revelation to officers and soldiers. They formed line rapidly and we started back to our boats."

The next half hour nearly ended Grant's participation in the war. With most of his men back on their ships after a retreat in which a thousand Union muskets were lost or thrown aside, Grant rode around near the shore to see that no one was left behind. At one point, coming through a cornfield, "I saw a body of troops marching past me not fifty yards away. I looked at them for a moment and then turned my horse toward the river and started back, first in a walk, and when I thought myself concealed from the enemy, as fast as my horse would carry me." The nearest Confederates had neither seen nor heard Grant, but from a different vantage point the Confederate general Polk, having crossed the river with reinforcements, spotted this lone Union officer. Polk said to his men, "There is a Yankee; you may try your marksmanship on him," but no one did.

Grant's problems did not end with that; reaching the steep riverbank, well above the water's edge, he found that every man of his expedition had hastily embarked, and the vessels had all cast off from the shore. "I was the only man of the National army between the rebels and our transports. The captain of a boat that had just pushed out recognized me and ordered the engineer not to start the engine; he then had a plank run out for me. My horse seemed to take in the situation . . . [He] put his fore feet over the bank without hesitation or urging, and with his hind feet well under him, slid down the bank and trotted aboard the boat twelve or fifteen feet away, over a single gang plank." Grant dismounted on the boat's deck and went into a cabin where he found a sofa. Lying down for a minute's rest, he then rose to go back out on deck. As he stepped away from the sofa, a Confederate musket ball came through the cabin wall and shattered the wooden frame of the sofa at the place where his head had just been.

On the way back up the river to Cairo, Grant sat by himself, clearly wishing to be left alone. He later admitted to his departmental surgeon that at one point in the battle he lost control of his forces, and he already knew he had not won the Union victory so many wanted. If he had, his men would not have been steaming back up the river but would have remained in possession of the positions they attacked. The Union losses proved to be 607 killed, wounded, or captured; the Confederate loss was 642. As soon as Grant reached his headquarters, he began sending off reports that made this large-scale raid sound better than it was and estimated that "the enemies [sic] loss must have been two or three times as great as ours." Then and later, he tried to clothe his attack as having been part of a larger strategic plan to forestall a Confederate advance; no one reading these first reports could have discerned that Grant acted entirely on his own. The following day, in his General Orders, Grant referred to himself and the battle in these words: "The General Comdg. this Military District, returns his thanks to the troops under his command . . . It has been his fortune, to have been in all the Battles fought in Mexico, by Genls. Scott and Taylor, save *Buena Vista*, and never saw one more hotly contested, or where troops behaved with more gallantry."

Starved for a victory, the Northern press decided that one had occurred. *The New York Herald* called the action at Belmont a success "as clear as ever warriors gained," and *The New York Times* said that "the success of the brilliant movement is due to Gen. Grant." In fact, the good news for the North was not the mixed results at Belmont. The first part of the good news was that the Union Army had a general who was eager to attack the enemy. The Battle of Belmont was the foundation of Grant's reputation for taking the war to the enemy, but it also demonstrated something subtler but equally important. Grant never escaped his image as a bluff soldier who smashed ahead when he could, but in him the Union had a man of the West, a man who had spent years near the Mississippi and viscerally understood the strategic importance of that river and its tributaries. In addition to that, while Grant was a general and not an admiral, he saw that he could work with warships and transport vessels to use a mighty river as part of a battlefield.

In Lancaster, Sherman's health and spirits began to improve during his twenty-day leave, but the press was not yet through with him. On December 9, *The New York Times* spoke of him as a general "whose disorders have removed him, perhaps permanently, from his command." The impact of this on Sherman, Ellen said in a letter she quickly wrote to his brother John, was that "it seemed to affect him more than anything that has hitherto appeared." Once again, Senator John Sherman went to work on behalf of his brother, this time having a long conversation with President Lincoln at the White House. Gently, Lincoln went through a recitation of facts, including, as John told Ellen, Lincoln's statement that "then came telegraphic dispatches from Cump that were unaccountable." John wrote that he himself was "well convinced that Cump made serious mistakes in Ky. . . . It is idle for him—for you or any of his friends to overlook the fact that his own fancies create enemies & difficulties where none exist." As for Lincoln's view of Sherman, John said that "the President evinced the kindest feelings for him & suggested that he come here on a visit," to which John had replied that Sherman was coming to the end of his leave back in Ohio and would be returning to Halleck's command in St. Louis.

Two days after the piece in *The New York Times* that so upset Sherman, the *Cincinnati Commercial* came out with an article on Sherman's time in Kentucky:

GENERAL WILLIAM T. SHERMAN INSANE

The painful intelligence reaches us, in such form that we are not at liberty to disclose it, that General William T. Sherman, late commander of the Department of the Cumberland, is insane. It appears that he was at the time while he was commanding in Kentucky, stark mad . . . He has of course been relieved altogether from command. The harsh criticisms that have been lavished on this gentleman, provoked by his strange conduct, will now give way to feelings of deepest sympathy for him in his great calamity. It seems providential that the country has not to mourn the loss of an army through the loss of mind of a general into whose hands was committed the vast responsibility of the command of Kentucky.

This eclipsed all that came before. From Lancaster, Ellen wrote John:

Nature will paint to your mind & heart what I felt when Tommy came in to us just now to say that a boy had told him that it was published in the paper that "Papa was Crazy."

I cannot persuade Cump to go to Washington. He is feeling terribly about this matter. If there were no kind of hereditary insanity in your family & if his feelings were not already in a marked state I would feel less concern about him but as it is I cannot bear to have him go back to St. Louis haunted by the spectre, dreading the effects of it in any apparent insubordination of officers and men.

The day after the *Commercial* story appeared, Sherman poured out his heart in a letter to his father-in-law and stepfather Thomas Ewing, who was on one of his frequent trips to Washington as one of the nation's most prominent lawyers. Sherman always felt both a great sense of obligation to the family who had taken him in when his father had died when he was nine, and a desire to impress Thomas Ewing in particular with whatever he could achieve. "Sir," he began, "Among the keenest feelings of my life is that arising from a consciousness that you will be mortified beyond measure at the disgrace which has befallen me—by the announcement in the Cincinnati Commercial that I am insane." Then followed twenty-five hundred words justifying his actions and pessimistic views of the war.

As Sherman headed back to St. Louis by himself to report for duty and face whatever awaited him there, other newspaper reports around the country started to echo the Cincinnati story. Ellen began to defend him like a tigress. She sent her lawyer brother Philemon, who thought that Sherman was "distressed almost to death," to Cincinnati to demand a retraction, possibly with the threat of a lawsuit. The *Commercial* published

Philemon's point-by-point denials and corrections of much that its article said, but the Ewing-Sherman counterattack was just getting under way. As Sherman began his new assignment in St. Louis—Halleck assigned him to train the thousands of recruits at nearby Benton Barracks, where he could keep an eye on him while Sherman served in a position in which he was not engaging the enemy—he received a letter from Ellen. It began, "I feel desolate in my room now, without you, dearest Cump," and was followed by one in which she laid out the reasons for a lawsuit against the *Commercial*: Ellen felt the family would win; the attendant publicity would frighten off other newspapers from writing similar stories; a successful verdict would clear his name with readers throughout the nation. In a letter she wrote three days later, Ellen emphasized that both her father and Sherman's senator brother agreed on the course to be taken. "So now my dearest Cump, as Father so strongly recommends a suit & John concurs with him in judgment you will not refuse it when you know it to be my most earnest wish. Not because I feel so vindictive against the miserable Editors but because I believe it will be a complete vindication of you & it will enable us to discover who is in the conspiracy against you. Do not let Halleck or anyone else (who is not affected by it & can thus treat it with great indifference) induce you to overlook the request of one who suffers keenly with you & for you." As for Halleck, she felt that he had no intention of ever giving her husband a post more significant than the training of recruits. Halleck was now writing people that Sherman had been exhausted, certainly not insane, but Ellen thought that his words displayed "true lawyer-like ambiguity."

Ellen was by no means done. While Sherman suffered in his humiliation and expressed his disapproval of a lawsuit that would inevitably bring repetitions of the insanity stories, Ellen wrote President Lincoln, beginning with, "Mr. Lincoln, Dear Sir," and pursued her idea of a conspiracy, asking, "Will you not defend him from the enemies who have combined against him, by removing him to the army of the East?" When she had no answer to this from Lincoln, she went to Washington with her father, who was trying an important case there.

By the time the two of them were ushered in to see Lincoln, things had changed in helpful ways. Secretary of War Cameron was going off in the ambassadorial position of minister to Russia, being replaced by one of Thomas Ewing's good friends, Edwin M. Stanton. In addition, Ellen's influential father was in Lincoln's particularly good graces because of his wise counsel in settling the recent *Trent* affair, in which the United States Navy had nearly brought England into the war on the Southern side by stopping a British vessel to seize two Confederate emissaries bound for London. The meeting went well: Ellen wrote her husband that Lincoln "seemed very anxious that we believe that he felt kindly towards you. He and Father are great friends just now." After a most successful tour of Washington society, during which Ellen spread the word that the president had confidence in her husband, she returned to Ohio, ready to acquiesce in Sherman's desire to drop all lawsuits.

Out in St. Louis, Halleck was balancing his own interests. As commander of the Department of the Missouri, he had mixed feelings about his subordinate Grant; when they had occasion to confer, both of them felt ill at ease. Halleck saw Grant as an able and aggressive general, but he wanted to bring the favorable attention of the nation's leaders to himself, not Grant, and win the position of top commander in all of the Western theater. With no special friends or influence to help him in Washington, Halleck was mindful of recent letters from Senator John Sherman and former senator and secretary of the interior Thomas Ewing, requesting that the man who was the senator's brother and Ewing's son-in-law be given a second chance. He was also aware that Lincoln had asked John Sherman's ally, Secretary of the Treasury Salmon P. Chase, to keep an eye on the political and military situation in the Western military departments that included Halleck's command.

While Halleck was considering whether Sherman could again be trusted with important responsibilities, Sherman was still trying to come to terms with his recent crisis. On New Year's Day of 1862 he wrote to his wife:

> Dearest Ellen:
>
> Again I have failed to write you as promised. Again have I neglected the almost only remaining chain of love & affection that binds me to earth. I have attempted to write you several times but feared to add to the feelings that already bear on you too heavily. Could I live over the last year I think I would do better, but my former associations with the South have rendered me almost crazy as one by one all links of hope [of averting war] were parted.

He went on to refer to "having so signally failed in Kentucky," and added, "I am about in the same health as at Lancaster but the idea of having brought disgrace on all associated with me is so horrible to contemplate that I cannot really endure it."

If he had ever been less than appreciative of Ellen, that was certainly no longer the case. "I will try & be more punctual in writing you my Dearest wife who has been true & noble and generous & comforting always. That She should thus be repaid is too bad—and our Dear Children—may God in his mercy keep them in his mind, and not let them Suffer for my faults . . . Bless you and keep you as their guide till they care for themselves."

Three days later, he wrote his brother John, "I am so sensible of my disgrace in having exaggerated the force of our enemy in Kentucky that I do think I Should have committed suicide were it not for my children. I do not think that I can again be entrusted with command." Sherman went on to say that he felt he could at least be useful as a high-level paymaster and added, "Suppose you see McClellan and ask him if I could not serve the Government better in such a capacity than the one I now hold."

While he slowly came out of his blackest moods, Sherman was doing an excellent job training twelve thousand soldiers at Benton Barracks. His idea of a uniform was sometimes worse than Grant's; during one of his frequent inspections of the camp, wearing a completely unmilitary coat and a tall civilian hat, he found a soldier beating a mule. When the man ignored Sherman's order to stop, Sherman asked, "Do you know who I am? I am General Sherman." The miscreant answered, "That's played out! Every man who comes along here with an old brown coat and stove-pipe hat, claims to be General Sherman."

Halleck slowly began to bring Sherman into a closer relationship, at one point sitting down with him to study a map and discuss the right line of advance for a future offensive, while still resisting Grant's efforts to resume just such a campaign. During this time, while Grant's wife and children were visiting him for several weeks at Cairo, with Julia nursing him through a number of migraine headaches, Grant suddenly had the chance to pull off a bold move. As with Belmont, where he attacked at just the moment between Frémont's departure and the arrival of the new commander Halleck, Grant found his opportunity in the form of Lincoln's President's General War Order No. 1 issued on January 27, 1862.

This unprecedented order represented the frustrated Lincoln's determination to make his commanders take the offensive on every front. Lincoln's order gave every department a month in which to start advancing on the enemy and made it clear that generals who failed to move might be replaced. Among the forces specifically named as being expected to act were "the Army and Flotilla at Cairo."

Grant saw the mandate he needed. Halleck had been holding him back since Belmont, and a previous suggestion by Grant in a meeting with Halleck that he try to capture Fort Henry, eighty miles southeast of Cairo up the waterway formed by the Ohio and then the Tennessee River, "was cut short as if my plan was preposterous." Now he pressed Halleck for specific authorization to make that attack, and three days after Lincoln's order became public, Halleck telegraphed, "Make your preparations to take and hold Fort Henry." Within seventy-two hours, Grant organized twenty-three regiments totaling seventeen thousand men and all the supplies to sustain them, assembled a fleet of transports led by seven of the navy's gunboats, and headed off at six in the morning.

Halleck's previous reluctance turned to panic, coupled with grim determination to support this effort at all costs. Arranging for reinforcements for Grant if he needed them, and redeploying units to replace those now on vessels steaming with Grant, during forty-eight hours Halleck sent or received twenty-two telegrams involving the general in chief McClellan in Washington and Sherman's successor Don Carlos Buell at Louisville.

Among the officers Halleck sent to support Grant was James B. McPherson, an exceptionally able lieutenant colonel of engineers who had graduated from West Point nine years before and whose peacetime Regular Army career had included fortifying Alcatraz Island in San Francisco Bay. Still worried about the old army stories of Grant's drinking, Halleck gave McPherson added instructions to report unusual behavior.

Halleck did something else: in his flurry of activity, moving officers here and there to cover the quickly developing situation, he ordered Sherman to turn over the training of recruits to someone else and proceed to the riverfront city of Paducah, ready to send men and supplies forward to Grant. It was the most important assignment Halleck would ever make.

4

GRANT MOVES FORWARD,
WITH SHERMAN IN A SUPPORTING ROLE

Boldness, good luck, and the United States Navy served Grant well at Fort Henry. As his convoy went east through the February night, going up the Ohio and turning southeast into the Tennessee River at Paducah, dark clouds made the ships hard to see. There were Confederate spies at both their embarkation point in Cairo and at Paducah, but no warning messages reached Fort Henry. The defenders first knew of the impending attack when Grant's fleet appeared downriver in the morning, and his troops began to disembark on both shores some distance away. Fully alert and by no means intimidated, the Confederate commander, Brigadier General Lloyd Tilghman, rushed his thirty-four hundred men into their entrenchments and telegraphed General Albert Sidney Johnston to the northeast at Bowling Green, Kentucky, saying, "If you can reinforce [us] strongly and quickly, there is a glorious chance to overwhelm the enemy."

No Confederate reinforcements were dispatched, but as Grant moved his troops ashore all day in pouring rain, preparing to fight the next morning, he had no idea of how many enemy troops he faced or how many more might soon arrive. He had come to a full appreciation of what naval vessels could do in this theater of war, and he had both the right ships for the day and the right commander. Four of the seven gunboats were "mud turtle" ironclads mounting thirteen cannon apiece: these maneuverable little paddle-wheelers, 150 feet long and 50 feet wide, with flat-bottomed hulls that required only six feet of water in which to operate, could go virtually anywhere on the Mississippi and its network of large and small tributaries.

The man leading the naval force was Commodore Samuel Foote, who had entered the navy forty years before and sailed every ocean, fighting on several. As aggressive as Grant, in 1856 he became involved in the Arrow War in China. During that conflict the United States was neutral, but when Foote came under fire from Chinese artillery while his squadron evacuated American soldiers stationed at the treaty port of Canton [Guangzhou], he fired back, giving much more than he got. Sending his United States Marines and other landing parties ashore, he captured four forts and 176 Chinese cannon.

On the eve of battle, Grant wrote Julia that despite the rainy weather, "The sight of our camp fires on both sides of the river is beautiful, and no doubt inspires the enemy with the idea that we have full 40,000 men. Tomorrow will come the tug of war. One side or the other must rest tomorrow night in possession of Fort Henry." He closed with, "Kiss the children for me. Kisses for yourself. Ulys."

Commodore Foote's action at Canton six years before was the perfect rehearsal for Fort Henry. At thirty minutes after noon on February 6, having told his ship captains, "It must be victory or death," Foote opened his barrage amid a downpour. Within an hour, his total of fifty-nine cannon destroyed all but four of the Confederate guns, and the enemy commander ran up the white flag. Because it all happened so quickly, while one wing of Grant's force was struggling through deep mud to take up positions and encircle Fort Henry, a large part of the Confederate force escaped.

As befitted what was an inland naval victory, Commodore Foote accepted Fort Henry's surrender, but Grant had to work out its details with a Confederate captain named Taylor, who sized up Grant as being "a modest, amiable, kind-hearted but resolute man." As they got down to business, Taylor described one of the things that happened: "[A Union] officer came in to report that he had not yet found any papers giving him information about our forces, and, to save him further looking, I informed him that I had destroyed all the papers on the subject, at which he seemed very wroth, fussily demanding, 'By what authority?' Did I not know that I laid myself open to further punishment, etc., etc. Before I could fully reply, General Grant quietly broke in with, 'I would be very much surprised and mortified if one of my subordinate officers should allow information which he could destroy to fall into the hands of the enemy.'"

This had been a real if not large victory, with Grant's troops occupying the enemy positions and staying there. The Northern press, still thirsty for signs of large-scale progress, read much into it, the *New York Tribune* saying that "a few more events such as the capture of Fort Henry, and the war will substantially be at an end."

Grant had been thinking ahead. Within hours of Fort Henry's surrender on February 6, he remarked that while his forces had momentum in that area he thought he would move right over to Fort Donelson, a far stronger Confederate position twelve miles to the east, on the Cumberland River, which roughly paralleled the Tennessee. That night, probably in an effort to forestall Halleck from slowing him down, he wired Halleck's headquarters that he could accomplish this within two days, but the relentless rains inundated the roads and reduced his effort the next day to making a reconnaissance of the fortress, approaching within a mile of it on horseback accompanied by some of his staff and a body of cavalry. Grant remained determined to attack Fort Donelson at the earliest moment, rather than waiting for more regiments to support him; knowing that the Confederates would be speeding reinforcements there, he said of his own force, "I felt that 15,000 men on the 8th would be more effective than 50,000 a month later."

Despite Grant's desire "to keep the ball moving as lively as possible"—an expression he was to repeat, and one that characterized his aggressive approach to warfare—floods swamped the terrain between Forts Henry and Donelson, and the extreme February temperatures, which at one point dropped to twelve degrees and had his troops camping in the snow, made an early attack impossible. "You have no conception of the amount of labor I have to perform," Grant wrote on February 9 from his temporary headquarters at Fort Henry to his sister Mary, in Covington, Kentucky. "An army of men all helpless looking to the commanding officer for every supply. Your plain brother has, as yet, had no reason to feel himself unequal to the task and fully believes that he will carry on a successful campaign against our rebel enemy. I do not speak boastfully but utter a presentiment." The next evening, Grant's son Fred, now twelve, arrived at his headquarters, sent there by Julia to share again in his father's military life. Writing Julia that night, Grant told her that "I will let him remain here for a few days but will not take him into danger," and soon sent him home.

Within three more days, Grant's force had Fort Donelson surrounded by land, and Foote's gunboats, having gone back down the Tennessee to come up the Cumberland, were on the way to take part in the attack. The stakes were high, with a strategic position in the balance and both North and South yearning for a major victory. After some probing attacks on February 13 during which the Union troops suffered heavy losses, the following morning Grant had his reinforced army of twenty-seven thousand men closing in around Fort Donelson, with Foote's flotilla still some distance down the river. Behind formidable fortifications on a bluff that at places rose 150 feet above the water, with cannon strongly emplaced on three levels, close to twenty thousand Confederates prepared to repulse attacks from any direction.

As for what had been going on at Halleck's headquarters in St. Louis between the fall of Fort Henry and this moment, Grant recalled that Halleck, who never did congratulate Grant on that victory, "did not approve or disapprove of my going to Fort Donelson. He said nothing whatever to me on the subject." While Halleck had his staff working diligently to fulfill Grant's requests for troops and supplies, he was also deep in his game of military politics. In response to McClellan's telegraphed suggestion that either Halleck or Don Carlos Buell take over Grant's expeditionary force with a view to expanding it into an effort to take not only Fort Donelson, but go on up the Cumberland to capture Nashville, Halleck responded that "I have no desire for any larger command than I have now." Then he undercut Buell, whom he considered to be a rival, by pointing out that Buell was outranked in seniority by Sherman. Completely reversing himself on Sherman, whom he had a few weeks earlier described to McClellan as being "entirely unfit to command," Halleck said that Sherman was ready for an important assignment. On the same day, Halleck telegraphed Secretary of War Edwin M. Stanton a suggested solution to some promotion problems that would have worked to his advantage and would also have placed the headstrong Grant under an additional level of command. Nothing came of Halleck's ideas, and soon after this he moved Sherman to the position in which he had the responsibility of forwarding men and supplies to Grant.

From the Northern point of view, the major attack on Fort Donelson began shockingly. At three in the afternoon of February 14, as Grant watched from high on a distant riverbank, Commodore Foote brought his ships up the Cumberland, but this swift-running stream, with a narrow channel that reduced the Union ships' maneuverability and made them easier to hit, was a different body of water from the slower-moving Tennessee at Fort Henry, and here the Confederate cannon were protected by earth-and-log parapets sixteen feet thick. As the gunboats came on, opening fire a mile away, cannonballs from Fort Donelson started tearing into them. During the first volleys, shots hit the pilothouse of Foote's flagship *St. Louis*, killing the man at the wheel, wounding the commodore in the ankle, and destroying all of the steering mechanism. Hit fifty-nine times, the *St. Louis* drifted downriver away from the fight, and the Confederate barrage soon disabled or forced the other Union ships to abandon the battle, killing or wounding fifty-four sailors while the defenders lost not a single cannon. Coupled with the casualties suffered in the probing land attacks the previous day, by nightfall the effort to capture Fort Donelson had thus far failed.

Responding to a request from the wounded Foote that he come to his anchorage seven miles downstream to confer, Grant left his headquarters the next morning at first light, having ordered his senior commanders to stay in their positions besieging the fortress and make no attacks until he returned. As he rode off for this meeting on a spirited stallion named Jack, Grant later said of that moment, "I had no idea that there would be any engagement on land unless I brought it on myself." When he came ashore at noon from his meeting with Commodore Foote aboard the shattered *St. Louis*, "I met Captain Hillyer of my staff, white with fear, not for his personal safety, but for the safety of the National troops. He said the enemy had come out of his lines in full force and attacked and scattered [Brigadier General John] McClernand's division, which was in full retreat."

If ever Grant needed his skills as a horseman, this was the moment, and in this crisis he had the right horse for his seven-mile gallop up icy, rutted roads to the scene of action. By the time Grant approached Fort Donelson, all the Confederates had fallen back inside their own lines after bloody fighting because of the resistance they met from two of his three divisions, but fifteen hundred Union soldiers had been killed or wounded. When he came to McClernand's division, "I saw the men standing in knots talking in the most excited manner. No officer seemed to be giving any directions. The soldiers had their muskets, but no ammunition, while there were tons of it at hand." As Grant calmed the men, some of them mentioned that when the Confederate soldiers had come out, they had their knapsacks strapped on and had haversacks containing rations slung over their shoulders. To Grant, this suggested that the enemy had attacked not so much to destroy the Union besiegers but to clear an avenue of escape for a days-long flight during which they would need blankets and food. Turning to a colonel of his staff, Grant said, "Some of our men are pretty badly demoralized, but the enemy must be more so, for he has attempted to force his way out, but has fallen back: the one who attacks now will be victorious." Grant rode on, calling out to the soldiers he saw, "Fill your cartridge-boxes, quick, and get into line; the enemy is trying to escape." Grant said that "this acted like a charm. The men only wanted some one to give them a command."

Riding up to the commander of his best division, Brigadier General Charles F. Smith, a Mexican War veteran of forty years' service who had a flowing white mustache, Grant ordered him to make this crucial attack up one of Fort Donelson's slopes. "The general was off in an incredibly short time." Grant had the right general: ordering his men to use only their bayonets and riding at the head of his lead regiment, the foot soldiers of the Second Iowa, Smith kept his horse abreast of the man carrying the regimental colors. As the troops advanced through a constant fusillade of musket balls, the man carrying the flag fell. Another soldier picked up the flag and was soon hit and went down, and a third man snatched up the flag and was hit. The fourth man to take the Iowa banner forward, a corporal with the unmilitary name of Voltaire P. Twombley, was struck by a musket ball fired from so far away that it had only the force to knock him down. Getting to his feet, Twombley continued to carry the flag forward beside Smith: together, the general on his horse and the corporal with the flag made a huge target as they led the Iowans up the slope through a defensive network of felled trees under everheavier fire. The fifty-seven-year-old Smith kept raising his cap high and calling over his shoulder, "No flinching now, my lads, this is the way— Come on!" Twombley stayed right beside his general, winning the Congressional Medal of Honor for his part in the attack. As the Iowans broke through the enemy rifle pits at the crest of the slope, Grant's other divisions also attacked. By nightfall, the total of Union killed, wounded, and missing climbed to twenty-eight hundred, but the defenders were penned inside the remaining inner parts of their bastion. Grant wrote of that evening, "General Smith with much of his division, bivouacked within the lines of the enemy. There was now no doubt that the enemy must surrender or be captured the next day."

During the night, more than three thousand of the defenders escaped, including the two most senior generals and a cavalry force under the brilliant and dashing but then little-known Colonel Nathan Bedford Forrest. (Ever resourceful, Forrest had a foot soldier swing up on a horse behind each of his seven hundred cavalrymen, thus doubling the number of troops he led away from certain capture.) Despite these escapes, Grant had demonstrated great military insight.

Before daylight, a Confederate soldier emerged from the enemy earthworks carrying a white flag of truce and a letter. It came from the general left in command, Simon Bolivar Buckner, who had been a year behind Grant at West Point, served with him in Mexico, and loaned him money when he passed through New York City in 1854, arriving by ship and out of funds after resigning from the army in California. In language that frequently governed arrangements for surrender, Buckner proposed "the appointment of Commissioners to agree upon terms of capitulation of the forces and fort under my command."

Grant replied with a sentence that made him famous: "No terms except an unconditional and immediate surrender can be accepted."

"Unconditional surrender." The North had been waiting to hear these iron words. U.S.: United States; U.S.: Unconditional Surrender Grant. The nation knew him now. As the day of surrender went on, the size of the victory became apparent: in addition to suffering greater casualties, fourteen thousand Confederate soldiers had been captured, along with twenty thousand muskets, sixty-five cannon, and more than two thousand horses. It was, as Grant wrote Julia during the day, "the largest capture I believe ever made on the continent." The numbers exceeded those surrendered by the British at Yorktown, and the victory had an effect greater than simply the destruction of one Confederate army. Northern confidence soared, *The New York Times* saying of the Confederacy that "the monster is already clutched in his death struggle," while the South realized that its heartland was suddenly vulnerable. Nashville was now within striking distance of these Union amphibious efforts, and between the actions at Fort Henry and Fort Donelson three of Foote's lighter gunboats had ventured up the Tennessee as far as Muscle Shoals, Alabama. The news reached Europe, dampening support for the Southern cause in England, a sympathy connected in good part to British textile mills' desperate need for the vast amounts of cotton usually exported by the South and now drastically reduced by the Union blockade.

As at Fort Henry, the details of the surrender had to be addressed. After Grant's harsh words to Buckner, which Buckner referred to in his forced letter of acceptance as "ungenerous and unchivalrous," when the two met later that morning the former cordiality between them returned. They agreed swiftly on procedures for burying the dead. Buckner explained that the cutoff fortress had run out of food, and Grant told him that his supply officers would provide rations for the thousands of captured Confederates. Seeing that Grant's staff was overwhelmed with everything involved in handling fourteen thousand prisoners, including many wounded men, Buckner had his own staff help assemble them to board Grant's ships that would take them downstream to be interned at Cairo.

Early in the day, as Grant and Buckner settled these matters, a Union Army surgeon, thinking there would be some formal surrender ceremony, with the Confederates parading up to a designated point to lay down their arms, asked when and where this would occur. Grant looked at him and said, "There will be nothing of the kind . . . We have the fort, the men, the guns. Why should we go through vain forms and mortify the spirit of brave men, who, after all, are our countrymen?" When it finally came time for Buckner to leave on a transport taking him as a prisoner to Cairo, Grant accompanied him down to the landing. Walking Buckner off to the side, Grant said, "You are separated from your people, and perhaps you need funds. My purse is at your disposal." Buckner declined the offer with thanks, afterward remarking that they both had in mind the time when Grant was penniless in Manhattan.

As Grant reorganized his victorious force at Fort Donelson, congratulations as well as administrative paperwork poured in from every direction, but during the battle and its aftermath Grant received two communications that particularly impressed him. At a time when Union generals everywhere were quarreling about seniority and promotions, a message came from Sherman, who was senior to Grant. Sending it from Paducah late on the afternoon of February 15, when the results of the battle were in doubt, Sherman told Grant that he was rushing an additional regiment up the river to support him, and added that he would "do everything in my power to hurry forward to you reinforcements and supplies, and if I could be of service myself would gladly come, without making any question of Rank." The same day Sherman telegraphed Grant, in reference to the enemy's potential to reinforce Fort Donelson,

"I feel anxious about you as I know the great facilities they have of con-centration by means of the River & R [rail] Road, but have faith in you — Command me in any way."

This began a flow of communications between Grant and Sherman. Grant later commented: "At that time he was my senior in rank and there was no authority of law to assign a junior to command a senior of the same grade. But every boat that came up with supplies or reinforce-ments brought a note of encouragement from Sherman, asking me to call on him for any assistance that he could render and saying that if he could be of service at the front I might send for him and he would waive rank."

Soon after this, and his promotion to major general of Volunteers that followed it, Grant, wanting to seize the opportunity to move up the river from Fort Donelson toward Nashville, told Sherman, "Send all rein-forcements up the Cumberland," and added, "I feel under many obliga-tions to you for the kind tone of your letter, and hope that should an opportunity occur you will win for yourself the promotion, which you are kind enough to say belongs to me. I care nothing for promotion so long as our arms are successful, and no political appointments are made." In saying that he cared "nothing for promotion," Grant was being disingenuous — four months before this, he had written Julia to see if she could do a little quiet lobbying with Congressman Washburne's wife to increase Washburne's interest in having him made major general, and the previous September he had engaged in a dispute over seniority with Brigadier General Benjamin M. Prentiss — but he was touched by Sher-man's eagerness to help in any capacity.

On the same day Grant thanked him for his congratulations, Sher-man wrote that he was sending up a riverboat loaded with grain, and told Grant he was coping with the many family members who were converg-ing on his headquarters, seeking news of men who had been in the bat-tle. He added, "Some of your wounded are here, and no efforts have been or will be spared to make them as comfortable as possible . . . The whole [surrounding] country is alive to the necessity of caring for the wounded." After explaining that he was forwarding some of the wounded on to hospitals as far away as Cincinnati, Sherman asked Grant to tell him what more he needed. "Do you wish surgeons, nurses — the wives of officers, Laundresses or any thing?" Even in the midst of everything else he had to do, Grant, a former supply officer, realized that this former supply officer down the river knew how to get things done and was even anticipating his needs.

As the news of the big victory came to St. Louis, Halleck telegraphed McClellan, "Make Buell, Grant and Pope major generals of Volunteers, and give me command in the West. I ask this in return for Forts Henry and Donelson." He claimed Grant's two victories as if he had won them himself and virtually demanded a far larger command that would keep Grant under him and place him above his rival Buell. Halleck never directly congratulated Grant, simply commending Grant and Commodore Foote strongly for their victories in his general orders while continuing to portray himself as the paramount leader in the West.

In Washington, Lincoln knew exactly who had won Fort Donelson. The day after the victory, when Secretary of War Stanton brought him the papers nominating Grant for promotion to major general of Volunteers, a rank still junior to Halleck's commission as a major general in the Regular Army, the president from Illinois signed them and said, "If the Southerners think that man for man they are better than our Illinois men, or Western men generally, they will discover themselves in a grievous mistake."

5

THE BOND FORGED AT SHILOH

With the fall of Fort Donelson, the South became vulnerable. Grant was eager to move on up the Cumberland and take Nashville. If he and other Union commanders could "keep the ball moving as lively as possible," pushing south into enemy territory, the Northern spearhead would cut the South's east-west railroad lines, which Jefferson Davis called "the vertebrae of the Confederacy." Unless the Confederate Army blocked the coming offensive, the way would be open for Union columns to march down to the Gulf of Mexico. The South could be split in two.

It became a race against time. Albert Sidney Johnston, the general entrusted by Jefferson Davis to organize the Confederate defense in the Western theater, was a handsome mustachioed West Pointer who had been a cadet with Robert E. Lee and Joseph E. Johnston. At the age of fifty-eight he was the South's oldest general, admired by his contemporaries and his troops. He had formed an east-to-west defensive line that Grant had now fractured. After Fort Donelson fell, Johnston's headquarters at Bowling Green, Kentucky, to the northeast, was in danger of being cut off from Confederate territory and armies to the south; Johnston abandoned Bowling Green and was managing to bring his troops, many of them sick from the winter weather, safely south in a difficult circuitous retreat, picking up additional brigades and regiments along the way until he had seventeen thousand men with him.

The Confederacy was throwing in its reserves. Lee, Stonewall Jackson, and Joseph E. Johnston continued as leaders in the Eastern theater of war, but Jefferson Davis sent Beauregard from Virginia to be Albert

Sidney Johnston's second in command. General Braxton Bragg was coming up from Mobile, Alabama, with ten thousand men, Leonidas K. Polk was retreating from Kentucky with another ten thousand, and five thousand were on the way from New Orleans. General John C. Breckinridge reported in to Johnston; William J. Hardee was already serving with him. There was a chance that as many as twenty thousand Confederate soldiers could come from Arkansas to join Johnston.

The potential number of Southern defenders was large, but Johnston knew that Grant could have as many as forty thousand men, confident after taking Forts Henry and Donelson, and Don Carlos Buell was marching slowly from Kentucky with more than twenty-five thousand. There would be an epic collision, somewhere. Speaking of the white population of the South, Johnston proclaimed to his soldiers that "the eyes and hopes of eight millions of people rest upon you." In the days to come, as more units from across the South rallied to him, Johnston decided to gather his forces at the important railway center of Corinth in northeastern Mississippi, just below the Tennessee border. The South could ill afford to lose this hub, which was a true railway crossroads: the Memphis and Charleston Railroad line, coming from Memphis eighty miles to the west, passed through Corinth going east to Decatur and Huntsville in Alabama, and on to Chattanooga; the north-south Mobile and Ohio line also ran through there, connecting trains moving to and from the Gulf of Mexico. If the Union Army seized Corinth, a critical portion of the South's railway system would fall under Northern control.

In times of peace, one of the ways that goods reached Corinth from the Tennessee River was by way of a road from Pittsburg Landing, a steamboat wharf twenty miles to the northeast. Three miles in from the high bluff above the wharf, on a ridge in the woods, stood a one-room Methodist church, a log meetinghouse with the biblical name of Shiloh—"Place of Peace."

As days passed and Johnston's force at Corinth grew, he had no idea of how much time he had to organize a defense, or whether he might possibly have enough time to prepare his green but eager troops to launch an offensive. It appeared that time was not on Johnston's side. If Grant could keep his momentum and take Tennessee's virtually undefended capital of Nashville, his next step would be to move his victorious army swiftly over from the Cumberland River to the Tennessee. If Grant could then steam up to Pittsburg Landing to attack and defeat Johnston's army at Corinth, and there might be no limit to how much farther federal columns could then penetrate the South.

Grant's superior General Henry Halleck gave Albert Sidney Johnston the time he needed. At a time of success, a time when Grant should indeed have been allowed to "keep the ball moving," Halleck hesitated to authorize Grant to do just that, partly for fear that Confederate General P.G.T. Beauregard might be able to cut north and retake either Fort Henry or Paducah, Kentucky; he even went so far as to call this doubtful threat "the crisis of the war in the west." As for the intellectual underpinnings of this caution, Halleck was following concepts he had put forth in his book *The Elements of Military Art and Science*, ideas based on those of the French military theorist Antoine Henri Jomini and others who believed that victories could best be won by maneuver and mass, rather than by aggressive frontal attacks. Halleck's ambition worked in tandem with his caution; intent on his vision of the war and his central place in it, Halleck felt he could delay any Union offensive long enough to bargain with Washington for the supreme command in the West.

In contrast to Halleck, Grant, who after Fort Donelson immediately sent naval and then land forces up the Cumberland to take unopposed possession of Clarksville, Tennessee, forty-five miles from Nashville, wired Halleck's headquarters for permission to go on, saying that he could "have Nashville" within ten days. To his surprise, Halleck shot back a telegram ordering him not to advance with his reorganized victorious army, which now numbered thirty-six thousand men, including twelve thousand just sent up to him by Sherman.

This was all difficult to fathom for men of action like Grant and Commodore Foote, but they were unaware of the extent to which Halleck's ambition was controlling the situation. Halleck apparently felt that he could dictate his terms for advancement and sent the Union general in chief McClellan a message saying that "I must have top command of the [Western] armies," adding that "hesitation and delay are losing us this golden opportunity," when in fact the only "hesitation and delay" were his own. He finished this with a peremptory, "Lay this before the President and Secretary of War. May I assume command? Answer quickly."

As Grant waited, McClellan refused to give Halleck what he wanted and refused to draw Lincoln and Stanton into the decision, but McClellan's reason for doing this showed yet another aspect to the game of military politics. McClellan, who at the time was giving lavish dinner parties in Washington while allowing the Confederates time to fortify and reinforce their positions in northern Virginia, and who was issuing few orders that would start a Union offensive in any theater of operations, wanted no rivals for his position as general in chief. In turning down Halleck, McClellan pointed out that Buell, who had replaced Sherman at Louisville and held equal rank with Halleck, was marching toward the scene of impending action. There was no reason to place Halleck above him.

Thwarted by McClellan, Halleck bypassed him, going out of the chain of command and approaching Secretary of War Stanton directly with a message that said, "One whole week has been lost by hesitation and delay. There was, and I think there still is, a golden opportunity to strike a fatal blow, but I can't do it unless I control Buell's army." When Stanton replied flatly, "The President does not think any change in the organization of the army or the military departments advisable," Halleck turned back to matters that were his to control and continued to hold Grant at Clarksville. (Writing to Halleck's chief of staff Brigadier General George W. Cullum, Grant said, "It is my impression that by following up our success Nashville would be an easy conquest," but he did not express the frustration shown by Commodore Foote, who wrote his wife that "I am disgusted that we were kept from going up and taking Nashville. It was jealousy on the part of McClellan and Halleck.")

While keeping Grant from seizing Nashville, Halleck continued to rehabilitate Sherman. Always mindful of Sherman's powerful connections in Washington and impressed by the job that he had done since his return from his breakdown and forced twenty-day leave, first in training thousands of troops and then in supporting Grant's efforts, Halleck told Sherman that he could start organizing various regiments into a division of his own, to lead in future battles.

Sherman went to work. As he wrote Ellen, "Learning some days past that the Confederates are simply abandoning Columbus [Kentucky, across the Mississippi River from Belmont, Missouri, the scene of Grant's first battle] I sent a party of cavalry to go as near as possible." In addition to this, Sherman embarked a regiment of nine hundred men on a large steamboat at Paducah and took them down there. At Columbus he encountered another example of Halleck's failure to take advantage of opportunities. Not only was the place empty, but Sherman found that the Confederates had been given so much time that "they carried off nearly all their [artillery] guns, and materials, burned their huts and some corn and provisions."

Landing and leaving troops to establish a garrison there, Sherman returned to Paducah. In one sense, it had been a venture with little military result, but for the first time in months he had been in active command of troops away from a headquarters, had planned the entire amphibious operation, and had executed it with precision. The man who had begged Lincoln to keep him always in a subordinate position had succeeded in an independent command, entirely on his own. Back at Paducah he continued to assemble his division, which soon numbered nine thousand men.

Earlier, tethered though Grant was, forty-five miles short of Nashville and with no idea of the reasons for being held there, he once again saw an opportunity to move Union troops forward. Don Carlos Buell had dispatched a division from Kentucky to support Grant at the time he was moving to attack Fort Donelson; now, as they arrived aboard a large fleet of paddle-wheelers a week after the battle, Grant realized that these thousands of men were not under Halleck's control. Telling their commander, Brigadier General William Nelson, not to bring his regiments ashore, he ordered him to proceed on with them and take Nashville, which they did, quickly and without bloodshed.

As a result of his quick thinking, Grant now had two commanders angry with him. Buell, who soon arrived at Nashville with the rest of his army, felt that Grant had commandeered some of his forces with an unauthorized order, robbing him of the chance to enter the city at the head of his troops. Halleck felt that Grant had broken the spirit if not the law of his own orders to stay where he was.

Next came a serious breakdown in communications between Grant and Halleck. This began when Grant, who had no orders to do so and normally should have stayed with his army, went to Nashville briefly to confer with Buell. Unknown to Grant or Halleck, the civilian telegraph operator at the Union headquarters in Cairo, Illinois, was certainly a Confederate sympathizer and possibly an active spy; in any event, the man threw away the messages and reports from Grant that were to be sent on to Halleck in St. Louis. As a result, Halleck heard nothing from

Grant for a week and did not know how much of that time Grant spent in Nashville, away from his command.

With no response to several requests that Grant send him various types of information, Halleck dispatched a complaining report of this to McClellan. Halleck said in part, "It is hard to censure a successful general immediately after a victory, but I think he richly deserves it . . . Satisfied with his victory, he sits down and enjoys it without any regard to the future. I am worn-out and tired with this neglect and inefficiency."

This time Halleck had McClellan on his side. Perhaps for a moment thinking of Grant only as the captain who had been on one of his "sprees" while outfitting McClellan's expedition to explore the Cascade Range in the Oregon Territory in 1853, and possibly with an eye to keeping in check a fast-rising potential rival, the general in chief replied the next day. "The future success of our cause demands that proceedings such as Grant's should at once be checked," McClellan told Halleck. "Generals must observe discipline as well as private soldiers. Do not hesitate to arrest him at once if the good of the service requires it." McClellan added that if Grant were removed, his replacement should be General Charles F. Smith, whose brave leadership storming a slope at Fort Donelson had done so much to win that day.

Halleck was still angry with Grant and thought that McClellan might have memories of the old army gossip about Grant's drinking. The next day, still delaying meaningful action fifteen days after the fall of Fort Donelson, Halleck spent still more time writing to McClellan: "A rumor has just reached me that since the taking of Fort Donelson General Grant has resumed his former bad habits. If so, it will account for his neglect of my oft-repeated orders. I do not deem it advisable to arrest him at present, but I have placed General Smith in command of the expedition up the Tennessee." Halleck was not replacing Grant, but leaving him in command in the rear while Smith, whom Grant greatly admired, was to come back down the Cumberland River where Fort Donelson stood and lead a portion of Grant's forces on amphibious raids up the Tennessee.

Until Grant received the order from Halleck spelling this out, he had no idea that Halleck was, as Sherman observed, "working himself into a passion" about his subordinate's seemingly defiant disregard of orders and command authority. He quickly replied to Halleck that he was sending Smith up the Tennessee as ordered and added that, based on intelligence reports, "Forces going . . . must go prepared to meet a force of 20,000 men. This will take all of my available troops." That last comment was an effort to regain lost momentum through others. Even though now tied by Halleck's order to his own rear headquarters, Grant was attempting to change the concept of a small raiding force into a major movement by his entire command. Then, addressing Halleck's overall complaint, he told Halleck, "I am not aware of ever having disobeyed any order from Head Quarters, and certainly never intended such a thing."

This exchange of communications between Halleck and Grant continued, with Halleck criticizing Grant while Grant forthrightly justified himself. As letters and telegrams moved back and forth, Grant told Halleck three times that if Halleck thought he was not doing his job properly, he should be allowed to resign his command. Grant was learning a bit about military politics himself: at one point he or his staff telegraphed a copy of one of Halleck's complaints, and Grant's answer, to Congressman Washburne in Washington. Washburne promptly went over to the White House and placed the matter before President Lincoln. This resulted in a message from Adjutant General Lorenzo Thomas to Halleck, informing him that the president was taking a personal interest in these allegations and requesting Halleck to spell out any charges of malfeasance and insubordination he had to make against Grant.

As this was going on, Sherman received the order to take his division up the Tennessee River as part of the expedition, led by General Smith, that Grant was being held back from commanding. On March 12 he wrote Ellen from Savannah, Tennessee, nine miles short of Pittsburg Landing on the opposite side of the Tennessee River.

Dearest Ellen,

Here we are up the Tennessee, near the Line [the South's railroad corridor from Memphis to Chattanooga] with about 50 boat loads of soldiers. I have the fifth Division composed mostly of Ohio Soldiers about 9000—but they are raw & Green . . . The object of the expedition is to cut the Line . . . along which are distributed the Enemy's forces.

Let what occur that may[,] you may rest assured that the devotion & affection you have exhibited in the past winter has endeared you more than ever, and that if it should so happen that I can regain my position and Self respect and should Peace ever be restored I will labor hard for you and for our children.

I am still of the opinion that although the blow at Fort Donelson was a terrible one to the Confederates they are still far from being defeated, and being in their own country they have great advantage . . . Today we shall move further up the river.

Despite the energetic and determined efforts made by Sherman and his men to cut the railway line east of Corinth, they were literally drowned out by rains that raised the Tennessee River fifteen feet in a single day. Sherman fell back down the river to Pittsburg Landing and began to prepare a vast encampment for the Union divisions that were to arrive there by ship and by road. At the moment, both Grant and Sherman thought of the place purely in terms of being a staging area for a march south to attack Corinth.

In the dispute between Grant and Halleck, by the time Halleck received the request from Lorenzo Thomas that he enumerate any formal charges he had to make against Grant, the entire high command of the Union Army had changed. Lincoln had vacated the position of general in chief by assigning McClellan to take active field command of the Eastern army, the Army of the Potomac. Although Lincoln was frustrated by McClellan's delay in mounting a major offensive into Virginia, and some in Washington were referring to McClellan as the Great American Tortoise, at this time there was no official censure of McClellan; "Little Mac" was simply being removed from overall command of the Union Army so that he could devote himself entirely to winning the war on the front south of Washington.

With this reorganization, Halleck was given what he wanted: command of all the forces in the Western theater, in a new entity called the Department of the Mississippi. Halleck was now equal to McClellan, with Buell as his subordinate; no Union officer stood higher than he. Rejoicing in his elevation and eager to dispose of what now seemed to him a minor matter involving Grant, Halleck took the position that the whole thing had been a misunderstanding. He had no intention of losing the services of his most successful subordinate and sent Grant a telegram assuring him that "instead of relieving you, I wish you, as soon as your new army is in the field, to assume the immediate command and lead it on to new victories."

Grant headed upstream on March 16 to take active field command of what was named the Army of the Tennessee. A month had passed since

Fort Donelson fell and the South had begun scrambling to repair its strategic position. Johnston would soon have forty thousand men at Corinth, with more trying to get there, and the Confederacy's leaders were in close touch with him. From the Confederate capital of Richmond, Virginia, Jefferson Davis wrote, "You have done wonderfully well, and now I can breathe easier." An equally encouraging but more specific message came from Johnston's old West Point classmate and friend Robert E. Lee, who knew that, in addition to the Union force under Sherman that was by this time ashore at Pittsburg Landing, and some other divisions coming up the river, Don Carlos Buell was slowly marching his army across country from Nashville to add to Grant's command. Lee wrote Johnston: "No one has sympathized with you in the troubles with which you are surrounded more sincerely than myself. I have watched your every movement, and know the difficulties with which you have had to contend . . . I need not urge you, when your army is united, to deal a blow at the enemy in your front, if possible before his rear gets up from Nashville. You have him divided, keep him so if you can."

When Grant took over the forces gathering at Pittsburg Landing, where Sherman had arrived with his division, he also became the superior officer of General Charles Smith, the hero of Fort Donelson, who had badly scraped his leg getting into a small boat; while still with the expedition he had led up the river, Smith was suffering from an infection that would cause his death five weeks later. Grant soon inspected the place where Sherman was continuing to prepare the encampment for the rapidly arriving Union divisions. Stretching inland from the wharf and seventy-foot-high bluff on the Tennessee River at Pittsburg Landing, the slightly rolling area combined dense woods and some open fields and orchards, with a number of streams running through it. Laid out like a huge triangle encompassing eighteen square miles, the river formed its eastern side, while the southern side had sentry posts facing in the direction of the enemy, twenty miles away at Corinth. The arriving regiments placed their tents along this line. The final leg of the triangle, nearly six miles long, led back from the inland corner of the encampment to the wharf and wooden warehouses at the landing.

Sherman's headquarters and sleeping tent stood next to the seldom-used Shiloh Church, at the triangle's inland corner, in the woods beyond a field serving as a parade ground. In a written report to Grant, who set up his headquarters nine miles downstream to the north at Savannah,

Sherman described the camp as being located on a "magnificent plain for camping and drilling, and a military point of great strength." Confederate cavalry patrols often came up from Corinth, approaching the camp through the woods, and there had been clashes with them.

Even though Halleck had given Grant the day-to-day command of these divisions, Halleck's slow, cautious nature constantly affected this situation. From hundreds of miles away, he ordered Grant "not to advance" until Buell's army joined him. Reverting to the philosophy of Jomini, who saw a campaign as being more of a chess game than an all-out attack, Halleck told Grant that "we must strike no blow till we are strong enough to admit no doubt of defeat" and added that until that time, no one was to bring on "an engagement." This stricture not only inhibited the Union responses to the increasingly frequent Confederate cavalry probes, but also created worries about what degree of response to *any* provocation might bring on "an engagement."

Although Grant wanted to move against Johnston's army at Corinth, he also wished to avoid another dispute with the over-bureaucratic Halleck. (An example of Halleck's overzealous attention to detail was a rebuke to Grant on March 24 for allegedly having as his departmental medical director a doctor who was not also an army officer. Halleck ordered Grant "to discharge him.") Confident after his victories at Fort Henry and Fort Donelson, Grant felt that in any case he held the initiative in the area. He was prepared to obey Halleck and wait for Buell's large reinforcements before marching the twenty miles to attack Corinth, but he wrote General Smith that "I am clearly of the opinion that the enemy are gathering strength at Corinth quite as rapidly as we are here, and the sooner we attack, the easier will be the task of taking the place." Still sure that it was he who would bring on the battle, Grant, who later said that he "had no expectation of needing fortifications" at Pittsburg Landing, built no earthworks.

Grant's letters to Julia reflected both his confidence and his certainty that his army would soon be in a major battle. He told her, "When you will hear of another great and important strike I can[']t tell you but it will be a big lick as far as numbers engaged is concerned. I have no misgivings myself as to the result and you must not feel the slightest alarm." In a later letter to her written on March 29, he mentioned that he had been suffering from what he spelled "Diaoreah" (as had Sherman and many others), and added that he thought the impending battle "will be the last in the West. This is all the time supposing that we will be successful which I never doubt for a moment." A few days later, referring to Halleck's long-distance restraint upon him, he told Julia, "Soon I hope to be permitted to move from here and when I do there will probably be the greatest battle fought of the War. I do not feel that there is the slightest doubt about the result . . . Knowing however that a terrible sacrifice of life must take place I feel conserned [sic] for my army and their friends at home."

While Grant felt in control of events, Sherman had a few moments of uncertainty. On one occasion, when he admitted to visiting war correspondents that the huge camp was vulnerable and they asked why he did not speak up about it, Sherman replied with a shrug, "Oh, they'd call me crazy again," but most of the time he thought as Grant did. Writing Ellen of the frequent sightings of enemy patrols in the woods just south of the encampment, Sherman told her, "We are constantly in the presence of the enemy's pickets, but I am satisfied that they will await our coming at Corinth."

The Confederates at Corinth had no intention of letting Grant decide the time and place of attack. While Grant waited for Buell's army to join him, Albert Sidney Johnston knew that his friend Robert E. Lee was right: he had to attack Grant before Buell reached Pittsburg Landing. At ten o'clock on the evening of Wednesday, April 2, 1862, Johnston's second in command, P.G.T. Beauregard, received an intelligence report telling him that the first regiments of Buell's column of twenty thousand or more men would reach Grant's army within the next few days. Beauregard immediately sent an orderly to Johnston's nearby headquarters, carrying this penciled message: "Now is the moment to advance and strike the enemy at Pittsburg Landing." By midnight, commanders throughout the forty-four-thousand-man Confederate camp started receiving orders: the army had to set out at dawn, march twenty miles during the day, and make a surprise attack on Grant's unfortified camp early on the following morning of Friday, April 4.

Johnston and his experienced Confederate generals liked the fact that forty thousand federal troops were packed into the triangle by the river at Pittsburg Landing. There were no defensive entrenchments; the hundreds of rows of tents placed just behind the picket line would hinder any swift effort to assemble and face the enemy, when the defenders finally saw thousands of Confederate infantrymen rush at them from the woods. After weeks of this strange situation—two opposing armies, twenty miles apart, spending most of their time carrying out the routine activities of camp life—Johnston's men had the chance to overrun the camp and drive Grant's army into the swamps away from Pittsburg Landing.

It was one thing to plan a swift march toward the enemy, and another to execute it. In what proved to be an unwise decision, Johnston decided to lead the march, while Beauregard would for the time being remain at the rear, directing the complicated sequence in which he wanted the different divisions to leave Corinth and move up to Pittsburg Landing. Beauregard told his corps commanders to start the march while his adjutant colonel, Thomas Jordan, began preparing written orders for the movement, but Confederate general William J. Hardee refused to start off before he had his orders in writing. Because Hardee's corps was the first major column in Beauregard's planned line of march, no large force left camp until past noon that day, April 3, and Beauregard postponed the attack from the morning of April 4 until sunrise on Saturday, April 5. On Friday night, a heavy cold rain began to fall on Johnston's army as his men tried to hurry forward to be ready for the dawn attack. The same kind of downpour that had raised the Tennessee River fifteen feet in a day now turned roads into bogs in which the Confederate artillery pieces and supply wagons sank to their axles. It became clear that no surprise attack could be made the next morning.

This same rain was also falling that night farther to the north, severely slowing Buell's effort to reach and support Grant. As for Grant, there had been so many minor skirmishes during the past two days that he did not go down the river by boat to spend the night at his headquarters at Savannah "until an hour when I felt there would be no further danger before the morning." Riding through the dark to learn more about a clash with some Confederate cavalry that had been reported to him in a quickly written report from Sherman, Grant had trouble.

> The night was one of impenetrable darkness, with rain pouring down in torrents; nothing was visible to the eye except as revealed by the frequent flashes of lightning. Under these circumstances I had to trust the horse, without guidance, to keep the road . . . On the way back to the boat my horse's feet slipped from under him, and he fell with my leg under his body. The extreme softness of the ground, from the excessive rains . . . no doubt saved me

from a severe injury and protracted lameness. As it was, my ankle was very much injured, so much so that my boot had to be cut off. For two or three days thereafter I was unable to walk except with crutches.

The next day, Saturday, April 5, able to be lifted onto a horse and with a crutch strapped to his saddle, Grant sent one of his now-frequent reports to Halleck. Not mentioning his injury, he told Halleck that the first division of Buell's column had arrived in the area of his headquarters downriver from Pittsburg Landing at Savannah, with the additional divisions expected "to-morrow and the next day." He continued, "I have scarsely [sic] the faintest idea of an attack (general one,) being made upon us but will be prepared should such a thing take place." He enclosed a two-part report made to him by Sherman, giving the details of the previous day's skirmish, with Sherman's additional comment that "the enemy is saucy, but got the worst of it yesterday, and will not press our pickets far—I will not be drawn out far unless with certainty of advantage, and I do not apprehend anything like an attack upon our position." Once again, Grant and Sherman were telling themselves, and each other, what they wanted to believe.

As these reports from Grant and Sherman went off to Halleck, General Albert Sidney Johnston was riding around among his advancing columns, trying to untangle another snarl in the forward movement of his army. Looking with disbelief at the confusion, he exclaimed, "This is puerile! This is not war!" Nonetheless, with the sun out and the day wearing on, thousands of Confederate soldiers neared the unsuspecting Union camp. The colonel of the Seventieth Ohio was conducting a review of his regiment, complete with its band playing: from a higher place in the woods beyond the camp, dozens of gray-clad Confederates stood quietly watching the parade. The afternoon sun caught the glint of the brass barrels of several Confederate cannon that had been brought forward through the trees and underbrush, but the Union sentries did not understand what they saw.

Even clearer warnings came: a Confederate patrol chased some federal troops from a house just a mile from Sherman's headquarters next to Shiloh Church, near the end of the encampment farthest from the river. Alarmed when his sentries reported unidentified men moving in the woods, Colonel Jesse J. Appler of the inexperienced Fifty-third Ohio sent a detachment to see what was out there. When he heard shots and his men came running back to report they had come under fire from "a line of men in butternut clothes," Appler had his musicians beat their drums to turn out his regiment under arms and ordered his quartermaster to report the situation to Sherman at his nearby headquarters. Sherman considered Colonel Appler to be a nervous, frightened old man. Within a few minutes, Appler's quartermaster reappeared with this message: "General Sherman says, 'Take your damned regiment back to Ohio. There is no enemy nearer than Corinth.'" Overhearing this, Appler's young soldiers laughed, broke ranks without being dismissed, and went back to looking for wild onions and turkey peas to add to their kettles for supper.

By dusk, the Confederates were moving into place within two miles of the Union encampment, where thousands of little fires and oil lanterns were lit. It was too late for an attack that day, and as it became dark Johnston held a council of war with his generals, gathering them around a campfire. They made quite a group. Among the men talking in the firelight and shadows with Johnston and Beauregard were Leonidas K. Polk, a graduate of West Point who had left the army and become a bishop of the Episcopal church, taking a Confederate commission when the Southern states seceded; John C. Breckinridge of Kentucky, an academy graduate who had served as vice president of the United States under President Buchanan; Braxton Bragg, another West Pointer, brevetted for bravery in Mexico, who had brought ten thousand soldiers up from along the Gulf Coast and now commanded 13,600 men; and William Hardee, a West Pointer who had later returned to the academy as commandant of cadets. One of Johnston's volunteer aides was Governor Isham Harris of Tennessee, who fled the state capital of Nashville before the Union Army marched in.

Of the twenty-six officers in Johnston's army who commanded divisions or brigades, ten were graduates of West Point, and eleven had fought in Mexico, several with distinction. They brought a wealth of military experience to the Southern side and had the cavalry leader Nathan Bedford Forrest providing them with a stream of accurate and timely reports from his scouts. A few miles from them were some interesting and able Union officers—including an untried brigadier from Ohio named James A. Garfield—but the leadership of the Union force lacked the experience possessed by the Confederate side. A year before this, Ulysses S.

Grant and William Tecumseh Sherman, captains who had resigned from the service and floundered in civilian life, had not yet returned to the army.

As Johnston's council of war got under way, Beauregard started to speak. Until recently he had hoped to mount a major offensive that would regain much that the South had lost in recent months, but now he began a litany of worries. The enemy had to know they were there: all afternoon, many untested soldiers, wondering if their powder was still dry after the heavy rain, had been firing their muskets. Along the line of march, there had been bugle calls and signals made by beating drums. Young troops had been shouting back and forth to each other; when a deer sprang from the woods beside the road, hundreds of youths cried out at the sight. The Union forces were not deaf, Beauregard argued: "Now they will be entrenched to the eyes!"

He had more to say. Because the march from Corinth had taken two days instead of one (due in good part to his unnecessarily complicated planning), the troops had used up their rations and would sleep hungry tonight and have to go into battle in the morning with empty stomachs. The men were also nervously exhausted by their long sleepless struggle through last night's rain and mud. The army was in no fit condition to make a do-or-die attack. To Johnston's amazement, the fiery Creole recommended that they all march back to Corinth and wait for a better opportunity.

Johnston listened patiently until Beauregard finished, and then, as Leonidas Polk wrote about the dramatic moment in the firelight, "remarked that this would never do." He replied to Beauregard that if the enemy knew they were there in great force, they would be under fire right then. Yes, the men were hungry, but the nearest food was in the Union camp, and the way to get it was to overrun the place in the morning. The meeting was over: "Gentlemen," Johnston told his generals, "we will attack at daylight tomorrow." As the leaders dispersed in the shadows to return to their commands, Johnston turned to an aide and said, "I would fight them if they were a million." He later added, "I mean to hammer 'em!"

That night, trying to sleep, Beauregard heard a drum beating nearby. Furious at this noise that was both keeping him awake and warning the Northern troops that something was up, he sent an aide to have it stopped. Within minutes the man returned to tell him that the drum was in the enemy camp. That was how close they were.

At three in the morning of Sunday, April 6, Grant was in bed at his headquarters in the house of a Union sympathizer downstream at Savannah, while Sherman slept in his headquarters tent next to the little log Shiloh meetinghouse. At that hour, Colonel Everett Peabody, a heavyset thirty-one-year-old Harvard graduate from a distinguished Massachusetts family, became worried about reports of Confederate activity in the woods in front of his brigade. He started assembling a force of three hundred men from one of his regiments, the Twenty-fifth Missouri. When they were ready, Peabody gave the major commanding them orders to take his troops forward and make a reconnaissance in force. Moving cautiously, at dawn the Missourians came to the edge of a clearing half a mile beyond the Union encampment and ran straight into a battalion of Confederates from Mississippi who were quietly coming the other way. As the two sides started firing at each other, thousands more Confederate soldiers began appearing out of the woods all along the six-mile-long edge of the Union camp that ran from the river to the vicinity of Shiloh Church. Jarred from sleep by the sounds of gunfire and the Union regimental drums beating the "long roll" signaling an attack, federal soldiers dashed out of their tents, grabbing their weapons and strapping on their equipment.

Just before all this started, at a meeting of the Confederate leaders in the woods, Beauregard had been telling Johnston once again that they should cancel the attack and take their army back to Corinth. As Johnston heard the crescendo of firing, he said to the officers around him, "The battle has opened, gentlemen; it is too late to change our dispositions now." At the edge of the Union camp, Colonel Appler of the Fifty-third Ohio, the target of Sherman's rebuke for being too apprehensive the previous day, saw a man of the Twenty-fifth Missouri come back from the direction of the firing with blood streaming from a wound on his arm; the man shouted, "Get into line—the Rebels are coming!" Sending word to Sherman, Appler turned out his men, only to receive a quick, caustic reply from Sherman, who was already up at his nearby headquarters but skeptical that this rattle of musketry was a major matter: the messenger said Sherman told him to say, "You must be badly scared over there." Appler, now seeing hundreds of men in gray coming straight at his right flank, shouted, "This is no place for us!" and led a retreat through his regiment's tents at the dead run, stopping on a ridge, where his men flung themselves down in the brush, pointing their muskets toward the advancing enemy.

At this point Sherman arrived, riding what he described as "a beautiful sorrel race mare that was fleet as a deer," and accompanied by his orderly, Private Thomas D. Holliday of the Second Illinois Cavalry, who always had a carbine ready to protect his general. As Sherman raised his field glasses to study the terrain in front of him, Confederate infantrymen sprang out of the bushes fifty yards to one side, their weapons at their shoulders. A Union lieutenant sprinted toward Sherman, yelling, "General, look to your right!" Sherman's head spun in that direction; as he shouted and threw up his right hand as if to ward off a bullet, a musket ball that he felt sure was meant for him killed his handsome young orderly, while some buckshot slashed open the third finger of Sherman's right hand. "Appler," he shouted to the colonel, "hold your position! I will support you!" Sherman wrapped a handkerchief around his bleeding hand and spurred off to organize the defense of the right end of the Union line. Looking in the direction of more firing, Sherman said that "I saw the rebel lines of battle in front coming down on us as far as the eye could reach." In the meantime, the first wave of Confederates had swept through Sherman's headquarters; his tent and the Shiloh meetinghouse were in enemy hands. In the gunfire, his two spare horses, tethered under a tree near his tent, were killed.

Some Union regiments fell back and formed lines to face the advancing enemy, while others, including Appler and most of his men, simply ran away toward the river. Johnston, who had told Beauregard to send men and supplies up from the rear while he directed the battle at the front, rode forward through the trees on his horse Fire-eater, thousands of his Confederate foot soldiers pressing through the woods to either side of him. As the sun rose and burnt through the morning mist, Johnston, convinced that his army would drive all the defenders right through their camp and sweep them into the swamps, said confidently, "Tonight we will water our horses in the Tennessee River."

Shortly before this—the hour is not recorded—Ulysses S. Grant was eating what he described as "a very early breakfast" at his headquarters nine miles down the river. He was hoping that Buell would arrive at the end of his long march so that he could confer with him, but now, as Grant put it, "heavy firing was heard in the direction of Pittsburg Landing." Getting up from the breakfast table, he hobbled out to the porch on his crutches, listened for a moment, and then said to his staff, "Gentlemen, the ball is in motion. Let us be off."

By the time Grant arrived at Pittsburg Landing aboard his headquarters paddle-wheeler *Tigress*, many hundreds of Union soldiers who had fled the battle were milling about aimlessly under the shelter of the steep bluffs along the shore. Riding to the hastily organized front, he came to the division in the center of the line, commanded by Brigadier General Benjamin M. Prentiss, with whom he had had the severest of disputes about seniority in rank the past autumn. Prentiss's men were falling back through their camps in the face of a Confederate bayonet charge. One of Prentiss's commanders, Colonel Everett Peabody, whom Prentiss had earlier accused of "bringing on this engagement" by sending forward three hundred men, was riding through the area of his regiment's tents, trying to rally his troops. Wounded four times as he kept trying to form them up to make a stand, a fifth musket ball hit him in the head, and he fell dead from his horse. Prentiss's regiments kept falling back, some of them turning to fire as they retreated, while others threw down their arms and ran to the river.

On the right end of the Union line, Sherman's men, nearly all of them in their first battle, were slowly falling back, but they kept their lines as they fired at the advancing enemy. Sherman, whose beautiful "race mare" had been wounded and then killed, was now using a horse he had taken over from one of his aides and was riding back and forth along his line, ignoring the danger as he calmly encouraged his men. He had dismounted and was standing, the handkerchief around his hand dark from drying blood, studying the situation and quietly giving orders, when one of Grant's aides came up to him and said that Grant was in the middle behind Prentiss's division and wanted to know how things were on this right end of the line. As enemy bullets and cannonballs flew past them, Sherman kept looking forward to where his men had started to hold fast, and said, "Tell Grant if he has any men to spare I can use them; if not, I will do the best I can. We are holding them pretty well just now—pretty well—but it's hot as hell."

In the center where Grant was, the hungry condition of the Confederate troops was giving the Union forces their first good development of the day. Many of the Southerners had not eaten for twenty-four hours or more; coming upon the deserted tent lines of Prentiss's division, they stopped to eat the food they found there and to loot the tents. By the time they resumed their forward movement, Prentiss had managed to round up a thousand of his soldiers and place them along a slightly sunken old wagon road running along a ridge a mile behind their abandoned camp. This indentation, hardly a trench, quickly became the defensive link between Sherman's sector on the right, and large federal units that were beginning to settle down to the left, in the area extending to the river.

Coming along the shaky defensive line from left to right, Grant reached Sherman's sector at ten in the morning, at a time when Sherman said that he and his men were "desperately engaged." Grant found that many of Sherman's soldiers had run off, but that the rest of them, despite being shot at for the first time, were holding where they were. Some had even begun to fight their way back toward the tents from which they had fled.

Here was Grant and Sherman's first meeting on a battlefield. Conferring under fire, Sherman told Grant that he needed more ammunition. Grant replied that wagonloads of it were on the way and expressed his admiration for the stand Sherman's division was making. As Sherman put it, "This gave him great satisfaction, and he told me that things did not look as well over on the left."

As Grant rode off, one of his aides remarked that things on Sherman's front looked "pretty squally." Grant answered, "Well, not so bad." Now he had seen Sherman in the midst of a battle, calmly handling everything as well as it could be done, and he liked what he saw. Speaking of the area near Shiloh Church, Grant later said that "this point was the key to our position and was held by Sherman. His division was wholly raw, no part of it ever having been in an engagement; but I thought this deficiency was more than made up by the superiority of the commander." He added, of this day:

> During the whole of Sunday I was continuously engaged in passing from one part of the field to another, giving directions to division commanders. In thus moving along the line, however, I never deemed it important to stay long with Sherman. Although his troops were then under fire for the first

time, their commander, by his constant presence with them, inspired a confidence in officers and men that enabled them to render services on that bloody battle-field worthy of the best of veterans.

As Grant indicated, he was not the only one who admired what Sherman did as the day went on. Men who had heard of this "crazy" general, frightened in Kentucky by the ghosts of nonexistent advancing columns, formed an impression that one man summed up this way: "All around him were excited orderlies and officers, but though his face was besmeared with powder and blood, battle seemed to have cooled his usually hot nerves." However, as he kept moving from place to place, exposing himself to fire as he received reports and gave orders, stopping at artillery batteries to coordinate their fire with that of other Union guns, Sherman did have "trouble keeping his cigar lit and he used up all of his matches and most of the men's." Smoking or not, he seemed to have a sense what was coming next, how to get ready for it, and when to use what was needed. When his right wing started falling back, Sherman grinned and said, "I was looking for that," and told an artillery officer to have his battery fire a prepared cannonade that stopped the Confederate charge in its tracks. The enemy responded to this repulse by sending a cavalry force thundering toward the Union guns in an effort to overrun and capture them, but the horsemen were suddenly blasted and thrown back by volleys from two infantry companies that Sherman had held in reserve for just such a contingency.

As the bloodiest battle fought thus far in the war roared on, Beauregard had come up to Shiloh Church and was using Sherman's tent as part of his headquarters. The Confederate attacks on the Union center encountered withering blasts from Prentiss's defenders along the old wagon road, who were able to lie with just their heads and arms and muskets above the edge of the indentation in the ground and fire from the prone position at the men running toward them. One Confederate, part of a force that came within ten yards of a brush-covered part of this makeshift trench before a desperate federal volley sent it reeling back, staggered up to a comrade and said, "It's a hornet's nest in there." As the Confederates started referring to that area of the Union line as "the Hornet's Nest" while the battle raged into the afternoon, both Johnston and Beauregard began calling for reserves to exploit advances at points along the front. They found they had none: all of the forty-four thousand men who had marched up from Corinth were already committed to the fighting along the six-mile line.

At a point near the old road stood a ten-acre peach orchard, in full bloom with pink petals on this April afternoon. The Union troops were defending a line along its front, and Johnston ordered one of Breckinridge's brigades to charge and break the line. The Confederates refused to make the attack. When Breckinridge, almost incoherent in his frustration, had to report this to Johnston, the fifty-eight-year-old commander said quietly, "Then I will help you. We can get them to make the charge."

Accompanied by his aide, Governor Isham Harris of Tennessee, Johnston began riding his horse Fire-eater slowly along the line of gray-clad infantrymen who were facing the peach orchard, in formation to attack, bayonets fixed on their muskets, but out of firing distance and unwilling to go forward. Reaching over from Fire-eater as he passed, he looked into his soldiers' eyes and touched their bayonets with a small tin cup that had been captured earlier in the day, saying with each click, "They will do the work." As he moved past man after man, Harris said, "The line was already thrilling and trembling." Coming to the middle of the line, Johnston, a commanding general whose traditional position would have been hundreds of yards to the rear, turned, drew his sword, and shouted, "I will lead you!" Screaming the high-pitched "yip-yip-yip!" of the Rebel Yell, the entire brigade dashed forward, with scores of men running right beside Johnston. Within minutes, they cleared the Union defenders out of the peach orchard.

When Governor Harris came riding up to Johnston after the successful attack, the general beamed at him and said, "Governor, they came near to putting me *hors de combat* in that charge." He raised his boot to show that its sole was flapping loose, cut from the rest of it by a musket ball. Johnston's gray uniform had been slashed by other shots, but he seemed unharmed and exultant. He gave Harris a message to take to another officer; when Harris came back, he found Johnston groggy and about to fall out of his saddle. What Harris did not know was that, in a duel fought in Texas twenty-five years before, a pistol ball cut the sciatic nerve in Johnston's right leg in a way that left it numb. During the attack just minutes past, a musket ball had severed an artery in that leg, but as he bled profusely, Johnston felt nothing, and in the excitement and confusion of the moment, no one else saw what was happening. By the time

Harris and others helped him to the ground, he was in critical condition. Even then the quick application of a tourniquet might have saved him—Johnston carried one in a pocket—but he had sent the nearest surgeon to help wounded men nearby, and none of his staff knew what to do. Within minutes, Albert Sidney Johnston died.

As soon as that news came to Beauregard, he ordered that Johnston's body be hidden and that no word of this loss should reach the men. Trying to weather the storm, Confederate brigadier Daniel Ruggles, a veteran twice brevetted for gallantry in the Mexican War, decided that piecemeal attacks by foot soldiers alone could not take the Hornet's Nest.

Moving to support Ruggles and the other Confederate infantry brigades in that area of the battle, Beauregard organized a counterattack. After an hour, fifty-three cannon, the largest concentration of guns to be put in line in any American battle fought to that day, had been brought up to face the center of the old road. When the massive Confederate attack struck in the late afternoon, the Northern line broke on both sides of the Hornet's Nest, and its defenders were surrounded. Some individual Union soldiers escaped the noose, and two Iowa regiments cut their way through to withdraw toward the river, but by six o'clock, after a long and gallant stand, Benjamin Prentiss and twenty-two hundred of his men had surrendered.

During these bloody hours, Grant had formed a new defensive line running in from the river at Pittsburg Landing, far behind the lines his army occupied at dawn. Now it was Grant's turn to assemble artillery, and he had fifty guns in place. All afternoon, units had been falling back, some in orderly fashion and some simply collapsing and making for the riverbank, where thousands of Union soldiers—some said five thousand or more—sat shocked and beaten. Grant had been everywhere all day, often placing small units personally during the confusion in the morning and trying to orchestrate the withdrawal to this last-ditch position in the afternoon. At one time, drawn up with his staff on horseback in an open space under fire, he became so intent on studying the battlefield that one of his staff told him that if they did not move from there, "We shall all be dead in five minutes." Grant came out of his trance, said "I guess that's so," and led them to a safer place. By late afternoon, every unit that Grant could rally, including those under Sherman that had finally fallen back from the right flank, was in place along this last defensive line. His artillery threw back one Confederate assault, and he was bracing his weakened army for a final Confederate attack that might drive them over the bluff and into the river.

It did not come. First, there was a delay while large numbers of Beauregard's men rounded up the twenty-two hundred prisoners from the Hornet's Nest and moved them some distance down the road to Corinth. But it was more than that: the Confederates were done for the day. The news of Johnston's death had reached many of them and, like the Union soldiers facing them, they had experienced hours of what one Confederate described thus: "It was an awful thing to hear no intermission in firing and hear the clatter of small arms and the whizing minny [minié musket] balls and rifle shot and the sing of grape shot the hum of cannon balls and the roar of the bomb shell and explosion of the same seaming to be a thousand every minute . . . O God forever keep me out of such a fight."

The ground at the front was strewn with corpses of men and horses, and from hundreds of thickets the wounded of both sides cried for help. The battle was not over; something had to happen tomorrow, but as the dusk of the spring evening closed on the square miles of battlefield and the Union and Confederate artillery continued to fire into each other's lines, no one had the strength to fight on foot.

Grant and Sherman had survived this Sunday by near miracles. After Sherman's fine "race mare" was shot from under him in the morning and he took his aide's horse, that mount too was killed as he rode it under fire from one position to another, and the horse he then borrowed from a surgeon was also killed later in the day. At one moment a spent bullet cut through the cloth of Sherman's uniform and bruised his shoulder without breaking the skin.

At twilight, Grant and Sherman nearly died within a minute of each other. Grant stood beside the road that led down to the wharf at Pittsburg Landing, watching with relief as some of Buell's reinforcements, ferried from across the river, landed and started marching up to the top of the bluff. An incoming cannonball tore the head off a captain who was standing beside him, cut off the top of a saddle on a horse just behind him, and went on to take off both legs of a soldier marching up from the river. As Sherman swung up into the saddle of the fourth horse he had used that day, with a major holding the horse's reins to make it easier for Sherman to mount because of his wounded hand, the horse pranced, and the reins became tangled around Sherman's neck. As Sherman bowed low above the horse's mane so that the major could lift the reins above his head to straighten them out, a cannonball cut through the reins two inches below the major's hands and slashed off the crown and back rim of Sherman's hat.

The first men coming up the bluff to reinforce Grant's army, marching with bayonets fixed on their muskets, were from a Kentucky brigade, troops who had been contemptuous of Sherman when he commanded them so timorously at Louisville. Now, as they reached the top of the bluff, they saw him sitting calmly on a horse in the evening light, a bloody bandage covering one hand, powder stains and crusted blood on his face, and his hat blown to rags. Putting their hats on their bayonets, they raised their muskets toward the sky and cheered him as they passed.

That night, Beauregard slept in Sherman's tent, while the battlefield echoed with the sound of United States Navy gunboats periodically firing at the enemy. At ten o'clock it started to rain in the torrential way it had when the Confederates were on the march from Corinth. Until late, Grant rode around his camp, watching as thousands of reinforcements marched in and talking with his commanders as he came to them. Perhaps remembering the moment at Fort Donelson when he sensed that whoever attacked next would win the battle, he said to one of his brigadiers, "Whichever side takes the initiative in the morning will make the other retire, and Beauregard will be mighty smart if he attacks before I do."

As the battlefield turned to mud—Grant said that, with their tents in enemy hands, "our troops were exposed to the storm without shelter"—agonized moans came from both the wounded who lay lost in the dark and the men being treated by overworked surgeons and medical orderlies. Grant described part of his experience during that night of suffering.

I made my headquarters under a tree a few hundred yards back from the riverbank. My ankle was so much swollen from the fall of my horse the Friday night preceding, and the bruise was so painful that I could get no rest. The drenching rain would have precluded the possibility of sleep without this additional cause.

Some time after midnight, growing restive under the storm and the continuous pain, I moved back to the log-house under the bank. This had been taken as a hospital, and all night wounded men were being brought in, their wounds dressed, a leg or an arm amputated as the case might require, and everything was being done to save life or alleviate suffering. The sight was more unendurable than encountering the enemy's fire, and I returned to my tree in the rain.

Sherman found him there, standing under that tree in the rainy night, supporting himself on a crutch. Grant had a lantern in one hand; his coat collar was up; rain dripped from the brim of his hat. He had a cigar clenched in his teeth. At this point Sherman, like Grant, knew that the incomplete casualty lists were already enormous: of the forty thousand Union soldiers who had started the day, ten thousand would be listed as killed, wounded, or missing, more than two thousand of the latter as prisoners. Thousands who had run from the battle still sat demoralized beside the river, huddled together in the rain, useless as soldiers.

During this bloody and terrible day, Grant had learned a lot about Sherman; now Sherman was about to encounter the essence of Grant. Sherman had reached the conclusion that the relentlessly courageous Confederate attacks had so shocked and battered the Union divisions that, even with the arrival of reinforcements, it would be best "to put the river between us and the enemy, and recuperate." He had sought out Grant to discuss how they could make such a withdrawal, from this bank of the river to the other side. Now, looking at Grant's strong, thoughtful face in the rain and lantern light, he was "moved by some wise and sudden instinct" not to mention retreat. Instead he said, "Well, Grant, we've had the devil's own day of it, haven't we?"

Grant said, "Yes," and remained silent for a minute as they stood together in the falling rain, Grant on one crutch and in pain from his injured ankle, and Sherman in pain from his wounded hand. Then Grant added, "Lick 'em tomorrow, though."

Dawn brought something close to the reverse image of the previous day. Beauregard, who the night before had sent Jefferson Davis a wire that included the triumphant words "A COMPLETE VICTORY," said of his frame of mind, "I thought I had General Grant just where I wanted him and could finish him up in the morning." He might have been less confident if he had received a report from Nathan Bedford Forrest, who dressed several of his cavalrymen in federal uniforms—had they been captured, they would have been executed as spies—and sent them to penetrate Grant's lines. They returned with reports of thousands of troops disembarking at Pittsburg Landing throughout the night, but when Forrest passed this on to Hardee, that general thought that the scouts had seen Grant evacuating his army. Beauregard never received any intelligence that Grant was being reinforced.

Grant intended to strike first, and hard. By three a.m. he had given orders to his commanders to push forward "heavy lines of skirmishers" at first light and keep moving until they encountered the enemy. Entire Union divisions were to march right behind the skirmishers, "to engage the enemy as soon as found." As usual, Grant left his commanders with great flexibility as to how to implement his plan: when Brigadier General Lew Wallace, commander of a newly arrived division of reinforcements (and the man who later wrote *Ben-Hur*) asked him where to start placing his troops for the dawn attack, Grant pointed toward Sherman's end of the line and said, "Move out that way." In answer to Wallace's question as to what formation he should adopt, Grant responded, "I leave that to your discretion." Wallace said of Grant's manner, "If he had studied to be undramatic, he could not have succeeded better."

The battle resumed at first light. A soldier of the Thirty-eighth Tennessee said, "At daybreak our pickets came rushing in under a murderous fire. The first thing we knew we were almost surrounded by six or seven regiments of Yankees." This time it was the Confederates who could not hold. Union brigadier Jacob Ammen, an old West Pointer who had been Grant's mathematics instructor at the academy, recorded that "the rebels fall back slowly, stubbornly, but they are losing ground." At places the Southerners counterattacked, but this morning Grant, with twenty thousand new if untried reinforcements, had the momentum and the same number of men as the day before, while Beauregard had only half his previous day's strength and no replacements on the way.

Once again, Grant, like Sherman, was in places of danger, and sometimes encountered trouble in places he did not consider hazardous. After a morning during which the outnumbered Confederates kept withdrawing but never broke, Grant and two members of his staff, Colonel James B. McPherson and a Major Hawkins, were, as Grant recalled, riding

along the northern edge of a clearing, very leisurely, toward the river above [upstream from] the landing. There did not appear to be an enemy to our right, until suddenly a battery with musketry opened upon us from the edge of the woods on the other side of the clearing. The shells and balls whistled about our ears very fast for about a minute. I do not think it took longer than that for us to get out of range and out of sight. In the sudden start we made, Major Hawkins lost his hat. He did not stop to pick it up.

When we arrived at a perfectly safe position we halted to take account of damages. McPherson's horse was panting as if ready to drop. On examination it was found that a ball had struck him forward of the flank just back of the saddle, and gone entirely through. In a few minutes the poor beast dropped dead; he had given no sign of injury until we came to a stop. A ball had struck the metal scabbard of my sword, just below the hilt, and broken it nearly off; before the battle was over it had broken off entirely. There were three of us: One had lost a horse, killed; one a hat and one a sword-scabbard. All were thankful that it was no worse.

It was two-thirty. Both sides had been fighting since dawn. The Confederates were still in the battle—one of Hardee's brigades, after a failed counterattack, had lost 1,000 killed and wounded, out of 2,750 who had started fighting the previous morning—but, as one of Beauregard's staff said, "The fire and animation had left our troops." Another of Beauregard's staff went to him and put the matter gently. "General," he said, "do you not think our troops are very much in the condition of lump sugar, thoroughly soaked with water, but yet preserving its original shape, though ready to dissolve? Would it not be judicious to get away with what we have?"

Beauregard, who was in all probability not thinking of his suffering army as a lump of sugar, replied, "I intend to withdraw in a few minutes." Some two thousand soldiers and twelve cannon were placed near Shiloh Church to cover the withdrawal, and within an hour the remnants of the late Albert Sidney Johnston's army began an orderly retreat back down the road to Corinth.

Here was the moment for Grant's divisions to strike them from the rear and vanquish them all, but the Union soldiers had nothing left. Men lay on the muddy ground, panting and vomiting from exhaustion; some fell asleep right where they were after these two days of battle; others who had been sick gave way to their illnesses. Grant said that he "wanted to pursue, but had not the heart to order the men who had fought desperately for two days, lying in the mud and rains whenever not fighting," and concluded that "my force was too much fatigued to pursue. Night closed in cloudy and with terrible rain, making the roads impracticable for artillery by the next morning."

This was not going to be as clear-cut a scene of victory as at Fort Donelson, with thousands of Confederate soldiers marching out of their earthworks to surrender. In fact, it was going to be far worse than that for the Southern soldiers. Within a mile or two down the Corinth road, men began staggering to the side of the road and falling, unable to go on. The retreating column was seven miles long, and a man with them described it.

> Here was a long line of wagons, loaded with wounded, piled in like bags of grain, groaning and cursing, while the mules plunged on in mud and water belly-deep, the water sometimes coming into the wagons. A cold, drizzling rain commenced about midnight and soon came harder and faster, then turned to pitiless blinding hail. This storm raged with unrelenting violence for three hours. I passed wagon trains filled with wounded and dying soldiers without even a blanket to shield them from the driving sleet and hail, which fell as large as partridge eggs, until it lay on the ground two inches deep.

The next morning, Sherman rode forward down the Corinth Road with four infantry brigades and some cavalry, not to resume the battle but to make sure that all the Confederates were in fact leaving the area. In some confusing terrain, he found himself and his staff separated from the screen of federal infantrymen who were supposed to be moving ahead of them. His sudden vulnerability was noticed by Nathan Bedford Forrest, who was guarding the rear of the Confederate column with 350 of his intrepid mounted troopers. Forrest shouted "Charge!" and came right for Sherman and the officers around him.

"I and my staff ingloriously fled pell mell through the mud," Sherman recalled. "I am sure that if Forrest had not emptied his pistols as he passed through the skirmish line, my career would have ended right there." Forrest's career also nearly ended: intent on trying to kill or capture what he rightly thought was a general and his staff, he spurred his horse through the underbrush straight into a line of federal foot soldiers he had not seen and was hit by a bullet that stopped in his back. With his unfailing presence of mind, Forrest grabbed a Union soldier up off the ground from where he stood, somehow placed the man behind him on his horse so that the soldier's comrades would not shoot at this human shield, and galloped away—the last man to shed blood at Shiloh.

When Sherman rode back into his original encampment around Shiloh Church at the head of the force he had taken down the Corinth Road, soldiers came from all directions to hail him. As the cheers went "rolling down the line," a man who was there said that "he rode slowly, his grizzled face beaming with animation, his tall form swaying from side to side, his arms waving." Halting on his horse in the middle of a crowd, his raspy voice rang out: "Boys, you have won a great victory. The enemy has retreated to Corinth."

For Grant and Sherman, there was the matter of telling their wives. Within hours of Sherman's triumphant return to camp, Grant wrote home.

> Dear Julia,
>
> Again another terrible battle has occurred in which our arms have been victorious. For the numbers engaged and the tenacity with which both parties held on for two days, it has no equal on this continent. The best troops of their rebels were engaged . . . and their ablest generals . . . The loss on both sides was heavy probably not less than 20,000 killed and wounded altogether.
>
> I got through all safe having but one shot which struck my sword but did not touch me.
>
> I am detaining a steamer to carry this and must cut it short.
>
> Give my love to all at home. Kiss the children for me. The same for yourself.
>
> Good night dear Julia.
>
> Ulys.

Three days later, Sherman told his wife of Shiloh in a long letter, the end of which is missing, which he began almost in the tone of an excited schoolboy.

> Dearest Ellen,
>
> Well we have had a big battle where they shot real bullets and I am safe, except a buckshot wound in the hand and a bruised shoulder from a spent ball . . . Beauregard, Bragg[,] Johnston, Breckinridge and all their Big men

were here, with their best soldiers and after the Battle was over I found
among the prisoners an old Louisiana Cadet named Barrow who sent for me
and told me all about the others, many of whom were here and Knew they
were fighting me. I gave him a pair of socks, drawers and Shirt and treated
him very kindly. I won[']t attempt to give an account of the Battle, but they
Say that I accomplished some important results, and Gen. Grant makes special mention of me in his report which he shew [sic] me . . .

The scenes on this field would have cured anybody of war. Mangled
bodies, dead, dying, in every conceivable shape, without heads, legs; and
horses! All I can say this was a Battle . . . I know you will read all accounts—
cut out paragraphs with my name for Willy's future Study—all Slurs you will
hide away, and gradually convince yourself that I am a soldier as famous as
Gen. [Nathanael, of the Revolution] Greene . . .

You ask for money—I have none, and am now without horse saddle bri-
dle, bed, or anything—The Rebels, Breckinridge had my Camp and cleaned
me out.

At about this time, Sherman began his appraisal of the performance
during the battle of his officers and men, and brought court-martial
charges against four of the twelve commanders of his infantry regiments.
Addressing the assembled officers and men of the Fifty-third Ohio, most
of whom had fled with their commander, Colonel Appler, to the riverbank early in the battle, with some trying to get themselves evacuated
with the wounded being taken downstream by boat, Sherman gave them
a verbal whipping that lasted for an hour. An Indiana soldier, a bystander
who heard it all, said Sherman told them that "they were a disgrace to
the nation and finally wound up by promising them that at the next battle they would be put in the foremost rank with a battery of Artillery immediately behind them and then if they attempted to run they would
open on them with grape and canister."

The Battle of Shiloh had ended, and the battle about what happened
there now began. For the South, the process was simple. After Beauregard's first triumphant telegram to Jefferson Davis, joy and relief pulsed
through the Confederacy. Then came the news of the death of Albert
Sidney Johnston, then the fact that some of the South's finest generals
had been forced to retreat back to Corinth, and then the casualty figures:
1,723 dead, 8,012 wounded, 959 missing. The Union losses were mar-

ginally higher, but, with no more talk about each Southern man being
able to beat three, four, five Yankees, it was clear that the South had lost
more blood than it could afford to lose. Strategically, so much had been
at stake for the Confederacy at Shiloh: it was the South's chance to reverse Grant's victories at Fort Henry and Fort Donelson, the opportunity
to fight back into Tennessee and Kentucky, the chance to become equal
again in the Mississippi River theater of operations that Grant and Sherman knew held the key to the war. All that was gone.

And there was morale. Memories of the rout inflicted on the North at
Bull Run were no comfort to Southerners hearing the realities of the
biggest battle fought on the North American continent to that date. The
Southern public learned of a pond in Shiloh's woods that turned red
from the blood of men killed and wounded around it, of fields you could
not walk across without stepping on bodies, of seven hundred Confederate soldiers buried in a mass grave, of the carcasses of five hundred
horses that had to be burnt where they fell, of a large and growing pile of
amputated arms and legs in the courtyard of the makeshift hospital
hastily set up in the Tishomingo Hotel at Corinth. The Confederacy was
determined to fight on, but as one of its soldiers, the writer George Washington Cable, said, "The South never smiled after Shiloh."

In the North, the first news was of a tremendous, categorically successful victory. Bells rang; Congress recessed for a day; President Lincoln
proclaimed a national day of prayerful thanksgiving. Beauregard had
called off the battle on Monday, April 7; on April 9, *The New York Times*,
The New York Herald, and the *New York Tribune* came out with headlines proclaiming a tremendous victory. Grant was the man of the hour.
The next day's edition of the *Tribune* was filled with stories of his courage
and skill: splendid, determined soldier; smokes cigars; man of few words
but magnificent results.

The next day's *Tribune* struck a different note, echoed in other cities
as reporters arrived at Pittsburg Landing and began buttonholing anyone
in sight for stories of the battle. Grant's army had been taken by surprise.
Why was he having breakfast nine miles away, when his men were attacked while they were asleep in their tents? Thousands of men ran away
from the battle. There was confusion everywhere, throughout the battle.

The casualty lists came in, with totals higher than the four previous
biggest battles of the war combined, and the journalistic hunt was on in
earnest. The Confederates had dashed into the Union soldiers' tents and

bayoneted them in their sleep. The *Tribune* came out with an editorial, "Let Us Have the Facts." There had been no preparations for anything at Pittsburg Landing. Grant was drunk. Was it a victory at all? A. K. McClure, a nationally prominent Republican supporter from Philadelphia, went to see Lincoln in the White House and urged him to remove this questionable figure Grant from command. Lincoln listened to everything McClure had to say, and thought for a while. Then he shook his head. "No, I can't do it. I can't lose this man. He fights."

Initially, the criticism focused on Grant. Then a paper in Ohio published an article by Lieutenant Governor Benjamin Stanton (no relation to Secretary of War Edwin M. Stanton). In a piece of political grandstanding, Stanton had arrived at Pittsburg Landing soon after the battle, bringing with him five thousand dollars to help troops from Ohio. Stanton heard the same inaccurate and exaggerated tales the reporters were getting and, when he met Sherman, the senior officer from Ohio present at Shiloh, never asked him a single question about the battle. Returning home, Stanton wrote a diatribe against the Union generals, not mentioning Sherman by name but referring to "the blundering stupidity and negligence" of Grant.

An incensed Sherman entered the fray with a letter to Stanton. Early in it he set the tone: "The accusatory part of your statement is all false, false in general, false in every particular, and I repeat, you could not have failed to know it false when you published that statement . . . Grant just fresh from the victory of Donelson, more rich in fruits than was Saratoga, Yorktown, or any other fought on this Continent, is yet held up to the people of Ohio . . . as one who in the opinion of intelligent cowards is worthy to be shot . . . Shame on You!" Sherman added that no colonel of the Ohio regiments had any idea of how Stanton spent the five thousand dollars that was supposed to benefit the soldiers from Ohio.

Stanton replied to this with another diatribe, this time mentioning Sherman by name, and the fight was on. The Sherman team went to work: Sherman's brother Senator John Sherman, Sherman's famous and well-regarded father-in-law, Thomas Ewing, and Ellen's brother, the influential lawyer Philemon Ewing, all Ohioans, filled newspaper columns with criticism of Stanton, with Stanton responding every time; before the storm ran its course, both sides were printing pamphlets setting forth their views.

As for Grant, he was heartsick about the slurs on his reputation, writing Julia that he had been "so shockingly abused" by the press, but he preferred to remain silent. Grant's father decided to go to the support of his son, releasing to the press a brief, entirely personal letter he received from Grant defending his conduct at Shiloh, and a long letter from Captain William S. Hillyer of Grant's staff that praised Grant's actions and asked the rhetorical question, "Is *success* a crime?" Grant complained to Julia that the publication of these letters "should never have occurred." His father made other efforts, which included two letters in the *Cincinnati Commercial* signed by a close friend of the elder Grant but possibly written by Jesse Grant himself. Grant's father also wrote Congressman Washburne to thank him for a speech he made on his son's behalf, and wrote the governor of Ohio, saying of Lieutenant Governor Stanton, "Shame on such a Demagogue."

Realizing that his father was defending him in part by criticizing the performance of other Union commanders, Grant sent him an angry letter saying that there was "not an enemy in the world who has done me so much injury as you in your efforts in my defense. I Require no defenders and for my sake leave me alone . . . Do nothing to correct what you have already done but for the future keep quiet on this subject." He closed the letter with, "My love to all at home. Ulys."

Julia, staying with her father-in-law in Covington, Kentucky, at the time Shiloh was fought, was reading a Cincinnati paper a few days after that. She had just finished an article that said Grant was at a "dance house" instead of being on the battlefield, when an unexpected guest arrived.

> A tall, handsome woman, clad in deepest mourning, entered the little parlour . . . Coming directly up to me, she said, "Mrs. Grant, I am an entire stranger to you, and I have come an entire day's journey out of my way to tell you this." She paused a moment, choking down a sob, and said, "I am the widow of Colonel Canfield. I have just lost my husband at Shiloh. I must tell you of your husband's kindness to me."

Feeling a premonition, this woman had managed somehow to get herself to Pittsburg Landing at the end of the first day of the Shiloh fighting. There at the waterfront she was told that her husband, Lieutenant Colonel Herman Canfield of the Seventy-second Ohio, had been wounded and was in a hospital down the river at a place she had passed on her way up to the battle. At that moment she saw Grant and his staff ride onto the pier and watched him being assisted off his horse because of his injured ankle and helped aboard his headquarters riverboat the *Tigress*, where he began writing dispatches. Warned by the sentries that she could not come aboard, she swept by them, found Grant, and explained her plight. Grant told her that she could stay on the *Tigress*, which would soon be taking the report he was writing downriver, and made out the appropriate passes to get her ashore and into the hospital where her husband lay.

Julia asked her, "Did you reach your husband in time, Mrs. Canfield?"

"Oh, no," she sobbed. "I was late, too late. I was conducted down the aisle between the cots in the hospital, and my escort paused and pointed to a cot, the blanket drawn up so as to cover the face. I knelt beside it and drew the covering down.

"He was dead—my husband, my beloved, my noble husband. I thrust my hand into his bosom. It was still warm, but his great heart had ceased beating.

"The blood was clotted on his beard and breast. I think he might have lived if I had been near," she sobbed. "I have determined to devote my time to the wounded soldiers during the war. My husband needed only the services of a kind nurse."

The ladies parted. Mrs. Canfield did indeed throw herself into nursing wounded soldiers, and Julia would see her next in three years, under supremely dramatic circumstances.

On April 11, four days after the large-scale fighting at Shiloh finished, Halleck arrived at Pittsburg Landing. Being senior, according to plan he took over Grant's field army and placed Grant in a meaningless role as his assistant commander of the armies of the West. It was in effect a demotion for Grant. At a time when Grant and Sherman felt that their army was sufficiently recovered from the battle to march the twenty miles to Corinth and finish the destruction of Beauregard's remaining force, Halleck began to plan his version of how to proceed. No one was going to catch Halleck by surprise or undermanned. Learning that Beauregard had received reinforcements at Corinth that brought the Confederate strength back up to sixty-six thousand, he did not begin his march until he had one hundred thousand men and two hundred cannon assembled at Pittsburg Landing. Then Halleck started the twenty-mile trek. He would march the army two or three miles in a day, stop and construct elaborate earthworks, camp there two or three days, and move on another few miles and do the same thing.

Although Sherman admired Halleck at this time, he recalled that his division "constructed seven distinct entrenched camps" in this astonishingly slow movement toward Corinth; a newspaper reporter who accompanied the Union force said that "Halleck crept forward at the rate of about three-quarters of a mile per day." (Grant, watching this without the authority to change a single detail, referred to it as a siege on the move.)

Halleck finally reached Corinth on May 28—seven and a half weeks after the last day of fighting at Shiloh—and the next day began a massive artillery bombardment of the town. From time to time trains moved in and out of Corinth, to the sounds of cheering, and Halleck saw this as proof of the arrival of more and more enemy troops and the Confederate determination to hold that essential transportation crossroads at all costs. In fact, Beauregard was evacuating his last units from Corinth, having the trains come back empty and ordering everyone left in the town to shout as they arrived. When Halleck finally sent foot soldiers into Corinth on May 30, they found not a single enemy soldier. (In a final touch, when Halleck rode in himself, his horse tripped over a telegraph wire hanging just above the ground.)

During this time, Grant remained with the army, doing nothing because as Halleck's second in command there was nothing for him to do. In addition to his feelings about the press controversy over his leadership at Shiloh, Grant was dismayed by the way Halleck so signally failed to exploit the battle in which so many men suffered and gave their lives. At one point during the ponderous march to Corinth, he wrote a formal request to Halleck, whose tent was only two hundred yards from his, asking that he be given a field command, or be relieved from further duty. Halleck, who recently had sent out a petulant order emphasizing that all letters on military matters *should relate to one matter only, and be properly folded*," refused to do either.

Sherman, whom Grant had praised and commended during and after the battle, stumbled upon the chance to save Grant for the Union cause. When Sherman paid a call on Halleck before leaving Corinth to see if it was possible to salvage some locomotives and railroad cars that had been abandoned in the swamps west of Chewalla, Tennessee, Halleck "casually mentioned to me that Grant was going away the next morning." When Sherman asked why, Halleck said that he did not know, but Grant had asked him for a thirty-day leave, and he had agreed.

Sherman knew why. For himself, he had reason to be grateful to Halleck, who had given him the division he led so well at Shiloh that he was now being promoted to major general, but he had seen Grant "chafing under the slights of his anomalous position" and decided to go and see him before he left for Chewalla. Arriving at Grant's headquarters, which "consisted of four or five tents, with a sapling railing around the front," he saw packing going on that indicated Grant was not simply going on leave, but leaving the army. Shown into Grant's tent, he found him "seated on a camp-stool, with papers on a rude camp-table," methodically sorting out letters and putting bundles of them aside.

> After passing the usual compliments, I inquired if it were true that he was going away. He said, "Yes." I then inquired the reason, and he said, "Sherman, you know. You know that I am in the way here. I have stood it as long as I can, and can endure it no longer." I inquired where he was going to, and he said, "St. Louis." I then asked him if he had any business there, and he said, "Not a bit." I then begged him to stay, illustrating his case by my own.
>
> Before the battle of Shiloh, I had been cast down by a mere newspaper assertion of "crazy," but that single battle had given me new life, and I was now in high feather; and I argued with him that, if he went away, events would go right along, and he would be left out; whereas, if he remained, some happy accident might restore him to favor and his true place.
>
> He certainly appreciated my friendly advice, and promised to wait awhile; at all events, not to go without seeing me again, or communicating with me.

That very evening, Grant wrote Julia that "Necessity however changes my plans, or the public service does, and I must yeald [sic]." The next day he wrote to Sherman at Chewalla, saying, as Sherman put it, "that he had reconsidered his intention and would remain." Later, Sherman could not find that note from Grant, but to the end of his life he kept a copy of the reply he sent Grant the same day. In it he went on to rail against the treatment they both had received from the press, but the part he chose to make public in his memoirs was this:

Chewalla, June 6, 1862

Major-General GRANT.

MY DEAR SIR: I have just received your note, and am rejoiced at your conclusion to remain; for you could not be quiet at home for a week when armies are moving, and rest could not relieve your mind of the gnawing sensation that injustice had been done you.

A letter from Grant that did survive is one he sent soon after that, not to Sherman but in reply to a letter from Ellen Sherman. Writing her from Memphis while Sherman was in another area, Grant tried to soothe the worries Ellen expressed concerning what Sherman described as "a touch of malarial fever" that he contracted during his inspection of the locomotives abandoned in the Tennessee swamps. In it he praised Sherman's "indefatigable zeal and energy" and assured Ellen that he had suggested Sherman take a leave to recover fully, but that "he would not listen to it." Grant said that "although Gen. Sherman's place would be hard to fill," he would again see "if he will consent to a leave," and went on to say, "There is nothing that he, or his friends for him, could do that I would not do [for him] if it were in my power. It is to him and some other brave men like himself that I have gained the little credit awarded me, and that our cause has triumphed to the extent it has."

The "happy accident" that Sherman hoped would happen for Grant occurred five weeks later: on July 11, after McClellan was completely outgeneralled by Robert E. Lee in the Seven Days' Battles in northern Virginia, Lincoln ordered Halleck to Washington to serve in the reconstituted post of general in chief. Henry Halleck now commanded the entire United States Army. There would eventually be questions of how wise a promotion that was for the Union cause as a whole, but Ulysses S. Grant took over from Halleck more or less by default, becoming the commander of two of the three Western armies. He thought that he could finally begin to fight the war in the Mississippi theater of operations the way he wanted to do it. Halleck was still Grant's superior, but he would be a thousand miles away, with much else on his mind.

The growing personal and military relationship between Grant and

Sherman had reached a somewhat paradoxical moment. From the time of Shiloh, Grant strongly praised Sherman. Writing his official report of the battle two days after it occurred, Grant singled out Sherman for his highest commendation: "I feel it a duty however to a gallant and able officer Brig Genl W T Sherman to make special mention. He not only was with his Command during the entire two days action, but displayed great judgment and skill in the management of his men. Altho severely wounded in the hand the first day, his place was never vacant. He was again wounded and had three horses killed under him."

Three weeks later he wrote to Julia, "In Gen. Sherman the country has an able and gallant defender and your husband a true friend," and in a later letter to her said, understating his role in Sherman's rise, "Although Gen. Sherman has been made a Maj. Gen. by the battle of Shiloh I have never done half justice by him. With green troops he was my standby during that trying day of Sunday, (there has been nothing like it on this continent—nor in history.) He kept his Division in place all day, and aided materially in keeping those to his right and left in place—He saw me frequently and received, and obeyed, my directions during that day."

In these five months that had brought them together, Sherman made varying estimates of Grant. After Fort Donelson, he wrote his brother John that "Grant's victory was most extraordinary and brilliant." After Shiloh, preferring to speak of Grant's earlier victory at Fort Donelson while controversy surrounded the recent battle, Sherman told Grant in a letter that "you obtained a just celebrity at Donelson, by a stroke of war more rich in consequences than was the battle of Saratoga." On the same day, he wrote Ellen that "he is brave as any man should be, has won several victories such as Donelson which ought to entitle him to universal praise . . ." but added, "He is not a brilliant man . . . but he is a good & brave soldier tried for years, is sober, very industrious, and as kind as a child."

By comparison, when Sherman learned that Halleck was ordered east as general in chief, with Grant to be his successor in the West, he wrote Halleck this:

I cannot express my heartfelt pain at hearing of your orders and intended departure . . . That success will attend you wherever you go I feel no doubt . . .

I attach more importance to the West than the East . . . The man who at the end of this war holds the military control of the Valley of the Mississippi will be the man. You should not be removed. I fear the consequences . . .

Instead of that calm, steady progress which has dismayed our enemy, I now fear alarms, hesitation, and doubt. You cannot be replaced out here.

This letter, written by Sherman to the man who gave him the chance to come back from disgrace to prominence and promotion, combined genuine admiration with flattery and perhaps a measure of duplicity. Telling Halleck, "You should not be removed" and "You cannot be replaced," indicated a preference for his leadership over that of Grant. To characterize Halleck's glacial military movements as "calm, steady progress" flew in the face of the squandered opportunities Sherman had witnessed. For Sherman to add that, with Halleck gone, he feared there would now be "hesitation, and doubt" was astonishing: those characteristics had marked all of Halleck's command in the West and were the antithesis of Grant's approach to war.

And yet it appeared, from other letters Sherman wrote, that he was not only grateful to Halleck but truly admired him, perhaps because of his orderly, methodical approach to so many matters. (Sherman was not alone in praising Halleck: many of Halleck's officers and men saw the fall of Corinth as an intelligently planned bloodless taking of an important strategic point and gave Halleck the nickname "Old Brains.") No one ever doubted that Sherman had a complicated personality: those words of praise for Halleck came from the same man who had supported Grant in every way since the war had brought them together. First, Sherman had done an outstanding job in forwarding men and supplies to Grant at Fort Donelson and in handling the movement of wounded and prisoners resulting from that battle. At Shiloh he outperformed Grant's other generals. Then Sherman, who had reason to avoid the attention of newspapers that could quickly remind their readers of his earlier failure of nerve in Kentucky, threw himself into defending Grant in his replies to an attack in which he himself was not initially named. Finally, as a newly promoted major general who could have regarded the departure from the army of another man of the same rank as an opportunity to advance himself, Sherman had talked Grant out of leaving the army.

What developed next between Grant and Sherman would influence both the military and political aspects of the war, and its results. Sherman had more to learn about Grant, and himself. Through conversations and letters they would, often without realizing it, teach each other about the nature of the war they were experiencing. Between them they would evolve a harsh and efficient philosophy of war that would affect the South's civilian population as well as its armies, and begin to apply those measures.

Fifteen months into the war, after Donelson, after Shiloh, at times both men thought it would soon end. Speaking of the Confederacy, Sherman wrote his brother that "the People are as bitter against us as ever, but the Leaders must now admit they are defeated." A week after Shiloh, Grant told Julia that he expected "one more fight and then easy sailing to the end of the war. I really will feel glad when this thing is over." Reflecting on the battle, Grant later concluded that "it is possible that the Southern man started in with a little more dash than his Northern brother; but he was correspondingly less enduring." In midsummer of 1862, both Grant and Sherman still had a lot to learn about Southern endurance, and about each other.

6

POLITICAL PROBLEMS, MILITARY CHALLENGES:
THE VICKSBURG CAMPAIGN DEVELOPS

When Halleck went east to take command of the entire United States Army, Grant inherited a situation in which Halleck had, as Sherman put it, "scattered" eighty thousand men into small garrisons all over northern Mississippi and western Tennessee. This was a time when the aggressive Grant and newly confident Sherman should have been able to take the experienced divisions that had fought at Shiloh and use them to seek out and attack the enemy. Instead, they found themselves inheriting a situation in which Halleck had tasked his forces with the duty of occupying cities and towns and guarding railroad lines, rather than engaging the Confederate generals they had recently defeated.

Sherman, newly promoted to major general of Volunteers, became the military governor of Memphis, then a city of twenty-three thousand, while his superior Grant made his headquarters at Corinth, a hundred miles to the east. Grant now commanded a military department that was geographically composed of northern Mississippi and the western parts of Tennessee and Kentucky; his spread-out field command consisted of the Army of the Tennessee, which he had commanded at Shiloh, and the army now under the command of Major General William Rosecrans that had been known as the Army of the Mississippi. In this period of flux, Grant was trying to rebuild a central striking force while dealing with Confederate guerrilla raids supported by the population of a large territory loyal to the Confederacy. He encouraged Sherman to send him not only military reports but also to "write freely and fully on all matters of public interest."

Memphis, initially paralyzed by the Union victory at Shiloh and the sight of long federal columns marching into the city, quickly provided Sherman with many such "matters of public interest." Soon after arriving at the city on the Mississippi in late July, he and his staff attended Sunday services at Calvary Episcopal Church. The intention was to appear quietly, in a neutral, peaceful setting. However, Sherman noticed that the clergyman omitted the prayer always said in prewar times for the president of the United States. Instantly, Sherman rose in the midst of the congregation and said the prayer in a loud, authoritative voice. The following day he decreed that the prayer would be offered at the next service, or the church would be closed. The prayer was duly said, and the church remained open.

That set the tone for Sherman's conduct of the Union occupation of Memphis and the city's response. Telling the mayor that "the Military for the time being must be superior to the Civil authority but does not therefore destroy it," he reorganized the police force, established order, reopened everything from schools to theaters and saloons, and encouraged the resumption of all commercial activity including local riverboat trade in nonmilitary items. The former California banker even organized a real estate department in which his quartermasters opened up buildings vacated due to the war, rented out the space, and held the profits to be paid out later to owners willing to declare their loyalty to the federal government. Even though, as one of his officers wrote home to his mother, "Sherman never utters a word to bring the blush to the cheek of a maiden," he let the city's famous bordellos, known as "parlor houses," continue their activities. Perhaps he learned of the attitudes of the many black prostitutes, who until then had only Southern white men as clients: a Union cavalryman said those women "felt loving towards us because they thought we were bringing them freedom, and they wouldn't charge us a cent."

During this time, Sherman and Grant had to develop ways to implement the evolving federal policies on the treatment of slaves. When the war began, the issue of secession, rather than the abolition of slavery, dominated the minds of most Northerners. Now, with Union armies controlling Southern communities, farms, and plantations, tens of thousands of slaves sought federal protection.

This reality—masses of slaves, many of them fugitives and all of them desperate for help and needing a new civil status—forced Abraham Lincoln to reconsider the issue. Although he wanted to free the slaves, he saw as his highest duty the preservation of the United States as one nation. At the war's outset, trying to keep the border states out of the Confederacy, he had skirted the slavery issue to avoid antagonizing the many slaveholding families in those states. As the war progressed, Lincoln still had hopes of negotiating an early end, and he hesitated to put the abolition of slavery foremost—the position of the Radical faction of the Republicans in Congress—while he explored the possibilities of reaching peace at a table with Confederate leaders. Now, however, there were slaves to be cared for, a negotiated peace seemed beyond reach, and yet there was no law protecting the freed slaves. Their status was in such a legal limbo that earlier in the year, in March, Congress had enacted an article of war expressly forbidding the Union Army to return fugitive slaves to their masters.

Apart from the ideal of ending slavery, the North began to see that freeing the slaves, hitherto considered to be their owners' legal property in the way that a horse or a house was, could be an economic weapon that also produced military advantages. Unpaid labor of slaves was an integral, vital part of the Southern economy, and to take slaves away from their owners would undermine the Confederacy's infrastructure in ways that would also reduce its ability to continue the fight. On July 17, four days before Sherman arrived to rule Memphis, Congress passed the Second Confiscation Act. Motivated by a combination of idealism and practicality, this law freed the slaves of "persons engaged in or assisting the rebellion" and also provided for the seizure and sale of other property belonging to those actively supporting the Southern cause. On September 22, while Sherman still governed Memphis, Lincoln issued the Preliminary Emancipation Proclamation. This statement declared that unless the Confederate states ceased their rebellion by the end of the year, as of January 1, 1863, *all* slaves in those states, not just those belonging to Confederate supporters, would be freed. In a continuing effort to keep the border states on the Union side, it did not declare that slaves outside of the seceded states would automatically be freed, but pledged a form of compensation for border states that adopted either immediate or gradual emancipation. (When the Emancipation Proclamation became operative, it also provided for the enlistment of black men in the army.)

The point came home to the entire white South: the war was now being waged not only on the military front; it would reach right into every part of the Confederacy's economic life. In addition to their many other duties and concerns, Grant and Sherman started turning law into reality. On August 17, Sherman wrote Grant, concerning the Southern whites' reaction, "Your orders about property and mine about niggers make them feel they can be hurt," and in a letter of October 16 he added this thought: "We cannot change the hearts of the people of the South but we can make war so terrible that they will realize the fact that however brave and gallant and devoted to their country, still they are mortal and should exhaust all peaceful remedies before they fly to war."

On the one hand, Sherman continued to see things as he had while living in prewar Louisiana. He thought that blacks were inferior beings, and, while regretting that slavery as an institution existed, he had no personal desire to force its abolition. Nonetheless, he was a soldier who intended to carry out his government's policy: the Southern civilian population must cooperate with federal rule, as most of the people of Memphis were doing. As for Grant, whose wife and Missouri in-laws still owned slaves, he had recently written his strongly antislavery father that "I have no hobby of my own with regard to the negro, either to effect his freedom or continue his bondage. If Congress pass any law and the President approves, I am willing to execute it." Echoing Sherman's view that the South was beginning to "hurt" in ways besides suffering militarily, he wrote this to his sister Mary: "Their *institution* [the slaves] are beginning to have ideas of their own and every time an expedition goes out more or less of them follow in the wake of the army and come into camp. I am using them as teamsters, Hospital attendants &c. thus saving soldiers to carry the musket. I don't know what is to become of these poor people in the end but it is weakning [sic] the enemy to take them from them."

By November of 1862, following the Preliminary Emancipation Proclamation, Grant began issuing specific orders concerning the treatment of "contrabands," as fugitive slaves were known. They were to be cared for and the men put to work except for "such men as are not fit for active field duty." For the first time in their lives, they would be paid. Grant set forth those conditions: "It will be the duty of the Superintendent of Contrabands to organize them into working parties in saving cotton, as pioneers [laborers assisting engineer troops] on railroads and steamboats, and in any way where their service can be made available . . . The negroes will be clothed, and in every way provided for, out of their earnings as far as practicable . . . In no case will negroes be forced into the service of the Government, or be enticed away from their homes except when it becomes a military necessity."

In addition to implementing federal policies concerning slaves, Sherman had a variety of experiences with white citizens of Memphis who tested Union vigilance. Smuggling became pervasive: military supplies for Confederate forces left the city in coffins and in the carcasses of slaughtered cattle and hogs. Salt, badly needed to preserve and flavor Confederate rations, went out in barrels labeled as being something else. Sherman classified the sending of salt, chloroform, and medicines to the enemy as a treasonable activity punishable by penalties including death, but he used discretion in these cases. When two women, one aged and the other pregnant, tried to sneak out of Memphis together, carrying banned goods as well as a trunk of their clothes and two dresses, Sherman ruled that "the commanding General directs that the goods except the trunk and two dresses be confiscated. The ladies will go home and not attempt this again."

Sherman and most of the citizens of Memphis continued on good terms, but outside the city, guerrillas wearing nothing that would identify them as Confederate soldiers fired on the Mississippi riverboats that Sherman allowed to move up and down the river. In late September, one of these vessels, not a Union gunboat or transport but a regular packet boat carrying nonmilitary goods and civilian passengers of both sides, was shot at near the rebellious small town of Randolph, Tennessee. Everyone in and near there got a taste of what Sherman was capable of doing. He had two infantry companies go into the town and burn it down. To make the scene of wreckage more dramatic, Sherman ordered that it be done "leaving one house," and reported to Grant's chief of staff John A. Rawlins that "the regiment has returned and Randolph is gone." (In a letter to Ellen, he disposed of the matter in twenty words: "The Boats coming down are occasionally fired on. I have just sent a party to destroy the town of Randolph.") Then he decreed that ten families would be expelled from Memphis every time such an attack occurred.

When a Confederate general sent him a letter under a flag of truce criticizing these actions, Sherman, never at a loss for words, replied that the general's protest "excites a smile" because he knew full well that the general himself would not countenance men "without uniform, without organization except on paper, wandering about the country pillaging friend and foe, firing on unarmed boats filled with women and children . . . always from ambush or where they have every advantage." The Confederate general persisted in the exchange, threatening to hang a captured Union officer. Sherman responded that when the guerrillas "fire on any boat, they are firing on their [own] Southern people, for such travel on every boat" and added that if a Union officer were hanged, "You initiate the game, and my word for it your people will regret it long after you pass from the earth." To Grant, Sherman wrote, "They cannot be made to love us, but may be made to fear us."

Other Northerners also were seeing the conflict in a harsher light. Senator John Sherman concurred with his brother, stating that "it is about time the North understood the truth, that the entire South, man woman & child is against us, armed and determined." Ellen's anger at the South took her even further than that. Two of her four brothers, Hugh and Charles, were now in the Union Army, with Charles serving as a lieutenant in one of Sherman's infantry regiments and Hugh a colonel of an infantry regiment in the Eastern theater of war. Worried about her husband and two brothers, she apparently began to see Southern whites as doing the work of the devil. Ellen wrote Sherman that "I hope this may be not only a war of emancipation but of extermination & that all under the influence of the foul fiend may be driven like the Swine into the Sea. May we carry fire & sword into their states till not one habitation is left standing."

While Sherman spent these months dealing with the complicated situation in and near Memphis, Grant remained at his headquarters in Corinth. Still without sufficient force to take the offensive while he tried to consolidate his units that had been "scattered" by Halleck, he was fortifying that railroad hub against an expected Confederate attack. None came, for the time being, and he sent for Julia and their four children to join him. She wrote of their arrival by train, at dusk.

> We found the General's ambulance awaiting us at the depot. The General and two or three of his staff officers accompanied us on horseback to headquarters. The General was so glad to see us and rode close beside the ambulance, stooping near and asking me if I was as glad to see him as he was to see me. He reached out and took my hand and gave it another and another warm pressure . . .
>
> As we entered the encampment, which extended from near the depot to

beyond the headquarters, the campfires were lighted, and I do not think I exaggerate when I say they numbered thousands. So it seemed to me. The men were singing "John Brown." It seemed as though a hundred or so sang the words and the whole army joined in the chorus ["Glory, Glory, Hallelujah!"]. Oh, how grand it was! And now when I hear "John Brown" sung, that weird night with its campfires and glorious anthem and my escort all come back to me.

For six weeks, Ulysses and Julia Grant and their children had an idyllic reunion. They all lived at his headquarters, which Julia described as being "in a handsome and very comfortable country house, situated in a magnificent oak grove of great extent. The house was a frame one, surrounded by wide piazzas, sheltered by some sweet odor-giving vine —Madeira vine, I think. On the grounds were plantain, mimosa, and magnolia trees."

A wide dirt walk surrounded the house and was kept neatly raked and sprinkled with water twice a day. The two youngest children, Nellie, seven, and Jess, four, loved to "make footprints with their little rosy feet in this freshly-raked earth." This led to an evening ritual.

Each day, as they were being bathed and dressed for the evening, the same petition came.

"Mamma, please let us make footprints. It is so cool and pleasant. Do, Mamma; it is such clean dirt. Let us, please." The General would answer, "Yes, you can. Why do you not ask me? I would always let you. It will not hurt you at all." The staff officers joined in the petition, so the little ones made the footprints and enjoyed it too.

Grant had much paperwork to attend to every day and worried about the amount of time the Confederate armies were being given to reorganize, but he thrived on Julia's attentions. Gaunt after Shiloh, he gained fifteen pounds during the weeks she was with him. She knew his tastes: Grant liked to eat small portions of plain food, unadorned by sauces or dressings. He enjoyed nibbling on fruit, and liked corn, pork, beans, and buckwheat cakes. Grant had a taste for oysters and clams; for breakfast he sometimes ate cucumbers with his coffee. (Once, when Julia remonstrated with him for mixing several of these different foods, a pickle among them, at one sitting, he said, "Let 'em fight it out down there," and continued to eat.)

Julia was at this time thirty-six, becoming stout but still with the lively personality and outgoing conversational manners of a Southern belle, a general's wife who was happy to sew on buttons for the officers working at headquarters, a woman who walked with a businesslike stride to visit officers and enlisted men in the hospital. Grant's staff, most of them younger than she, enjoyed her company and were devoted to her. Not only did Julia and the children bring with them a taste of the family life they missed, but she accomplished something far more important: when she was with her husband, he never drank.

Grant's susceptibility to alcohol was the subject of many different stories, during the war and ever after. They ranged from statements by those close to him that he drank not a drop during the war years, to accounts that had him crawling on his hands and knees, vomiting, and being carried to bed unconscious. His staff threw a nearly impenetrable mantle over the subject. There are internal headquarters letters on this subject to Grant from John A. Rawlins, the roughhewn lawyer from Galena who was the chief of Grant's headquarters staff and a man Grant particularly trusted. A visitor characterized Rawlins as having "very little respect for persons and a rough style of conversation," even cursing at Grant on occasion, but Rawlins used a respectful, even solemn, tone in his letters reproaching Grant for drinking incidents and warning him of the dangers of repeating them. Many years later, Sherman wrote, "We all knew at the time that Genl Grant would occasionally drink too much—he always encouraged me to talk freely of this & other things and I always noticed that he could with an hour's sleep wake up perfectly sober & bright—and when any thing was pending, he was invariably abstinent of drink." (According to a famous story, when Lincoln received a complaint about Grant's drinking, he told the person making the complaint to find out the brand of liquor Grant drank, so that he could send barrels of it to his other generals. When asked about this, Lincoln, who loved a good story, replied that he wished he'd said it, but hadn't.)

In other matters, unusual things happened to Julia at Corinth. A day or two before her visit ended, she was sitting near Grant, writing a letter, when she looked up and saw a young man in civilian clothes whom she did not know near the door and peering into the room. When he seemed startled to see her, Julia calmly penned this on the margin of the letter she was writing, and handed it to her husband: "Who is this strange young man? He is much interested in what is going on here. I am sure he must be a spy." Without a word, Grant wrote his answer on the same piece of paper and handed it back to her. "You are right. He is in our employ." (It was later discovered that this spy, free to come and go, was in fact a double agent, working at different times for both sides.)

During this same autumn of 1862, Sherman was joined in Memphis by Ellen and their six-year-old son Tommy. She wrote that at the age of forty-two, her husband looked "thin & worn being more wrinkled than most men of sixty," but found him "cheerful & well." Tommy had a wonderful time. Occasionally he was allowed to take a blanket and sleep with some of Sherman's soldiers in their tents. The men liked him, and a company tailor made him a uniform that had corporal's stripes on its sleeves. Writing a letter that began "My Dear Children," after Ellen and Tommy left for home with Tommy proudly wearing his uniform, Sherman told them that Tommy "thinks he is a Real soldier with a leave of absence for 7 years until he becomes fourteen when he must join his Company. No body can tell what may happen in the next seven years and therefore Tommy was very prudent in getting a seven years absence."

Soon after Julia and their children left Grant at Corinth, the relatively quiet and indecisive nature of the post-Shiloh military situation changed. Halleck had given Grant two of the three Union armies operating in the Western theater of war—the names and structure of the Western forces were changed several times—and the third was under the separate command of Don Carlos Buell. With Grant and Sherman's forces still heavily committed to occupation duty in the Mississippi River area, the same Southern generals whom Grant and Sherman defeated at Shiloh decided to attack Buell. Moving slowly as usual, Buell was well east of the Mississippi, marching his army toward Confederate-held Chattanooga. Braxton Bragg came up with the plan: link up with Edward Kirby Smith's army of eighteen thousand men who were in eastern Tennessee, smash Buell, recapture Nashville, get back into Kentucky, take Louisville, and move up the Ohio River to Cincinnati. (Beauregard loved the idea, citing the words of the French revolutionary leader Danton's slogan that translate as, "Audacity, more audacity, always audacity.")

· Bragg completely outmaneuvered Buell, retaking central Tennessee without fighting a battle, and Halleck was forced to strip Grant of nearly half his men and have their commanders rush them north. Grant had no choice but to remain on the defensive, far south of the new and very real Confederate threat to territory that had come under Union control many months before. It seemed entirely possible that, five months after Shiloh, the Confederate Army might take the war right back to, and in some cases beyond, its original boundaries. Buell was falling back toward Louisville, with Bragg marching to the same destination on a parallel route. Kirby Smith rode into Lexington, Kentucky, just 80 miles south of Cincinnati and the Ohio River. In the East, Robert E. Lee won decisively at Second Manassas and crossed the Potomac heading north into Maryland. Days after that, Stonewall Jackson occupied Frederick, Maryland, and went on to take the United States Arsenal at Harpers Ferry, capturing twelve thousand men and vast stores of weapons and equipment. The North found itself in the worst situation since the war began.

Grant's contribution to turning the tide came when the Confederate high command underestimated what he could do with the reduced Union forces he still had with him. At the time when Bragg was writing his subordinate Sterling Price that he was sure that Price and Confederate general Earl Van Dorn could "dispose of" Sherman and Union general William Rosecrans, after which they should move north "and we shall confidently meet you on the Ohio," Grant once again decided to attack first. The result was the battle on September 19 at Iuka, twenty-five miles east of Corinth. An indecisive engagement in itself, it stopped Price's move north to reinforce Bragg. At that point Price and Van Dorn threw their twenty-two thousand soldiers at Corinth, where Rosecrans had twenty thousand men waiting for them in the entrenchments Grant had built. In fierce fighting, the Confederates were repulsed. Tears coming down his cheeks, Price watched his torn-to-pieces divisions march away from Corinth: of the twenty-two thousand men who attacked in the morning, by late afternoon five thousand were dead, wounded, or missing.

The Confederate Army's position also deteriorated in the East, and then in Kentucky. In a tremendous clash at Antietam Creek in Maryland on September 17, Lee, with forty thousand men, outmaneuvered and initially held his own against McClellan's field army of seventy-five thousand—a force nearly twice that of the Confederates. Although Lee had to withdraw, it was the clearest possible demonstration of his skill as well as the ability and determination of his Army of Northern Virginia, and McClellan's failure to pursue Lee brought about the end of his military career, but the losses on both sides, totaling 23,500, exceeded those at Shiloh. Despite his performance at Antietam, Lee's effort to carry the war into the North had failed, with casualties the Confederacy could not afford, and he withdrew his divisions south across the Potomac. (Lincoln had been waiting for some good news to strengthen the Union's political position and war aims, and, although Lee's failed offensive was short of the kind of victory he hoped for, he issued the Preliminary Emancipation Proclamation five days later.) At Perryville, Kentucky, an outnumbered Braxton Bragg nearly prevailed over Don Carlos Buell, but Bragg's momentum and strength were spent, and he had to retreat to Tennessee. (A result of the Battle of Antietam was that Ellen Sherman's brother Hugh, after distinguishing himself there, was promoted to brigadier general and sent west to serve under Sherman.)

Ever since the Confederates' costly failure at Corinth, Grant had been planning to take the offensive, with Sherman ceasing his role as military governor of Memphis to lead Union forces in battle. It would require many weeks to get the massive and complicated movement organized and under way, but Grant intended to take the immensely strong Southern bastion of Vicksburg, Mississippi, 200 miles down the Mississippi River from Memphis. Across the river from this stronghold was the bayou country of Louisiana. Known as the "Gibraltar of the Confederacy," Vicksburg was a small city, but it was surrounded by a lethal defensive network that would make the whole complex singularly hard to capture. One Union observer described its combination of natural defenses and fortifications filled with artillery as being part of "an ugly place, with its line of bluffs commanding the channel for fully seven miles, and battery piled above battery all the way." (Some of those bluffs rose two hundred feet above the river.)

Before making any attack, points near the city would have to be approached through a cleverly defended maze of waterways, many miles of which looped through the flat marshland of the bayous. Grant saw that it would require an entire campaign, fighting a number of battles, simply to reach the places from which to launch a final offensive: as he put it in a letter to Julia, "Heretofore I have had nothing to do but fight the enemy. This time I have to overcome obsticles [sic] to reach him."

Even more than before, Grant needed the men and ships of the

United States Navy to work with him, as they had at Belmont, Fort Henry, Fort Donelson, and Shiloh. The senior Union naval officer in the area, Acting Rear Admiral David Dixon Porter, had recently taken command of the Mississippi Squadron. Once again, Grant had an exceptionally able naval officer ready to support him, a man eager to play his part in amphibious operations. The forty-nine-year-old Porter was the son of a United States Navy commander who had been a hero in the War of 1812; the elder Porter had also adopted a boy named David Farragut. In 1826, Porter's father resigned from the United States Navy to become commander in chief of the Mexican Navy. At the age of fourteen, young Porter went to sea aboard his father's flagship; after training he was transferred to other vessels, and before his fifteenth birthday was captured in a battle with a Spanish warship and spent six months as a prisoner of war in a prison ship in Havana Harbor. Making his way back to the United States when he was sixteen, he became a midshipman aboard the famous American man-o'-war *Constellation*, and in the next twenty years he rose to command his own ship, the USS *Spitfire*, during the Mexican War.

Now, in the Civil War, Porter was coming into his own. In the capture of New Orleans, which took place in the weeks just after Shiloh, he had served with distinction under his foster brother David Farragut, but it was Porter who conceived of the amphibious operation that seized the city. (Farragut would become immortal with his "Damn the torpedoes! Full speed ahead!" uttered at Mobile Bay, but Porter endeared himself to many with his remark, "A ship without Marines is like a garment without buttons.") Five feet six inches in height, Porter was an outspoken, ambitious officer with a considerable ego. He lived well. A man who loved to ride when he was ashore, Porter maintained two fine horses aboard his side-wheeler flag boat *Blackhawk*, along with cows to supply milk and butter for his highly regarded officer's mess.

Porter described in his diary his awkward first meeting with Grant. On an evening in early December of 1862, the nattily uniformed Porter was at a dinner party aboard an army quartermaster's riverboat at Cairo, Illinois, feasting on roast duck and champagne. His army host was called away from the table and returned with an unexpected guest, a travel-stained man wearing a wrinkled brown civilian coat and gray trousers. "Admiral Porter," the army quartermaster said, "meet General Grant."

Possessing a measure of the mistrust that often existed between the navy and army, and worried that Grant might think of him more as a bon vivant than the fighting sailor he was, Porter soon found that the plain-spoken Grant had only one thing on his mind: Vicksburg, and how to take it. As if the roast duck and champagne did not exist, for twenty minutes Grant spoke earnestly with Porter, indicating his need for all the help Porter and his ships could give him in the forthcoming campaign and telling him something of his plans. Impressed by Grant's "determination" and "calm, imperturbable face," Porter pledged him his fullest cooperation, and Grant walked off the ship.

In his journal, Porter also recorded the details of his first meeting with Sherman, some days later in Memphis. He first described Sherman's headquarters in the Gayoso House, the best hotel in town. Struck by the sparse furnishings and the intense concentration displayed by the men at every desk, he noted that the officers were "bronzed and weather-beaten," and dressed in the simplest of uniforms. Despite the air of efficiency, this naval officer on whom much depended had to wait an hour before Sherman appeared from an inner room. "He seemed surprised to see me, when I introduced myself, and informed me that he did not know I was there." Having said that, Sherman began conferring with one of his quartermasters as if Porter were not present. "I was not, I must confess, much impressed with Gen'l Sherman's courtesy." Then Sherman finished his conversation. "He turned to me in the most pleasant way, poked up the fire, and talked as if he had known me all his life . . . He told me all he had done, what he was doing, and what he intended to do, jumping up every three minutes to send a message to someone." Porter walked out of the Gayoso House liking Sherman and feeling confident that they would work well together. (They remained friends for the rest of their lives and died in the same month in 1891.)

As 1862 ended, Grant continued preparing for the Vicksburg campaign. More than ever, he came to rely on Sherman, and the tone of his letters differed from those he sent his other generals. Writing Sherman at a time when their headquarters were only ten miles apart, he enclosed a letter from Halleck, in which the general in chief, mistakenly informed that Grenada, Mississippi, had been captured by Union forces, told Grant that this "may change our plans in regard to Vicksburg." Grant, thinking that the report was accurate, said to Sherman, "I wish you would come over this evening and stay to-night, or come over in the morning. I would like to talk with you about this matter." Grant described two courses of action and added, "Of the two plans I look most favorably on the former." He closed with, "Come over and we will talk this matter over."

At a time when Grant needed to concentrate on military plans, more problems with "matters of public interest" arose. The first involved illegal trading in cotton, the South's great export crop and the one so widely grown in the area of Grant's and Sherman's responsibility. Before the war, Southern cotton had flowed in constant great quantities to Northern textile mills and to those in England. Now, a year and a half into the war, that trade was disrupted. Not only did the Union naval blockade of the Southern ports and the seizure of New Orleans the past April deprive the English mills of virtually all of the commodity on which their production depended, but the war produced a similar crisis on the American scene. With armies fighting in the areas between the Northern mills and their source of supply, Northern textile operators were themselves desperate for cotton. The North needed cotton, and the South needed Northern-manufactured goods, as well as cash that was not in the form of the already depreciated Confederate dollar. Some patriotic Southern planters burnt their cotton to deny it to the North, but others stored hundreds of thousands of bales and waited to see what would happen.

A lot did. For months, speculators had been coming south, offering the highest prices for cotton that the South had seen in sixty years. In Washington, the Treasury Department thought that restoring the cotton trade would in effect bribe many Southerners back into loyalty to the Union, and indeed the case could be made that the Union Army itself needed shirts, bandages, tents, and other items made from cotton. The War Department, however, felt that these Northern dollars would end up financing the Confederate Army and took a negative view of this trade between enemies. A compromise reached within the federal government produced instructions to Grant. He was to permit the activities of these Northern traders, as long as they held permits, did not go into enemy territory, and did not offer gold to the cotton sellers. This became meaningless: the speculators attached themselves to Union army regiments, avoided any kind of regulation, and handsomely bribed federal officers and men to look the other way while they dealt with anyone they chose, in any way that suited them.

Both Grant and Sherman found the situation infuriating: as they saw it, while brave Northern boys died, profiteers poured into the South to

trade with the enemy, ruining discipline in Union Army camps and creating bad feeling among the men, some of whom were making little fortunes working with the speculators, while others were not. (In Memphis,
Sherman believed there was even some treasonous barter in which cotton was traded for Northern pistols and chemicals that were used for ex-
plosives.) Sherman tried to have Northern cotton buyers pay in such a
way that the profits would be held by his quartermasters until the rebellion was over, thus guaranteeing that no Northern money could aid the
Confederate military, but Halleck sent on to him a federal government
order to desist. Sherman grudgingly obeyed, but he wrote: "Commerce
must follow the flag, but in truth commerce supplies our enemy with the
means to follow the [enemy] flag and the Government whose emblem
it is."

Grant wanted the speculators banned from his army's camps and decided that the only way to accomplish that was to ban them from the entire area of his military department. Most of the speculators were not
Jews, but a good number were, and both Grant and Sherman began to
characterize them all as being Jewish. Sherman's position on the "cotton
order" was that he had tried to regulate the Memphis economy by forcing all Southerners to trade with one another in Confederate dollars, but
this speculation in cotton with freelance Northern brokers was a different
matter. He had already written Ellen that Memphis was "full of Jews &
speculators buying cotton for gold & [federal] treasury notes, the very
things the Confederates wanted, money. I am satisfied the [Confederate]
army got enough money & supplies from this Quarter to last a year." He
also wrote an angry letter to the army's adjutant general Lorenzo
Thomas, who a year before had characterized Sherman's attitude while
in Louisville as being "insane" but now treated carefully the man who
had done so much at Shiloh and been promoted to major general. In it,
Sherman said: "If the policy of this government demands cotton, order
us to seize it . . . This cotton order is worse to us than a defeat. The [sur-
rounding] country will swarm with dishonest Jews who will smuggle
powder, pistols, percussion caps, etc. in spite of all the guards and pre-
cautions we can give."

Grant soon had a personally embittering experience involving Jews,
in which he felt that his father betrayed him. A year before, when his father had asked him to use his influence in getting an army contract for
the manufacture of harnesses, Grant told him firmly in a letter that "I

cannot take an active part in securing contracts" and tried to explain to his dyed-in-the-wool businessman father the concept of conflict of interest. Now his father arrived at his headquarters at Holly Springs, Mississippi, to visit him, accompanied by three of his business acquaintances, brothers from Cincinnati named Henry, Harmon, and Simon Mack. The Macks, who were Jewish, had a trading enterprise known as Mack and Brothers, and had entered into some form of partnership with Jesse Grant.

Grant was happy to see his father, with whom he had always had a difficult relationship, and thought that he had come on a purely personal visit. He soon saw how wrong he was: his father and the Macks wanted him to use his influence to get them one of the prized permits to buy cotton and ship it north. Grant had the three Macks put on the next train going north, and Jesse Grant also left.

Just how much this confrontation influenced Grant cannot be said, but on December 17, 1862, Grant had published for the guidance of his entire military department his General Orders No. 11.

> The Jews, as a class, violating every regulation of trade established by the Treasury Department, and also [War] Department orders, are hereby expelled from the Department.
>
> Within twenty-four hours from the receipt of this order by Post Commanders, they will see that all of this class of people are furnished with passes and required to leave, and anyone remaining after such notification, will be arrested and held in confinement until an opportunity occurs of sending them out as prisoners unless furnished with permits from these headquarters.

It took a while for the contents of this order to reach the North, but apparently the first person to act was Cesar J. Kaskel, of Paducah, who led a delegation of Jews to Washington and met with President Lincoln. Hearing what his visitors had to say, Lincoln commented, "And so the children of Israel were driven from the happy land of Canaan?" Kaskel replied, "Yes, and that is why we have come to Father Abraham's bosom, asking protection." Lincoln said, "And this protection they shall have at once."

Halleck had the duty of informing Grant that the order must be "immediately revoked" and later explained to him that "the President has no objection to your expelling traders and Jew pedlars, which I suppose was the object of your order, but [because] it prescribed an entire class, some of whom are fighting in our ranks, the President deemed it necessary to revoke it." Grant grudgingly complied, but the story reached the press: *The New York Times*, which previously had praised Grant, now condemned him for issuing an order in "the spirit of the medieval age." Congress voted along party lines on a measure to censure Grant. The Democrats nearly prevailed in the House, losing by three votes, fifty-three to fifty-six, but the Republican-controlled Senate defeated it thirty to seven.

This controversy faded; now Grant became involved in a new chapter of the old story of political appointments of military officers. Readying his army for the campaign down the Mississippi toward Vicksburg, with Sherman slated for a vital role in that, Grant learned of the recent activities of Major General John McClernand. It was McClernand's division that had collapsed and run at Fort Donelson, requiring Grant to restore order in its ranks, and at Shiloh, while McClernand stood his ground, Grant formed an opinion he later expressed that McClernand was "incompetent."

As was the case with so many political appointments, McClernand was entirely unqualified to be a general. An influential lawyer and politician from Illinois, he had served three terms in the Illinois legislature and had represented Abraham Lincoln's congressional district in the House of Representatives. His sole military experience prior to being commissioned as a brigadier general of Illinois Volunteers at the outset of the Civil War had been three months' service, thirty years before, as a private during the campaign against the Sac and Fox Indians in northern Illinois known as the Black Hawk War. During that conflict, he had displayed courage and resourcefulness in taking a dispatch through a hundred miles of territory held by hostile warriors, but he had never commanded even a squad of soldiers. Although a Democrat, McClernand had been a good friend to Lincoln, who always took care of those from the state in which he rose politically. Second in seniority within Grant's military department because of his date of rank, in September McClernand had taken a long leave, which Grant was happy to approve. Now Grant learned that McClernand had gone to Washington and talked his friends Lincoln and Secretary of War Stanton into letting him go to Iowa, Illinois, and Indiana, to recruit a large and entirely separate force of Volunteers, which he would then command, taking over the expedition to capture Vicksburg. McClernand was now successfully raising those many regiments. Grant and Sherman, who were emerging as the two ablest officers in the Western theater and were presumably the leaders who might be able to take Vicksburg if it could be done at all, had been told none of this.

It had been a stunning lapse on the part of Abraham Lincoln, whose military judgment was frequently better than that of his generals. In this case, politics completely dominated the president's thinking. Lincoln had needed the support of McClernand and other prominent "War Democrats" to support his Republican administration's decision to enter the war. In the recent midterm congressional elections, the Republicans had fared badly. As the casualty lists lengthened, the war was becoming increasingly unpopular in just those Democrat-leaning states in which McClernand proposed to enlist many thousands of needed soldiers. The vision that McClernand had presented to Lincoln and Stanton was of himself, a well-known Democrat from Lincoln's home state, giving a bi partisan flavor to a new Western army that he would lead to an enormously significant victory that would strengthen Lincoln's position as well as his own.

Halleck had the most serious reservations about McClernand and the entire scheme, and shared Grant and Sherman's dismay about it. For a change, Halleck moved swiftly. He saw the thorniest part of the problem: incompetent or not, McClernand was senior to Sherman, and McClernand's letter from the president authorizing his actions could give him the power to override Grant. If McClernand got down to Memphis with what was in effect a newly raised private army, it would be hard to stop him from wrecking Grant's plans. He might be able to proceed to Vicksburg and lose every man of his untrained force. The stage was set for a spectacular disaster.

Having demonstrated after Shiloh that he could hogtie Grant, Halleck now used his bureaucratic skills to stop McClernand. First, he told Grant to set up his headquarters in Memphis, where he was reunited with Sherman. Then, as McClernand's newly recruited regiments came into being, Halleck ordered each new unit to report to Grant in Memphis. Grant asked Halleck if these troops, and Sherman, were under his command or "reserved for some special service." The moment that Halleck replied that every soldier now within the boundaries of Grant's department was his to command, Grant understood the unspoken part of the message: start yourself, Sherman, and all the troops down the river heading for Vicksburg, before McClernand comes down and starts trying to take over.

Pleased, Grant, then in La Grange, Tennessee, wrote Sherman, "The mysterious rumors of McClernand's command left me in doubt as to what I should do . . . I therefore telegraphed Halleck . . . He replied that all troops sent into the Department would be under my controll [*sic*]. Fight the enemy in my own way . . . I think it advisable to move on the enemy as soon as you can leave Memphis."

When McClernand eventually did show up along the Mississippi and began using Lincoln's letter of authorization in an effort to take command of the four divisions under Sherman, Halleck gave Grant the backing he had so often and so conspicuously withheld: "You are hereby authorized to relieve McClernand from command of the expedition against Vicksburg, giving it to the next in rank [Sherman] or taking it yourself."

Grant decided to take command himself, working closely with Sherman as he had always planned to do. When McClernand insisted that the matter receive Lincoln's personal review, Grant agreed. Someone, surely Halleck and perhaps others, must have been talking to Lincoln and Stanton: the president quickly told McClernand to obey the orders given him by his military superiors and to be satisfied with the command of the army corps Grant assigned to him for the expedition.

During the closing months of 1862, Grant and Sherman had worked their way through a political and administrative labyrinth. Grant had emerged as the undisputed leader in the Western theater and had consolidated the units of his command that Halleck had, as Sherman put it, "scattered" across a wide area. As for Sherman, he had come to Memphis as its military governor, found the city in a chaotic condition, and in four months had brought it to a state of law-abiding prosperity. Placed back in command of troops by Grant, he was relieved of his role as military governor.

Now, in December, Grant and Sherman began to face the realities of the most formidable military challenge any Union generals had thus far confronted. This was the attempt to capture Vicksburg, the immensely strong bastion located deep in the Confederacy, which was defended by an army of brave and determined men. Sherman was to describe it as the strongest defensive position he ever saw. Grant offered his appraisal of the fortified riverfront city that had a peacetime population of forty-five hundred, the second largest city in Mississippi after Natchez: "Admirable for defense. On the north it is about two hundred feet above the Mississippi River at the highest point and very much cut up with cane and underbrush by the washing rains; the ravines were grown up with cane and underbrush while the sides and tops were covered with a dense forest."

It would be suicide to come across the river from the Louisiana side and land directly under the cliffs of this fortress on the east bank of the Mississippi, but simply getting near enough to attack from any other direction presented enormous problems. The approach on land from the north led through marshes, and to approach from the east would extend supply lines already being successfully raided by Nathan Bedford Forrest.

The approach from the south, which was Vicksburg's least defensible area, presented great advantages, coupled with a possibly insurmountable problem. If an army could get into place on dry land below Vicksburg on the eastern side of the river, it could move north up to the city and begin extending itself around its defenses, to the point where it might be possible to have it encircled on three sides, with the other side being the river. Then there would be no escape for the besieged defenders. The problem was how to get such a large force there. For many more weeks, the entire low-lying Louisiana side of the river opposite Vicksburg would be so inundated with winter and spring rains that Sherman could not march his forces south along that bank. Men on foot and on horseback might possibly struggle through the bayous, but it was impassable for artillery pieces and wagonloads of supplies. That left only the extraordinarily hazardous choice of moving down the river itself. If Grant tried to use ships to move his troops, heavy equipment, and supplies, they would have to pass within range of the Confederate cannon placed on the bluffs above the water.

In the months to come, Grant would make seven different efforts to reach Vicksburg and surround it. In the first move, Grant led forty thousand men south toward Jackson, Mississippi, east of Vicksburg, in a diversionary maneuver intended to pull Confederate forces out of that stronghold while Sherman brought thirty-two thousand men down the river aboard ships in an amphibious operation and landed them at Chickasaw Bluffs, just north of the city. Grant's overland approach was thwarted by the movements of thirty-five hundred Confederate cavalry-men under Major General Earl Van Dorn. Sherman's four divisions were decisively repulsed when enormous rains, at places raising the Mississippi twelve feet above its usual level, literally washed them out of their attacking positions as the defenders inflicted on them casualties of 208 killed, 1,005 wounded, and 563 missing. (Sherman wrote Ellen, "Well we have been to Vicksburg and it was too much for us and we have backed out.")

Immediately after his withdrawal, Sherman told Grant that "I assume responsibility and attach fault to no one." Then he saw an article written by Thomas W. Knox, the leading war correspondent of *The New York Herald.* Ignoring Sherman's order that no journalists could accompany the recent failed expedition, Knox had been aboard a transport. Relying on interviews he conducted more or less at random with some of the participants immediately after the retreat, Knox wrote that Sherman had bungled the attack and failed to take adequate care of his wounded. He termed Sherman's behavior "unaccountable"—a word that echoed the reports by journalists a year before that Sherman was insane.

Learning of Sherman's anger about his article and the way the information was obtained, Knox wrote a conciliatory note to Sherman, telling him that he had since learned from the battle reports many mitigating facts about the recent failure at Chickasaw and offering to retract his story. In no way mollified, Sherman had him arrested and held for trial by a court-martial. When they met face-to-face, Knox managed to worsen the situation. Clearly referring to the manner in which Sherman had banned reporters from his camps as long ago as his time in Kentucky, he said, "Of course, General Sherman, I have no feeling against you personally, but you are regarded as the enemy of our set [the press] and we must in self-defense write you down." Sherman had Knox charged with violating his order excluding nonmilitary personnel from the recent expedition and of breaking a War Department rule that no information concerning military activities could be printed without permission from the commanding officer of the area in which they took place.

The matter quickly became a cause célèbre. It was the first time in American history that a journalist had faced a military court. While Sherman went on to win an engagement at Arkansas Post, up the river from Vicksburg, and then to save Admiral Porter's fleet from destruction when it was surrounded by enemy forces in the swamps at Steele's Bayou, much of the press concentrated on condemning him both as a commander and as an enemy of free speech. Sherman remained resolute, informing the court-martial judges that he considered Knox to be in effect a spy who gave the enemy information about Union movements and losses that Knox had broken a specific ban to acquire. As the trial continued for several weeks, Sherman gathered statements from officers who had been at Chickasaw, including Porter: the naval leader observed that two of Sherman's generals had handled their divisions in a deficient manner but said of Sherman's leadership of the force as a whole, "Sherman managed his men most beautifully . . . He did nobly until the rain drowned his army out of the swamps." Sherman had all this copied and sent to his senator brother in Washington, to his wife and her lawyer brother Philemon, and to Grant. He wrote Ellen that if the press was running the war, he might as well resign from the army.

Grant, who was trying to conduct an extraordinarily complex military campaign, had through his own experience come to dislike and mistrust the press, although his overall handling of journalists was far better than Sherman's. When the verdict came down, ordering that Knox be banished from Grant's military department and would be arrested if he returned, Grant knew that President Lincoln was under intense pressure to use his discretionary power to overturn the decision. Knox went to Washington and saw the president, who needed the goodwill of the Northern press. Lincoln told him that he would revoke the sentence if the commanding general of the department, Grant, would agree to that.

Grant had no doubt as to what his commander in chief wanted him to do. He made his decision in the form of a letter to Knox. Grant told the reporter that he had in fact violated Sherman's order against traveling with the amphibious force, that he had attacked Sherman's reputation and had suggested that Sherman was insane. He went on to say that "General Sherman is one of the ablest soldiers and purest men in this country . . . Whilst I would conform to the slightest wish of the President where it is formed upon a fair representation of both sides of any question, my respect for General Sherman is such that in this case I must decline, unless General Sherman first gives his consent to your remaining."

Sherman had the last word. He wrote Knox, "Come with a sword or musket in your hand, prepared to share with us our fate in sunshine and storm, in prosperity and adversity, in plenty and scarcity, and I will welcome you as a brother and associate; but come as you do now, expecting me to ally the reputation and honor of my country and my fellow-soldiers with you as the representative of the press which you yourself say makes so slight a difference between truth and falsehood and my answer is Never!"

Grant had risked his standing in Washington for Sherman; later in the campaign, Sherman would demonstrate his willingness to sacrifice his military reputation for Grant. When Grant considered having Sherman distract the enemy's attention by making a large and deceptive thrust toward Hayne's Bluff, up the river from Vicksburg, he wrote Sherman that he thought the feint would be "good," but added that he was "loth" to order it, in effect because the public might consider it to be another defeat such as Sherman had suffered at Chickasaw Bluffs. Sherman stoutly replied that Grant had "good reason to divert attention . . . That is sufficient for me and it shall be done." Recognizing that there were journalists who were after both of them (as the unsuccessful efforts to approach Vicksburg dragged on, one editor characterized Grant as a "foolish, drunken, stupid . . . ass"), Sherman added, "As to the Reports in newspapers we must scorn them, else they will ruin us and the country. They are as much enemies to Good Government as the sesech [secessionists] and between the two I like the sesech best, because they are a brave open enemy & not a set of sneaking croaking scoundrels. I believe a diversion at Haines [sic] Bluff is proper right and will make it, let whatever reports of *Repulse* be made."

When Sherman's troops were landed at Hayne's Bluff and then withdrawn, the amphibious maneuver proved to be not particularly useful in distracting Confederate attention from Union movements farther south, but the significance of the correspondence about the operation between Grant and Sherman was that it underscored the way they were supporting each other. As for their problems with the press, Grant needed no encouragement from Sherman regarding what both believed to be their right to conduct their operations free from newspaper reports that might assist the enemy. From Milliken's Bend, Louisiana, twenty miles above Vicksburg on the Mississippi, Grant fired off a message to Major General Stephen Hurlbut, who had replaced Sherman as military governor of Memphis: "Suppress the entire press of Memphis for giving aid and comfort to the enemy by publishing in their columns every move made here by troops and every [engineering] work commenced. Arrest the Editor of the Bulliten [sic] and send him here a prisoner, under guard, for his publication of present plans."

Although they saw the press in the same light, there were times when Sherman disagreed with Grant's strategy, and said so. At one point, when nothing seemed to be working, he strongly recommended that Grant take most of his army back to Memphis and move south again by an untried route. After advocating his plan in a conversation with Grant, he set it forth in a seven-point memorandum to Grant's chief of staff John Rawlins, but closed with this: "I make these suggestions, with the request that General Grant will read them and give them, as I know he will, a share of his thoughts. I should prefer that he should not answer this letter, but merely give it as much or as little weight as it deserves. Whatever plan of action he may adopt will receive from me the same zealous cooperation and energetic support as though conceived by myself."

Grant had a plan entirely different from Sherman's, possibly the boldest idea of the war. He had already tried approaching Vicksburg by digging canals through the bayous and blowing the levee at Yazoo Pass to create a flood, and had been defeated by combinations of weather and terrain, long, vulnerable supply lines, and Confederates moving within their home territory to counter him at every point. Now he decided to do what the enemy, and everyone else, considered to be impossible. Placing his faith in his naval colleague Porter, he would run some gunboats and transports down the river under the murderous rows of artillery batteries placed on the miles of bluffs at Vicksburg, but do it at night.

The plan had enormous risks: not only would the ships have to run the gauntlet of enemy cannon, but in the darkness they could easily collide or run aground. If they got through, however, passing the swamps to their west, Grant's men could go ashore on solid ground on the Louisiana side of the Mississippi below Vicksburg. This could then be used as a staging area for both the soldiers on the transports and the many thousands of Sherman's troops who would soon be able march down the west bank in drier weather to join them there. If all this worked, Grant could ferry his army across the river to the east bank, below Vicksburg, and attack from the side of the city the enemy was least prepared to defend.

The news of what Grant intended to do reached the authorities in

Washington. For a change, the secret was kept, but many who knew of it thought that Grant was about to lose Porter's ships and a good part of his army. Stanton disapproved; Halleck was worried. But months had passed, the press and the public wanted action, Grant had tried so many strategies, and nothing else had worked. No one told Grant to stop, but the stage was set for a shocked disavowal of Grant's actions if news of a disaster came north.

Before embarking on this dramatic move, Grant wanted to make a reconnaissance up the Yazoo River, a tributary of the Mississippi above Vicksburg. At just this point a journalist of a different sort arrived at Grant's headquarters, a man with far more power than the officers there at first realized he had. He was Charles A. Dana, the former managing editor of Horace Greeley's *New York Tribune*. At forty-three, he still had decades ahead of him in one of the great American journalistic careers of the nineteenth century. He came now not as a reporter but as a "special commissioner" appointed by Secretary of War Stanton to investigate the practices of paymasters in the Mississippi theater of war.

Dana was a brilliant, mercurial man, a descendant of a galaxy of New England figures that included Abigail Adams. He had studied at Harvard until his eyes failed him; he had combined physical labor with intellectual activity living at the Brook Farm experiment started by Ralph Waldo Emerson and other Transcendentalists; he had covered the revolutions of 1848 in Paris as a foreign correspondent. Going to Germany later that year, he met Karl Marx, who a few months earlier had published the *Communist Manifesto*. Three years later, Dana arranged for Marx to write letters on European affairs for the *Tribune*, and when Dana and the *Tribune*'s literary editor George Ripley began what became the sixteen-volume *American Cyclopaedia*, Karl Marx and Friedrich Engels started contributing what became a total of eighty-one articles on politics and military affairs. Four years before the war began, while continuing all his journalistic activities, Dana founded and edited the widely read and profitable *Household Book of Poetry*, which broadened the readership of such writers as Edgar Allan Poe. One of Lincoln's strong journalistic supporters, he knew both Lincoln and Stanton. Dana had met Grant in Memphis the previous year and recalled "the pleasant impression Grant made—that of a man of simple manners, straightforward, cordial, and unpretending." A combination of intellectual and outdoorsman, the athletic Dana had been unable to join the military because of his bad eyes and wanted to help the Union cause.

"Special Commissioner" Dana was in fact a spy of sorts, sent by Lincoln and Stanton to stay at Grant's headquarters and make his own estimate of Grant's behavior and ability. Tales continued to reach Washington of Grant's occasionally being drunk, and General McClernand, while outwardly reconciled to his position subordinate to Grant, had been sending his allies in the capital incessant criticism of Grant's performance as a general. The campaign intended to capture Vicksburg had been under way for four months; Lincoln wanted to continue to believe in Grant, but at the moment found it hard to do. The War Department had set up a special cipher for Dana to use in sending his reports to Stanton, a code known only by those close to Stanton, who would take the messages to Lincoln. If Dana sent back negative reports, Grant might well be relieved of command.

Before Dana arrived, two officers close to Grant met to talk about handling this man who was clearly going to be more than a visitor. Grant's rough-tongued chief of staff Rawlins sat down with Grant's inspector general, Lieutenant Colonel James Harrison Wilson. When Wilson first reported in at Grant's headquarters, two months before, Rawlins had come right to the point. According to Wilson, Rawlins said, "I'm glad you've come, you're an Illinois man and so am I. I need you here. Now I want you to know what kind of man we are serving. He's a goddamned drunkard, and he's surrounded by a set of Goddamned scalawags who pander to his weakness. Now for all that, he is a good man, and a nice man, and I want you to help me in an offensive alliance against the Goddamned sons-of-bitches."

Now, with Dana about to start living in their midst, Rawlins and Wilson came to an agreement. Describing their talk, Wilson said that "it was finally decided that he was to have access to everything, favorable and unfavorable, official or personal . . . With plenty of enemies about to bring him both truth and exaggerations, the worst tactics would be to arouse his suspicions by attempted concealment. A wise decision and fully endorsed by Grant."

A situation arose that tested Dana's loyalty and judgment. Included on a reconnaissance trip Grant made up the Yazoo River aboard Admiral Porter's flag boat *Blackhawk*, he later described the General's behavior: "Grant wound up going on board a steamer . . . and getting so stupidly drunk as the immortal nature of man would allow; but the next day he came out fresh as a rose, without any trace of the spree he had just passed through. So it was on two or three occasions of the sort and when it was all over, no outsider would have suspected such things had been."

So there it was. If Dana did what he had been sent to do, a telegram in secret code intended for President Lincoln would be on its way as soon as the *Blackhawk* landed, and Grant's career might be finished.

Dana did nothing. A professional journalist who had witnessed a sensational story, the thing he wanted most was for the North to win the war, and he had decided to invest his faith in this man Grant. (While Dana had recorded of their first meeting that Grant had "simple manners" and was "straightforward, cordial, and unpretending," he expanded his estimate to "Grant was an uncommon fellow—the most modest, most disinterested, and the most honest man I ever knew, with a temper that nothing could disturb, and a judgment that was judicial in its comprehensiveness and wisdom.") Dana was to be present at another time Grant drank, with equally bad results, but he waited for this campaign to end before he wrote Stanton that, when necessary, Rawlins could control Grant's drinking.

Perhaps as a result of the episode on the Yazoo, Julia Grant appeared, accompanied by their children. It was unusual for Julia to be allowed to come this close to the actual fighting, but she soon witnessed one of the most spectacular scenes of the war. On the evening of April 16, she and two other ladies went aboard the riverboat *Henry von Phul.* Julia said that "we dined on board with many officers"—Grant, Sherman, and Admiral Porter were not among them—"and when quite dark we silently dropped down the river."

Grant was about to send selected ships of the fleet past Vicksburg's miles of cannon. Aboard his warship *Benton*, Admiral Porter would be leading six other gunboats, including one named the *Henry Clay*, followed by three transports loaded with thousands of men. The transports were towing a total of ten barges loaded with coal that would be needed for the future operations down the river if they got through. Far astern of the others, by herself to minimize damage if she were hit and exploded, was a barge loaded with ammunition that would also be needed if this fleet survived.

Grant and Porter had worked hard, trying to anticipate everything as they prepared for this gamble. As Grant put it:

The great essential was to protect the boilers from the enemy's shot, and to conceal the fires under the boilers from view. This he [Porter] accomplished by loading the steamers, between the guards and boilers on the boiler deck up to the deck above, with bales of hay and cotton, and the deck in front of the boilers in the same way, adding sacks of grain. The hay and grain would be wanted below [down the river], and could not be supported in sufficient quantity by the muddy roads over which we expected to march.

Before this I had been collecting, from St. Louis and Chicago, yawls and barges to be used as ferries when we got below . . . Men were stationed in the holds of the transports to partially stop with cotton shot-holes that might be made in the hulls.

Ready to observe what he called this "perilous trip," Grant was accompanied by Dana and his staff aboard a vessel in the middle of the river upstream of the enemy's gun emplacements. On the river below the city, Sherman was well out from the shore on one of the yawls Grant had acquired. On Sherman's orders, his men had hauled four of these sailing craft across the swamps and placed them in the Mississippi below Vicksburg: ready for a disaster, Sherman had "manned them with soldiers, ready to pick up any of the disabled wrecks as they passed by."

Everyone was ready. Watching from the deck of the *Henry von Phul*, Julia noticed that it was so quiet that she could hear the frogs and katydids along the dark riverbank singing "their summer songs." Standing near Grant on the river transport, Dana saw it begin.

Just before ten o'clock . . . the squadron cut loose its moorings. It was a strange scene. First a mass of black things detached itself from the shore, and we saw it float out towards the middle of the stream. There was nothing to be seen except this big black mass, which dropped slowly down the river. Soon another black mass detached itself, and another, then another . . . They floated down the Mississippi darkly and silently, showing neither steam nor light, save occasionally a signal astern, where the enemy could not see it.

Julia, Dana, and Sherman saw what happened next from their different places on the river. Julia said, "All was going well when a red flare flashed up from the Vicksburg shore and the flotilla of gunboats and transports and our own boats were made visible." Dana recalled that the gunboats "were immediately under the guns of nearly all the Confederate batteries, when there was a flash from the upper forts." Down the river, Sherman had the best view:

> As soon as the rebel gunners detected the Benton, which was in the lead, they opened on her, and on the others, in quick succession, with shot and shell; houses on the Vicksburg side and the opposite shore were set on fire [something in fact done by Confederates who crossed over] which lighted up the whole river; and the roar of cannon, the bursting of shells, and finally the burning of the Henry Clay, drifting with the current, made a picture of the terrible not often seen. Each gunboat returned the fire as she passed the town, while the transports hugged the opposite shore.

The sound of the cannon was heard sixty miles away, and the firing lit up the sky for more than three hours; Julia said that, even at her vantage point upstream, "The air was full of sulphurous smoke."

As the ships came down the river, Sherman was everywhere. He managed to pull alongside the *Benton* and "had a few words with" Admiral Porter, checked on the gunboat *Tuscumbia* as she towed the "transport Forest Queen into the bank out of the range of fire," and reported that "the Henry Clay was set on fire by bursting shells and burned up; one of my yawls picked up her pilot floating on a piece of wreck, and the bulk of her crew escaped in their own yawl-boat to the shore."

Grant's idea had worked. One ship had been lost and any number of shells had hit the gunboats, but to everyone's amazement, not a single soldier or sailor was killed. Grant had ships and men and supplies below Vicksburg, ready to cross the river and approach the city from its most vulnerable side. (Union soldiers later learned that on this night, many citizens of Vicksburg began the evening by dancing at a "gala ball" to celebrate the invincibility of their "Gibraltar"—a confidence shaken when the sudden sound and physical thud of cannon fire interrupted the music.)

Grant was always a man to follow up on an opportunity. When Dana was with him the next day, he decided to send massive loads of supplies down the river on ships, running the same gauntlet while his columns of soldiers started making their way down the bank of the river to a point below Vicksburg where they could be ferried across. Dana said that Grant "ordered that six transport steamers, each loaded with one hundred thousand rations and forty days' coal, should be made ready to run the Vicksburg batteries . . . The transports were manned throughout, officers, pilots, and deck hands, by volunteers from the army . . . This dangerous service was sought with great eagerness, and experienced men found for every post. If ten thousand men had been wanted instead of one hundred and fifty, they would have engaged with zeal in the venture."

This second effort to pass down the river went off on the night of April 26. The transports got through, but Grant's headquarters steamer *Tigress* was hit and sunk, although Grant was not aboard. Sherman, who was right there in a yawl, helping those aboard the *Tigress* get to shore, explained what these two night operations had accomplished: "Thus General Grant's army had below Vicksburg an abundance of stores, and boats with which to cross the river."

Grant had done it. His army still had to make its way down the east side of the river to be ferried across, but the willingness of his soldiers to volunteer for hazardous duty such as manning the ships during the second run past Vicksburg's batteries showed that his men believed in him as never before. There were sighs of relief in the White House and the War Department. Against all odds, with Sherman's help Grant was placing his army where he wanted it to be, but the great fortress still had to be taken.

7

THE SIEGE OF VICKSBURG

On April 30, Grant ferried his troops across the river from the Louisiana shore to Bruinsburg, Mississippi, and the first columns began an eight-mile march east to the inland town of Port Gibson, twenty-five miles south of Vicksburg. On May 1, the Union troops fought an all-day battle against two Confederate brigades, throwing them back and taking Port Gibson.

Grant had orders telling him that, before he moved against Vicksburg, he must first march south to join Major General Nathaniel Banks in an effort to capture Port Hudson, Louisiana, but he learned from Banks that he was still clearing the west side of the Mississippi in his Red River campaign and could not move on Fort Hudson for another month. At the same time, Grant received intelligence that Confederate General Joseph E. Johnston, a master of maneuver, was assembling forces that would total twenty-four thousand men in Alabama and eastern Mississippi. Johnston's plan was to have these men follow him as soon as possible to Jackson, forty-five miles east of Vicksburg, where there were six thousand Confederate soldiers in place. The situation was fluid: Johnston hoped to find a way to link up with the Confederate defenders now inside Vicksburg and either defeat Grant in the field or in some other way relieve the pressure that was sure to be brought on that city.

Johnston's movements prompted Grant to make a decision that was in its way as bold as the one to run down the river in front of Vicksburg. Disobeying orders and against the advice of his subordinates, Grant took forty thousand men and headed straight for Jackson. His plan was to intercept Johnston and defeat or throw him back, and then to turn and give his full attention to Vicksburg.

As Grant's troops began this move, news came of the brilliant Confederate victory over the far larger Union forces under General Joseph Hooker at Chancellorsville in Virginia. (Three weeks before Robert E. Lee's victory over Hooker, Sherman had written Ellen, "I know Hooker well, and tremble to think of his handling 100,000 men in the presence of Lee.") The news of the Confederate success was accompanied by the details of the death of Stonewall Jackson. Just after the supreme triumphal moment of the war's other great military partnership, that between Lee and Jackson, Lee learned that Jackson, wounded by musket balls mistakenly fired at him by his own men, had been operated on and his left arm removed. In sending a chaplain to Jackson with a message of "affectionate regards" and wishes for a speedy recovery, Lee said to the clergyman, "He has lost his left arm, but I have lost my right." Now Jackson was dead—a great blow to Confederate arms, but Lee's tour de force at Chancellorsville left the North once again disheartened and thirsting for news of a success. (When Lincoln heard of the Union defeat, he exclaimed, "My God! What will the country say! What will the country say!")

Whatever the reverses in the East, Grant kept up his characteristic momentum in the West. Assisted by a bold diversionary cavalry raid through east Mississippi made by Colonel Benjamin Grierson and seventeen hundred men, in the next two weeks he brought off a military masterpiece. Marching 180 miles, he changed directions twice and fought five large-scale engagements against different forces whose total numbers were greater than his own. At the end of this sweep, his troops had killed, wounded, or taken prisoner more than seven thousand Confederates while suffering less than half that many casualties themselves. The results were impressive, and they demonstrated Grant's mastery. As far back as the Battle of Belmont in November 1861, he had shown that he understood how to work with rivers and ships, but this was the first time that he had shown his ability to employ cavalry in the strategy of a campaign. Always, he kept the ball moving—on May 7, as Sherman began a march from Grand Gulf, Mississippi, south of Vicksburg, to join him, Grant wrote, "It's unnecessary for me to remind you of the overwhelming importance of celerity in your movements."

Sherman understood the need to keep things going, but worried about traffic jams of wagons along Grant's lines of supply, he wrote Grant on May 9 urging him to "stop all troops" until the wagons could bring up supplies for the marching men. Grant had other ideas: he decided to leave the supply wagons behind as he marched to attack and briefly hold Jackson.

The manner in which Grant made this march again showed boldness and originality similar to that which enabled him to get his army past the guns of Vicksburg. Breaking with the traditional ideas of feeding an army from its existing supplies, Grant told Sherman that he intended to start off with "what rations of hard bread, coffee & salt we can and make the country furnish the balance." They would live off the land, foraging for fruit, corn, and other vegetables, while slaughtering whatever chickens, sheep, hogs, and cattle they found, and let nothing slow them down. Sherman learned from what he was about to see.

By now, Grant understood Halleck. Saying later of the general in chief in Washington that "I knew well that Halleck's caution would lead him to disapprove this course" and hold him in place if he knew that he intended to cut loose from his slow-moving supply lines, Grant simply sent Halleck a brief telegram whose last words were, "You may not hear from me for several days." By the time Grant's next communication arrived in Washington, he had taken Jackson. His advance was so swift that Joseph E. Johnston, now in overall command of the Confederate forces in Mississippi, after some initial fighting on the outskirts quickly ordered the six thousand men defending the city to evacuate it, and the Union forces took it without further bloodshed.

Sherman had come up quickly from Grand Gulf, and on the morning of May 14 his corps was one of the two that had swept away the last of the Confederate resistance. Jackson had fallen so swiftly that when Grant and Sherman rode into the city together during a driving rainstorm, many of the residents did not know anything had happened. The two men entered a textile mill where women workers were weaving cloth to make tents for the Confederate Army. None of the women looked up from working at their looms. After watching for a while, Grant finally told Sherman he thought they had done work enough. He later described how "the operatives were told they could leave and take with them what cloth they could carry." As they left, Grant, knowing that he would soon give up the city to turn back toward Vicksburg, ordered Sherman to burn down the factory and to wreck everything else that could be useful to the Southern cause—the city's other factories and machine shops, its foundries and railroad facilities, and the state arsenal.

Grant's army had two interesting supernumeraries riding with it: Charles Dana, and Grant's son Fred, now thirteen, whom Julia had willingly left behind when she returned north after her visit. Dana and Fred, riding together, seemed to have an ability to get near the action without bothering anyone or being hurt. Grant remembered seeing them on the quick forced march to Jackson, "mounted on two enormous horses grown white with age." Fred, wearing the sword his father never used, got into Jackson ahead of the main Union force and had to duck down a side street to avoid a company of Confederate foot soldiers who were hurrying out of the city as part of Joseph E. Johnston's evacuation. A few minutes later he saw an advance party raise the Stars and Stripes above the statehouse and was there to greet his father when he and Sherman rode into the city at the head of the main force.

Grant's presence in Jackson made Johnston think he had a great opportunity. Assuming that if a good part of Grant's army was in Jackson, it meant that there were supply lines stretched all along the more than forty miles back to the Mississippi River, Johnston ordered the general commanding Vicksburg to come out of the city and strike at those wagon trains. As Johnston saw it, any combination of advantages could be gained: supplies could be destroyed or captured, and if Grant headed back the way he had come—from the west—to protect the route under attack, Johnston could take reinforcements around to the north side of Vicksburg and add to its defenses.

The Confederate leader who brought a large part of the defending force out in response to Johnston's order was General John C. Pemberton, a West Pointer from Pennsylvania who, as a young lieutenant, had been involved in the same attack on Mexico City's Garita San Cosme during which Grant had managed to place a mountain howitzer in a church's belfry and lob cannonballs behind enemy lines. At least partly influenced by his beautiful Virginian wife and her family, Pemberton had chosen to fight for the South. Now, at the head of twenty-three thousand men, Pemberton started moving back and forth in the area south of Vicksburg, looking for convoys of wagons that did not exist because of Grant's new policy of living off the land.

By now, Grant was already moving back toward Vicksburg, not with convoys but with regiments of combat infantrymen. A spy gave Grant information about Pemberton's movements; coming forward to reconnoiter as fast as he could, Grant saw that he had been given a chance to fight many of the defenders of Vicksburg in open country, rather than having to attack all of them when they were behind the strongest entrenchments either side had built during the war. His intuition, correct as it so often proved to be, was that there was an opportunity to fight Pemberton without having to fight Johnston at the same time. Grant ordered Sherman, still back in Jackson destroying that city's warmaking capacity, to bring his corps to join him with the same "celerity" he had previously asked for, and prepared to strike the bewildered Pemberton.

The result was the crucial battle of Champion's Hill, eighteen miles east of Vicksburg. Starting at seven in the morning on May 16, with Sherman still on the way from Jackson, Grant moved his thirty thousand men into action, and the opposing armies began grappling with each other. By ten o'clock, Pemberton had his forces placed in good defensive positions atop an L-shaped ridgeline. At the corner of this forested higher ground was Pemberton's defensive anchor, Champion's Hill, 140 feet high. With the capricious and controversial Union general McClernand inexplicably failing to move his corps forward as ordered, and Sherman still not there despite a remarkably swift march, Grant fought the battle with the three divisions available to him on the right side of his line. After bloody attacks and counterattacks, at two-thirty in the afternoon, with McClernand still not putting pressure on the enemy and Sherman six miles away, the Confederates poured down the side of Champion's Hill toward Grant's men, threatening to scatter them.

In a situation where everything seemed to be going against him, Grant was smoking a cigar while he quickly organized all his available artillery. He turned to one of his generals who was just now bringing some fresh troops up to the battle. Matter-of-factly, Grant said that he was ready to make his last stand then and there, and sensed that the enemy "is not in good plight himself. If we can go in there again and make a little showing, I think he will give way." As he spoke, an enlisted man was struck by his calm, recalling that "I was close enough to see his features. Earnest they were, but signs of inward movement there were none."

Grant let loose a blast from his artillery, stopping the Confederate advance. He had only two depleted regiments ready to counterattack, totaling five hundred men, but he threw them in. The Seventeenth Iowa and

Tenth Missouri charged forward against larger numbers, but in those few minutes they changed the tide of battle. The Confederates began an orderly withdrawal, with Grant finding and sending in more units to add to the momentum of his counterattack. By four in the afternoon, Grant's men had Champion's Hill, and Pemberton's Confederates were headed back in the direction of Vicksburg.

Unlike many battles, in which both sides knew in advance roughly where the fighting would take place, this action had materialized and been fought within twenty-four hours. Pemberton's force had lost 3,840 men killed, wounded, or missing, in contrast with a Union loss of 2,441. The disparity in numbers was less important than the fact that Grant, deep in Confederate territory and opposing forces under the overall command of the greatly admired Joseph E. Johnston, was outmaneuvering the enemy at every turn and winning every engagement.

Within hours of losing at Champion's Hill, Pemberton fell back that night to the Big Black River, which was the last natural defensive position he could hold outside of Vicksburg. The next day, a spirited Union attack forced Pemberton's men back across the Big Black River Bridge, eight miles east of Vicksburg, but the Confederates destroyed it before making the final march of their retreat into the fortress city. Pemberton had lost 1,751 men and twenty-seven cannon at the Big Black River, while Union losses were just 200.

That night, Sherman was back with Grant's main force. The immediate problem was to get across the river at the place where the Confederates had destroyed the bridge; once that was accomplished, the entrenchments of Vicksburg lay just eight miles ahead, and the job of taking the great bastion could at last begin. Grant came toward Sherman in the dark, and Sherman described the quietly dramatic time they had together:

> A pontoon-bridge was at once begun, finished by night, and the troops began the passage. After dark, the whole scene was lit up by fires of pitch-pine. General Grant joined me there, and we sat on a log, looking at the passage of those troops by the light of those fires; the bridge swayed to and fro under the passing feet, and made a fine war-picture.

The essence of this moment was not that two increasingly important generals were watching their forces cross a river by night, but that two soldiers who had been under fire together were sitting on a log in friendly comradeship. Undoubtedly Sherman was talking and Grant was as usual politely listening, but the bond between them was that known only to those who become friends at a time when they know that any day may be their last.

The next day, May 18, 1863, Sherman marched his corps around to the north of Vicksburg. Johnston had remained away from the city, and Grant achieved the objective he had sought for months: Union troops surrounded the enemy bastion on three sides, and on its west side, Porter's warships controlled the waters of the Mississippi. In the final movement that sealed the ring around the city and its miles of defenses, Grant and Sherman rode among the advanced skirmishers at the front of one of their columns to secure the entrenchments atop Chickasaw Bluffs, which Sherman had been unable to take during the torrential rains five months earlier. Grant described what happened as they approached the hostile trenches: "These were still occupied by the enemy, or else the garrison from Haines' [sic] Bluff had not all got past on their way to Vicksburg. At all events the bullets of the enemy flew thick and fast for a short time. In a few minutes Sherman had the pleasure of looking down from the spot coveted so much by him the December before where his command had lain so helpless for offensive action."

As they rode into the just-abandoned Confederate positions, all was quiet. Now the muddy slopes of the previous December were dry, and the swamps below them bore no resemblance to the boiling brown rapids that had tumbled Sherman's men back when they tried to advance under fire. The cannon still in the enemy trenches had been spiked by the withdrawing defenders to make them useless.

Sherman, usually talkative, remained silent for a while. It was he who had at one point urged Grant to take his army back to Memphis and strike out again for Vicksburg on a different route. Along with virtually everyone else, including Halleck and Stanton, Sherman had disapproved of Grant's plan to move ships down the river under Vicksburg's guns. When he had urged Grant to wait for his supply trains to catch up to the troops, Grant had cut loose, foraged off the country, and, movement upon movement, seized every opportunity to bring them to this day. Now he and Grant were sitting on their horses, atop strategically placed heights that looked toward the great Southern bastion that their army had surrounded. Inside Vicksburg were thirty thousand men who, no matter how well they defended their positions, could no longer get out to help other Confederate forces.

Turning to Grant, Sherman spoke words that combined apology and admiration: "Until this moment I never thought your expedition a success. I never could see the end clearly until now. But this is a campaign; this is a success if we never take the town."

As for taking the town, Grant intended to do that as soon as he could. Pemberton was behind his fortifications with about thirty-three thousand soldiers, with another six thousand civilians of all ages and both sexes in the little city, but Grant knew that Pemberton was not the only enemy general he needed to think about. After his clashes with Grant, Joseph E. Johnston had decided not to come into Vicksburg but stay out in the open country to the east of Jackson for the time being, and, as Grant put it, "was in my rear, only fifty miles away, with an army not much inferior in number to the one I had with me, and I knew he was being reinforced. There was danger of his coming to the assistance of Pemberton." Grant hoped to take Vicksburg at one stroke, so that he could be free to turn east again and face Johnston, if need be. Accordingly, at two in the afternoon on May 19, the day after he and Sherman were under fire together at Chickasaw Bluffs, Grant struck the Vicksburg defenses with all three of his corps: the one led by Sherman, the one belonging to the self-promoting McClernand, and the one under young Major General James B. McPherson, who had moved swiftly upward through the officer ranks and was a man both Grant and Sherman saw as a great future leader in the war.

As they headed up the slopes of the Vicksburg defenses, Grant's men ran into a wall of fire. Grant later minimized the extent of that repulse, saying that the overall attack "resulted in securing more advanced positions for all our troops," but Sherman wrote Ellen a more accurate appraisal: "The heads of Colums [sic] are swept away as Chaff thrown from the hand on a windy day." Echoing Grant's concerns, he added, "We must work smartly as Joe Johnston is collecting the shattered forces, those we beat at Jackson and Champion's Hill, and may get reinforcements from Bragg . . . and come pouncing down on our Rear." Telling Ellen about one of her younger brothers, Captain Charles Ewing, he said, "Charley was very conspicuous in the 1st assault, and brought off

the colors of the Battalion which are now in front of my tent[,] the Staff
1/4 cut away by a ball that took with it a part of his finger." Summing up
the campaign to this moment, he said, "Grant[']s movement was the
most hazardous, but thus far the most successful of the war. He is entitled to all the Credit, for I would not have advised it." As for his fellow
corps commanders, "McPherson is a noble fellow, but McClernand a
dirty dog."

Three days after this, on May 22, Grant once again threw all three of
his corps at the slopes of Vicksburg. This time it was another of Ellen
Sherman's younger brothers, Brigadier General Hugh Ewing, who was
in the thick of the action. Lying with his men in a ditch just down the
slope below an enemy parapet from which they had fallen back under a
withering fire, he handed a captain a new regimental battle flag and
said, "I want this planted on the top." The captain took the banner forward and was killed; the man's younger brother managed to spike the
flag's staff into the earth at the crest and rolled back down the slope to
safety. The area became a no-man's-land; for hours, the Confederates
kept trying to rush that battle flag at the edge of their defenses and capture it, while Ewing's men exposed themselves to rise up and shoot them
down. At dusk, when orders came to withdraw down the slope, a private
volunteered to crawl forward and yank the flag out of the earth. He succeeded, and Ewing and his men came down the slope with their flag.

It had been an afternoon of brave fighting on both sides, but of this
effort, which cost him more than three thousand casualties, Grant said,
"The attack was gallant, and portions of each of the three corps succeeded in getting up to the very parapets of the enemy and placing their
battle flags upon them; but at no place were we able to enter." As for an
additional attack during the afternoon that Grant ordered in deference to
repeated messages from McClernand that he was about to break through
and needed additional support from Sherman on his right and McPherson on his left, Grant commented, "This last attack only served to increase our casualties without giving any benefit whatever."

As aggressive a general as the Civil War produced, Grant nevertheless
realized that more frontal attacks would be futile. "I now determined
upon a regular siege—to 'out-camp the enemy,' as it were, and to incur
no more losses. The experience of the 22d convinced officers and men
that this was best, and they went to work on the defenses and approaches
with a will. With the navy holding the river, the investment of Vicksburg

was complete. As long as we could hold our position the enemy was limited in supplies of food, men and munitions of war to what they had on hand. These could not last always."

While Grant's army began its various entrenchments, building earthworks to house the artillery that would ceaselessly slam away at the "Gibraltar of the Confederacy," visitors, some of them coming down the river on ships that disembarked them above the city, arrived to visit their family members who were soldiers and to see where the next act of the long Vicksburg drama would take place. Grant found himself amused by the sight of families of soldiers bringing the men "a dozen or two of poultry." Unaware that in living off the land, Grant's troops had wrung the necks of any number of chickens, ducks, and turkeys, hastily cooking them and often eating them while they marched, "They did not know how little the gift would be appreciated . . . the sight of poultry . . . almost took away their appetite. But the intention was good."

Not only the families of soldiers came to see besieged Vicksburg. Grant described one of the most important visitors and how Sherman refused to take any credit for the success of the campaign to date and directed it all toward Grant.

Among the earliest arrivals was the Governor of Illinois, with most of the State officers. I . . . took them to Sherman's headquarters and presented them. [Fifteen of Sherman's fifty regiments were from Illinois.] Before starting out to look at the lines—possibly while Sherman's horse was being saddled—there were many questions asked about the late campaign . . . There was a little knot around Sherman and another around me, and I heard Sherman repeating, in the most animated manner, what he had said to me when we first looked down . . . upon the land below on the 18th of May, adding: "Grant is entitled to every bit of the credit for the campaign; I opposed it . . ."

But for this speech it is not likely that Sherman's opposition would ever have been heard of. His untiring energy and great efficiency during the campaign entitle him to a full share of all the credit due for its success. He could not have done more if the plan were his own.

The siege began. From out in the river, the 100 cannon aboard Admiral Porter's gunboats began firing shells into the fortified city at all hours, and Grant's 220 cannon and mortars opened up on the inland side. The Confederate defenders responded with artillery fire from more

than 170 guns, and sharpshooters from both armies began firing at anything that moved. As the siege continued, Grant received reinforcements
from Memphis; eventually he had between seventy and eighty thousand
men, and used half of them to guard his rear against Joseph E. Johnston,
who at times had thirty thousand men under his command in the area
east of the city but had almost no artillery with him and little in the way
of a supply line. (On May 29, Johnston tried to get a letter through the
lines to Pemberton. In it he said, "I am too weak to save Vicksburg," and
held out only the hope that he might be able to "save you and your garrison" if Pemberton could "cooperate" in an effort to break out of the city
and link up with him. The impossibility of that was shown by the fact
that the Union lines were so closely drawn around the city that the letter
could not be sneaked through to Pemberton until sixteen days later.)

As the daily bombardments and sniping continued, attackers and defenders had an enormous variety of experiences. Within the besieged
area, Henry Ginder, a civilian construction engineer who was continuing to improve the already-formidable Confederate fortifications, wrote
an account of the dangers he faced and of a shell that was a dud and did
not explode.

> Not a day passed but in riding back and forth from my labors the shells burst
> around my path and minié balls whiz past my ears. Last night I was on foot
> returning from the scene of my labors, and I heard a 13-inch shell coming
> but couldn't see it; it came nearer and nearer until I thought it would light
> on my head, when splosh! it went into the earth a few feet to my left, throw-
> ing the dirt into my face with such force as to sting me for some time after-
> wards. The Lord kept it from exploding . . . Otherwise it would have singed
> the hair off my head and blown me to pieces into the bargain.

Many of the civilians inside Vicksburg began spending much of their
time in caves, to avoid being hurt in the bombardment. What could happen to a house was recorded in her diary by Dora Richards, the young
wife of a lawyer: "I was just within the door when the crash came that
threw me to the floor . . . Shaken and deafened I picked myself up . . .
The candles were useless in the dense smoke, and it was many minutes
before we could see. Then we found the entire side of the room torn
out." The defenders started to cope with shortages: running out of
newsprint, the defiant Vicksburg *Citizen* and the Vicksburg *Whig*, both

assuring their readers that Joseph E. Johnston was on his way to break the siege and save the city, put out their editions in a small format printed on one side of cut-up wallpaper.

Among the besiegers, various unlooked-for things occurred. "Old Abe," the American bald eagle that accompanied the Eighth Wisconsin as its mascot, was wounded by the defenders' fire but survived. Captain J. J. Kellogg of the 113th Illinois, a company commander in the brigade led by Ellen Sherman's brother Hugh Ewing, a few days earlier had seen Grant and Sherman looking at him through their field glasses as he led a charge up to the parapet on which a Union battle flag was finally planted. Now, redeployed with his company to a seemingly far safer position beside a bayou on an approach to Vicksburg, he started to put up a sleeping tent and encountered a surprise.

> When I was driving stakes for my new home, a great green-headed alligator poked his nozzle above the surface of the bayou waters and smiled at me.
>
> Upon examination of the ground along the bayou shore, I discovered alligator tracks where they had waltzed around under the beautiful light of the moon on a very recent occasion, so I built my bunk high enough to enable me to roost out of reach of these hideous creatures.
>
> Though I had built high enough to escape the prowling alligators I had not built high enough to get above the deadly malaria distilled by that cantankerous bayou.

On one of the early days of the siege, a private of the Fourth Minnesota saw an older Union soldier in a rumpled uniform standing at the top of an observation tower near the front, looking toward the entrenchments on the enemy-held slopes, and shouted, "Say! You old bastard, you better keep down from there or you will get shot!"

The man paid no attention, but when the Minnesotan started to shout again, his captain grabbed him and said, "That's General Grant!"

While he was stationed on the northern end of besieged Vicksburg, Sherman kept up with developments in Washington. He learned that the federal government, needing ever-greater numbers of soldiers to add to the dwindling number of volunteers for the Union Army, intended to introduce conscription and draft three hundred thousand men into the service. Sherman saw that as necessity, but the plan for how these new troops were to be used shocked him. One hundred thousand would be trained and sent forward to fill up the ranks of the existing regiments, many of which by now had an excellent level of combat experience shared by veteran officers and men, but the remaining two hundred thousand were to be formed into entirely new regiments. This was to be 1861 all over again: new colonels would be commissioned from civilian life by political appointment, and recruits with a few weeks' training would march to unnecessary deaths in a military version of the blind leading the blind. Any of the experienced, proud old regiments whose casualties had caused their numbers to fall below three hundred were arbitrarily to be consolidated with other old regiments, instead of receiving recruits who could fill their ranks and immediately profit from the experience to be gained by serving with combat veterans.

As a man with a penchant for order who frequently found the workings of a democracy incompatible with the realities of raising an efficient army and fighting successful campaigns, Sherman was appalled by the prospect of having more "political colonels" and sending into battle more than a hundred untried regiments. On June 2, he wrote Grant a letter on the subject.

Dear General:

I would most respectfully suggest that you use your personal influence with President Lincoln to accomplish a result on which it may be, the Ultimate Peace and Security of our Country depends.

. . . All who deal with troops in fact instead of theory, know that the knowledge of the little details of Camp Life is absolutely necessary to keep men alive. New Regiments for want of this knowledge have measles, mumps, Diarrhea and the whole Catalogue of Infantile diseases, whereas the same number of men distributed among the older Regiments would learn from the Sergeants, and Corporals and Privates the art of taking care of themselves . . . Also recruits distributed among older Companies catch up, from close and intimate contact, a knowledge of drill, the care and use of arms, and all the instructions which otherwise it would take months to impart.

. . . I am assured by many that the President does actually wish to support & sustain the Army, and that he desires to know the wishes and opinions of the officers who serve in the woods instead of the "Salon." If so you would be listened to . . . I have several Regiments who have lost . . . more than half their original men . . . Fill up our present ranks, and there is not an Officer or man of this Army, but would feel renewed hope and courage to meet the struggles before us.

I regard this matter as more important, than any other that could possibly arrest the attention of President Lincoln and it is for this reason, that I ask you to urge it upon him at the auspicious time.

Grant forwarded Sherman's letter to Lincoln, along with his own letter endorsing Sherman's facts and reasoning, and told the president, "I would add that our old regiments, all that remains of them, are veterans equaling regulars in discipline . . . A recruit added to them would become an old soldier, from the very contact, before he was aware of it." He went on to point out that the existing regiments already had their encampments, garrison equipment, and supply trains, and that in addition to considerations of military efficiency and morale, it would cost the government far less to put new recruits into existing regiments than to buy and construct everything necessary to organize new ones. What he and Sherman got for their trouble was a letter to Grant from Halleck, saying that, as planned, two hundred thousand men would go into new regiments. Lincoln was still making military appointments as political favors.

As the siege went on, new developments in other matters continued to occur. On June 7, an unusual battle took place at Milliken's Bend. Four understrength and outnumbered regiments of virtually untrained black Union Army soldiers, newly freed slaves from Louisiana and Mississippi who had volunteered only since the siege began, using obsolete Belgian muskets and supported by one of Admiral Porter's gunboats, drove off a Confederate brigade that was trying to raid a Union supply line. There were reports that the Confederates murdered some of the black soldiers they captured, and two of the white Union officers who led the men were apparently also executed.

This battle, though brief and small in size, changed the minds of many Union Army commanders concerning blacks' willingness and ability to fight. Charles Dana, still with Grant's army, had this to say:

A force of some two thousand Confederates engaged about a thousand negro troops defending Milliken's Bend. This engagement at Milliken's Bend became famous from the conduct of the colored troops. General E. S. Dennis, who saw the battle, told me that it was the hardest fought engagement he had ever seen. It was fought mainly hand to hand. After it was over many men were found dead with bayonet stabs, and others with their skulls broken open by the butts of muskets. "It is impossible," said General Dennis, "for men to show greater gallantry than the negro troops in that fight."

The bravery of the blacks at Milliken's Bend completely revolutionized the sentiment of the army with regard to the employment of negro troops. I heard prominent officers who formerly in private had sneered at the idea of negroes fighting express themselves after that as heartily in favor of it.

Grant was among those persuaded. In a letter to Adjutant General Lorenzo Thomas, he made this comment on a reorganization of the black regiments that were coming into existence: "I am anxious to get as many of these negro regiments as possible and to have them full and completely equipped." In a letter to Halleck he said, "The negro troops are easier to preserve discipline among than our White troops and I doubt not will prove equally good for garrison duty. All that have been tried have fought bravely."

Sherman had a different attitude. Months after Milliken's Bend, he was still expressing his mistrust of the abilities of black troops. In a letter to Ellen he told her, "I would prefer to have this a white man's war and provide for the negroes after the time has passed, but we are in revolution and I must not pretend to judge. With my opinions of negroes and my experience, yes, prejudice, I cannot trust them yet."

At this point in the siege, Grant's nemesis, alcohol, reentered the picture. The evidence on the point is an in-headquarters letter to Grant from his chief of staff Rawlins. Saying that his motivation was "the great solicitude I feel for the safety of this army," Rawlins made reference to a report that Grant had been drinking with a military surgeon at Sherman's headquarters "a few days ago," but concentrated on this: "Tonight when you should, because of the condition of your health if nothing else, have been in bed, I find you where the wine bottle has just been emptied, in company with those who drink and urge you to do likewise, and the lack of your usual promptness and clearness in expressing yourself in writing tended to confirm my suspicions."

Years later, Charles Dana said that he had been present when Rawlins "delivered that admirable communication. It was a dull period in the campaign. The siege of Vicksburg was progressing with regularity. No surprise from within the city or from without was to be apprehended;

and when Grant started out in drinking, the fact could not imperil the situation of the army or any member of it except himself." At the time, Dana clearly maintained the policy he had adopted: this incident was just what he was supposed to report to Washington in the special tele-graphic code devised for him to use, but he believed in Grant and in-tended to tell Lincoln and Stanton about Grant's drinking only after the critical situation at Vicksburg came to its end.

As if Grant did not have problems enough, his ambitious, fractious subordinate McClernand now sought to further his own reputation with the public in a way that broke army regulations and was guaranteed to anger Grant and Sherman, both of whom had suffered at the hands of the press. McClernand was already on the thinnest of ice: through Dana, Secretary of War Stanton had recently passed the word to Grant that he was free to relieve McClernand at any time and send him north for reas-signment. Unbeknownst to Grant, on May 30 McClernand had written what he called his "General Orders 72." Ostensibly a document congrat-ulating his troops for their bravery, it was in fact an astonishing piece of self-promotion, which, in addition to being circulated among his units, McClernand had sent to St. Louis to be published in the *Missouri Democrat*. He presented himself as the hero of the failed attacks on May 22 that he had in fact made worse by urging an additional attack— the one that Grant said "only served to increase our casualties without giving any benefit whatever." The text of McClernand's order implied that Sherman, on his right, and McPherson, on his left, had failed to support him, and that Grant, by not sending him reinforcements, had lost the opportunity to take Vicksburg that day.

The piece appeared in St. Louis on June 10, and a copy arrived at Grant's headquarters three days later, where its effect was symbolically like that of an incoming Confederate salvo. Not only did it misrepresent the costly support McClernand had received—Grant said that it "did great injustice to the other troops engaged in the campaign"—but its publication violated the rules of both the War Department and Grant's military department, which required that no document of this sort could appear in the press without the permission of the departmental com-mander, i.e., Grant.

Even now, Grant postponed action on the matter; a letter he sent to McClernand two days later speaks only of troop movements. On June 17, Sherman saw a copy of the *Memphis Evening Bulletin* reprinting McClernand's orders and sent a blazing letter to Rawlins. Sherman said that on May 22 McClernand had lied about the extent of his advance in order to convince Grant to support a further effort, and when that effort was made, swiftly and fully, "we lost, needlessly, many of our best officers and men." One account had it that Sherman also appeared at Grant's headquarters, holding the Memphis newspaper and so angry that he could not speak for several minutes. On that day, Grant wrote a peremptory note to McClernand demanding that he either confirm or deny that Orders 72 was his work. The following day, McClernand telegraphed that he had written it, and stood by it, but thought it had been sent to Grant before being published. Within hours, Grant relieved McClernand of command; in a telegram to Halleck telling him of his action, Grant said, "I should have relieved him long ago for general unfitness for his position." The next day, Dana wired Stanton that Grant's most pressing reason for removing McClernand was that, if Grant were incapacitated, McClernand would outrank both Sherman and McPherson, which would have "most pernicious consequences to the cause."

Other than saying of McClernand's departure that "not an officer or soldier here but rejoices he is gone away," Sherman had little time to think about anything but the campaign at hand. Worried about reports that the elusive and skillful Joseph E. Johnston was leading an army of thirty thousand toward Vicksburg from the east, Grant ordered Sherman to ready himself to move out immediately toward the Big Black River. Feeling that Sherman was the best man to find Johnston and oppose him, he used characteristically concise instructions: "You will go and command the entire force." To Admiral Porter, Grant wrote of Johnston, as if he could see it happening, "I have given all the necessary orders to meet him twenty-five miles out, Sherman commanding."

Grant knew that Sherman had hoped to take part in a final victorious entry into the city, riding in at the head of his troops and perhaps becoming for a time the military governor of Vicksburg, as he had been of Memphis. Disappointed, Sherman would write Ellen, "I did hope Grant would have given me Vicksburg and let some one else follow up the enemy inland." But he obeyed without discussion and headed out with thirty-four thousand men. (Describing the closeness of his relationship with Grant, Sherman wrote his brother John that "with him I am as a second self. We are personal and official friends.")

As Sherman moved his force through the countryside, seeking to find and engage Johnston, Grant expressed the strength of his support for him in a letter he wrote on June 23. Grant referred to troops he had with him at a place "Near Vicksburg" and others at nearby Young's Point, and listed the units he had ready to reinforce Sherman if he should need them, closing with, "Use all the forces indicated above as you deem most advantageous, and should more be required, call on me and they will be furnished to the last man here and at Young's Point."

The shortage of food caused by the siege began to take its toll on Vicksburg's defenders. Meat sold as beef was described in one account as "very often oxen killed by the enemy's shells, and picked up by the butchers." A bitter Vicksburg resident invented a fictional "Hotel de Vicksburg" and wrote out a bill of fare that began with Mule Tail Soup, offered Mule Rump Stuffed with Rice as a roast, and included among its entrées Mule Spare Ribs Plain and Mule Liver Hashed. Sergeant William H. Tunnard of the Third Louisiana Infantry, entrenched in the besieged city, recorded this in his regimental history: "How the other troops felt, we know not, but the boys of the Third Regiment were *always hungry.*" Soon dogs and cats began disappearing; the city's stoic citizens and soldiers made jokes about "What's become of Fido?" but no one doubted the animals' fate. While Vicksburg was running out of food, hundreds of tons of ammunition were available.

As the Union forces pushed nearer the town's defenses, constantly digging trenches that snaked toward the enemy positions, the proximity of the two armies produced both sniping at closer range and occasional impromptu truces. In places, the Union and Confederate soldiers were so near each other that, although they kept their heads down, they needed only to raise their voices slightly to communicate across the narrowing no-man's-land. Blackberry bushes grew in profusion between the opposing trenches; the troops of both sides suffered from diarrhea and knew that blackberries helped to cure it, so quick conferences produced agreements that allowed men from both sides to go out and pick the berries.

One hot June day, after hours of desultory sniping, a private of the Eleventh Wisconsin said to his comrades, "I'm going down into the ravine and shake hands with them Rebs!" and he did just that. More men from both sides came out, shaking hands with their enemies, until hundreds of men were milling about in the no-man's-land of this ravine. They talked about everything: how hot it was, the kind of illnesses they had, what they thought of their generals. Union soldiers traded rations of coffee for Confederate tobacco. Farmboys swapped knives and chatted about their hometowns, and some soldiers even pulled out tintypes of their wives and sweethearts to show to men who had been shooting at them an hour before. A young Confederate, talking with some Wisconsin boys, suddenly blurted out, "I want to see my ma," and went off to sit by himself on a fallen tree trunk.

A Union officer came walking into the middle of this friendly gathering and began berating the men of both sides for all this fraternization. The young men fell silent, said good-bye to one another, slowly walked back up the slopes to their respective trenches, and soon began shooting at one another again.

While the siege continued, with Sherman maneuvering to the east and finding that the wily Johnston had no intention of fighting his superior force unless he could catch the Union regiments by surprise, Grant dealt with a variety of matters at his headquarters. On June 25, he wrote to Lorenzo Thomas in Washington, asking for the speedy assignment of Ely S. Parker to his army as an assistant adjutant general. "I am personally acquainted with Mr. Parker," Grant said, "and think [him] eminently qualified for the position. He is a full blooded Indian but highly qualified and very accomplished. He is a Civil Engineer of conciderable [sic] eminence and served the Government some years in superintending the building of Marine Hospitals and Custom Houses on the upper Miss. river."

Parker was duly assigned to Grant and quickly proved to be one of the ablest members of his staff. A thirty-five-year-old Seneca who had been educated at Rensselaer Polytechnic Institute, Parker had at first been rejected by the army because he was not a white man. The recommendation from Grant, who had known Parker when he worked on projects at Galena, brought him into Grant's "military family" as a captain, and he stayed with Grant through all that lay ahead.

Unknown to Grant, at just this time something extraordinary occurred in Kentucky. In parts of that state, Confederate raiders moved about periodically and unpredictably. Julia Grant's sister Emma, who had first met Grant thirty years before, when, "pretty as a doll," young Lieutenant Grant rode into the yard at White Haven to call on the Dent family, was now living near Caseyville, Kentucky, close to the Ohio

River. She was the wife of James F. Casey, who in the custom of the day she always referred to as "Mr. Casey." Emma described the situation in her area in these terms: "There were a good many bands of guerrillas prowling about the country at this time, as well as several other bands of irregular Confederate soldiers, but, as they never molested us, we were scarcely aware of their presence."

Grant's son Fred, who had been with him during much of the Vicksburg siege, had begun feeling sick, and to improve his health Grant had sent him north from Mississippi to Kentucky to stay for a week or more at Emma's house. She said of her thirteen-year-old nephew, "He was very fond of us, and we of him." On a morning when Fred rode into nearby Caseyville with his uncle, Emma recounted what happened:

A man dressed in the tattered uniform of a Confederate officer rode into the yard and [after dismounting] asked me for a drink of water. I gave it to him, and as he lifted the cup to his lips he said, casually:

"I guess Fred Grant is visiting you, isn't he?"

Instantly a cold suspicion struck me like a dart through the heart, and I answered him as casually as he had questioned me:

"Why, no."

"OH!" he said. "Isn't he?"

"No, he's gone."

"Gone, has he? Is that so?" He looked at me with a smile slowly breaking out over his face. "Surely, he has," he said again, as if speaking to himself. Then he remounted his horse, took off his hat, made me a sweeping bow, and rode away. I did not lose a moment, but as quick as one of the horses could be caught out of the pasture, I put a black boy on his back and sent him to find my husband. I sent Mr. Casey word to put Fred on a coal boat and get him down the river to Cairo as fast as ever he could. I also suggested that if he could communicate with a gunboat on the river it might be very well.

Later, Emma reported:

A squad of eight hard-riding, grim-looking, and tattered cavalrymen rode up to the gate. One of them, heavily armed, and looking as fierce as a Greek bandit, came up to the porch.

"Is this Mr. Casey's?" he asked, politely. I told him that it was.

"Isn't there a boy visiting here?"

"No, he has gone back to his mother, at Cairo."

"Are you sure?"

"Yes. And I think there is likely to be some gunboats coming up the river very shortly, looking for some one. Perhaps you gentlemen will be interested in seeing them."

The fierce-looking bandit laughed pleasantly, said that it was a nice day, and rejoined his companions at the gate. They talked in low voices for a while, then sprang on their horses and rode away.

As Emma put it, had they captured Fred, "It is mere speculation to consider what effect this might have had on the cause of the Union." What needs no speculation is what Grant thought later in the war, when he was presented with a plan to abduct Jefferson Davis and bring him north as a prisoner. Grant cut off the discussion with the observation that he and his men were not kidnappers.

Inside Vicksburg, on June 28, Confederate general Pemberton received a letter that was signed, "Many Soldiers." Down to a quarter of their normal rations a day—and in some units there was less than that to eat—his men told him, "If you can't feed us, you had better surrender, horrible as the idea is, than suffer this noble army to disgrace themselves by desertion . . . This army is now ripe for mutiny, unless it can be fed." Pemberton could see that, strong as the besieging Union forces were, hunger was even stronger and that he would probably soon have to surrender to both that and Grant.

The following day, in a letter to Julia, Grant, whom the Caseys had not told about his son's near capture, wrote her that "Fred. Has returned from his uncle[']s. He does not look very well but is not willing to go back until Vicksburg falls." Grant added that Joseph E. Johnston was "still hovering beyond the Black River." Johnston faced a painful choice. As things stood, if he did not act, the city's defense seemed certain to collapse. If he took the risk of fighting Sherman's larger numbers in an effort to break through to Vicksburg, he might be defeated in the area east of the city. Then Vicksburg would have to surrender in any case, and Johnston would lose additional thousands of his own men who could otherwise be used in future campaigns. Grant told Julia that he thought

Johnston would feel compelled to advance and fight Sherman, but that, either way, Vicksburg would have to surrender within a week. With his often uncanny sense of a military situation, he told Julia that "Saturday or Sunday next [July 4 or July 5] I set for the fall of Vicksburg." As usual, he closed his letter with, "Kiss the children for me. Ulys."

At ten in the morning of July 3, white flags began appearing along the crest of Vicksburg's fortified slopes. Confederate major general John Bowen came riding out of the Confederate lines, sent by Pemberton and accompanied by one of Pemberton's staff. Taken to Union headquarters, Bowen handed one of Grant's staff a letter to Grant in which Pemberton said that he wanted to arrange "terms for the capitulation of Vicksburg." Grant soon composed a letter in which he told Pemberton that there would be no discussion of terms other than unconditional surrender. He did, however, add, "Men who have shown so much endurance and courage as those now in Vicksburg, will always challenge the respect of an adversary, and I can assure you will always be treated with all the respect due to prisoners of war." Bowen asked to speak directly with Grant but was told that if Pemberton wanted to meet Grant face-to-face, he was welcome to come out of Vicksburg for that purpose "at any hour in the afternoon which Pemberton might appoint."

At three that afternoon, Pemberton, along with Bowen and several other officers, rode out to meet with Grant. They found him standing on a slope near their entrenchments, accompanied by a number of Union officers. His son Fred and Charles Dana were also there. With Sherman still absent while he blocked Johnston from making any last-minute advance, the next officer in seniority to Grant was Sherman's fellow corps commander, young Major General James B. McPherson. For a time it appeared that there could be no agreement on accepting or softening Grant's "unconditional surrender" statement and that the fighting would resume. Dana noted that "Pemberton was much excited, and was impatient in his answers to Grant." Then Bowen suggested that he and McPherson try to talk things through between themselves.

What happened next was ironic: during a formal, intense half-hour session, Bowen and McPherson got nowhere, while Grant and Pemberton stood under a stunted oak tree and exchanged reminiscences about their experiences in the Mexican War. By the time Bowen and McPherson came back and grimly announced that they could find no grounds for a mutually acceptable mode of surrender, Grant informed them that he would have something worked out by ten o'clock that night that might satisfy both Pemberton and himself. The two groups parted, Pemberton going back into the besieged city and Grant returning to his headquarters.

The problem involved not the fact of surrender but whether these more than thirty thousand Confederate soldiers were to be sent to prison camps in the North or be paroled. Pemberton wanted his men to be paroled and, as a matter of honor, be allowed to march out of Vicksburg with their flags flying, before laying down their arms. In the strict sense, this would not be the "unconditional surrender" first insisted upon by Grant, but that evening Grant sent a letter to Pemberton agreeing to these conditions and sat in his tent awaiting an answer. Grant had decided to parole the enemy troops because it would immediately free all the men and ships at his command to continue combat operations rather than having to furnish guards and transportation to take the multitude of defeated Confederates to prison camps.

In Grant's tent that night, young Fred later wrote that he was "sitting on my little cot, and feeling restless, but scarcely knowing why." He went on:

> Presently a messenger handed father a note. He opened it, gave a sigh of relief, and said calmly, "Vicksburg has surrendered."
>
> I was thus the first to hear the news officially, announcing the fall of the Gibraltar of America, and, filled with enthusiasm, I ran out to spread the glad tidings. Officers rapidly assembled and there was a general rejoicing.

At ten in the morning of the Fourth of July, 1863, ending the forty-seven-day siege, Pemberton had the Confederate Stars and Bars lowered from the highest point in Vicksburg's defenses, and at his command the Stars and Stripes was raised. White flags appeared everywhere along the enemy entrenchments. The hungry, tattered Confederate regiments came marching out as if on parade, muskets on their shoulders, with their bands playing and battle flags flying. They halted, and the men laid down their weapons, in some places stacking them right on the parapets of Union trenches that had been dug forward to within a few yards of the defensive slopes. Quietly watching this, Grant observed of his own men,

"Not a cheer went up, not a remark was made that would cause pain." Union troops moved forward among the disarmed Confederates to give them food and share their kettles of coffee. A soldier from Wisconsin later said, "It was good to see them eat . . . We could never remember anything that gave us greater pleasure than the eagerness of the rebels to get a drink of coffee . . . [Later, that night,] many of us did not sleep at all, talking with the prisoners."

As the day of surrender continued, the dimensions of the victory became even clearer. Grant said, "At Vicksburg, 31,600 prisoners were surrendered, together with 172 cannon, about 60,000 muskets and a large amount of ammunition." As a result of reports that groups of Union soldiers had entered the city without authorization and were looting it, the Forty-fifth Illinois was ordered to be the first unit of its division to march in, set up advance headquarters at the courthouse, and begin to impose order on everyone.

Inside the city, Lida Lord, the young daughter of the minister of Christ Episcopal Church, a man who was also chaplain of the First Mississippi Brigade, came out of the cave in which she and her mother and brothers and sisters had stayed for safety during the siege. They encountered some disarmed Confederates who apparently had been sent back into the city to await the formalities of signing their paroles.

> We met group after group of soldiers and stopped to shake hands with them all. We were crying like babies, while tears ran down their dusty cheeks, and eyes that had fearlessly looked into the cannon's mouth fell before our heartbroken glances.
>
> "Ladies, we would have fought for you forever. Nothing but starvation whipped us," muttered the poor fellows, and one man told us that he had [to avoid surrendering it] wrapped his torn battle-flag around his body under his clothes.

A Southern woman who had endured the siege, watching for nearly two months the daily deterioration of the soldiers and civilians in Vicksburg, saw the Union soldiers march in. "What a contrast to the suffering creatures we had seen so long were these stalwart, well-fed men, so splendidly set up and accoutered. Sleek horses, polished arms, bright plumes—this was the pride and panoply of war. Civilization, discipline, and order seemed to enter with the measured tramp of these marching feet."

Colonel Robert S. Bevier, a lawyer from Russellville, Kentucky, who was serving with the First Missouri Confederate Brigade, dismissed his surrendered troops and watched the surrender unfold.

> I rode into the city to see the vast Federal fleet come down to the landing, with pinions and streamers fluttering, and blaring music and blowing whistles . . .
>
> When returning to camp I was politely accosted by an officer in blue, who overtook me. We had some conversation, chiefly complimentary on his part, to the stubborn bravery of the troops, when, noticing that a large staff followed him, which I had not observed before, as the road was crowded with equestrians, I looked at him closely and found that it was General Grant himself.

At the waterfront, Grant went aboard Admiral Porter's flagship, the *Blackhawk*, which had pennants flying from all its rigging, and its crew turned out in white dress pants and navy blue jackets. Grant joined in the victory party for a few minutes but then walked off to sit by himself. Porter said of the moment, "No one, to see him sitting there with that calm exterior amid all the jollity . . . would ever have taken him for the great general who had accomplished one of the most stupendous military feats on record."

Throughout these climactic days, Grant and Sherman remained in constant communication. The day before the surrender took place, Grant had sent Sherman a telegram that began, "I judge, Johnston is not coming to Vicksburg, he must be watched though." The challenge for Sherman now was to hold himself ready to come to Grant's side if the negotiations failed at the last moment and a final massive attack became necessary, while not letting Johnston slip away as he often had. Informing Sherman that negotiations for Vicksburg's surrender were under way, Grant made the assumption that the city would soon be in Union hands and gave Sherman his usual kind of directive—one that stressed action and momentum, and left it to the commander on the scene to work out the details: "When we go in [to Vicksburg], I want you to drive Johnston from the Mississippi Central Rail Road,—destroy bridges as far as Grenada with your cavalry, and do the enemy all the harm possible—

You can make your own arrangements and have all the troops of my Command, except one Corps."

Later the same day, Grant sent Sherman another message, explaining how the negotiations then stood and, thinking of the situation facing Sherman out in the country well to the east of Vicksburg, adding, "I want Johnston broken up as effectively as possible, and [rail]roads destroyed. I cannot say where you will find the most effective point to strike."

Thinking that by the time he answered the city might have surrendered, Sherman fired off a telegram that said in part, "If you are in Vicksburg Glory Hallelujah the best Fourth of July since 1776," and assured Grant that he was ready to move. While waiting for Pemberton's answer that evening, Grant telegraphed Sherman yet again, saying, "There is but little doubt, but the enemy will surrender to night or in the morning—make your calculations to attack Johnston." Continuing the exchange of messages, Sherman assured Grant that he had some units already on the move and would throw everything else forward at Johnston as soon as he knew that Grant would not need his support in taking Vicksburg if the surrender negotiations failed. Sherman wrote Grant's chief of staff Rawlins a detailed plan of what he intended to do and said of Vicksburg, "The news is so good I can hardly believe it."

Finally, Sherman got definite reports that the surrender was taking place. He sat down at his advanced "Camp at Bear Creek" to write Grant: "I can hardly contain myself . . . Did I not know the honesty, modesty, and purity of your nature, I would be tempted to follow the example of my standard enemies of the press in indulging in wanton flattery; but as a man and a soldier, and ardent friend of yours, I warn you against the incense of flattery that will fill our land from one extreme to the other. Be natural and yourself, and this glittering flattery will be as the passing breeze of the sea on a warm summer day." After referring to the spirit of Grant's treatment of Pemberton and his men as "the delicacy with which you have treated a brave but deluded enemy," Sherman said, "This is a day of jubilee," and assured Grant that he now had all his units moving to find Johnston. "Already are my orders out to give one big huzza and sling the knapsack for new fields."

Still on the day of his greatest victory, thinking of momentum and not of mutual congratulations and praise, Grant sent Sherman yet another letter on purely operational matters. After telling Sherman that his decision as to which corps to use as a reserve "is just right" and asking to be told instantly about any new reports of Johnston's movements, he closed with another example of his keep-the-ball-moving spirit: "I have no orders or suggestions to give. I want you to drive Johnston out in your own way, and inflict on the enemy all the punishment you can. I will support you to the last man that can be spared."

It seemed hardly possible that any news could rival that of Vicksburg's capture, but on July 3, the day before the city surrendered, Robert E. Lee's Army of Northern Virginia was thrown back decisively at Gettysburg by the Union's Army of the Potomac under General George Gordon Meade. Lee's invasion of Pennsylvania, which had threatened Washington, Baltimore, Philadelphia, and Harrisburg, came to a bloody end. During the massive three-day battle, just under ninety thousand federal troops fought seventy-five thousand Confederates, but the Confederate losses in killed, wounded, captured, and missing were larger, totaling twenty-eight thousand compared to the Union's twenty-three thousand, and, as before, the North had a greater capacity to replace its losses.

A most important victory had been won, but the greatest drama of Gettysburg centered on one man: Robert E. Lee, in whom the hopes of the South were so profoundly embodied. Grant's old friend, Julia's cousin Confederate general James Longstreet, had done everything in his power to avoid having to execute the final failed uphill attack that Lee ordered, the gallant doomed effort that came to be known as Pickett's Charge. When Lee saw the shocked and wounded survivors of the fifteen thousand men he had sent up Cemetery Ridge come staggering back down the slope on the afternoon of July 3, he immediately took the entire blame upon himself. To his despondent general Cadmus Marcellus Wilcox, who had been a groomsman at Grant and Julia's wedding, he said, "Never mind, General, all this has been my fault—it is I that have lost this fight, and you must help me out of it the best way you can." When he saw Pickett, whose division had just been slaughtered, Lee told him, "It's all my fault. I thought my men were invincible."

As aggressive a general as Grant, at the end of this day of defeat Lee showed yet another aspect of what he was. It was later described by a Union soldier who said of himself, "I had been a most bitter anti-Southman, and fought and cursed the Confederates desperately." A musket ball had shattered the man's left leg, and as he lay on the ground near Cemetery Ridge, Lee and his officers came by, starting their retreat.

> As they came along I recognized him, and, though faint from exposure and loss of blood, I raised up my hands, looked Lee in the face, and shouted as loud as I could, "Hurrah for the Union!"
>
> The General heard me, looked, stopped his horse, dismounted, and came toward me. I confess that at first I thought he meant to kill me. But as he came up he looked at me with such a sad expression upon his face that I wondered what he was about. He extended his hand to me, and grasping mine firmly and looking right into my eyes, said, "My son, I hope you will soon be well."
>
> If I live a thousand years I shall never forget the expression on General Lee's face. There he was, defeated, retiring from a field that had cost him and his cause almost their last hope, and yet he stopped to say words like those to a soldier of the opposition who had taunted him as he passed by! As soon as the General had left me I cried myself to sleep there upon the bloody ground!

While Pemberton's troops marched out of their fortifications at Vicksburg to surrender to Grant's men, Lee's shattered army was retreating to Virginia, with Meade failing to pursue them as closely as Lincoln hoped he would. Deep in the details of the immediate aftermath of the Vicksburg surrender and planning his next moves, Grant simply noted in a letter to General Nathaniel Banks that he had received a telegram from Washington "stating that Meade had whipped Lee badly," but Sherman, while concentrating on the Vicksburg victory in a letter to Ellen, said that "the news from the Potomac . . . appears so favorable that I sometimes begin to think that the Secech will have to give in and submit." (It was in this letter that Sherman wrote Ellen, "I want to hear from you after you hear of the fall of Vicksburg. I have bet you will get tight on the occasion, à la fashion of Green Street California.")

In the North, the victory at Gettysburg naturally resonated far more loudly than the news from far-off Vicksburg. *The Philadelphia Inquirer* announced in a headline, "Victory! Waterloo Eclipsed!" The Northern church bells rang for Gettysburg, but the men of the West understood what the removal of the great Southern bastion on the Mississippi River meant. When Port Hudson, the last remaining Confederate fortress be-

tween Vicksburg and New Orleans, surrendered a few days later as a result of Vicksburg's loss, the Confederacy was cut vertically in half, and
Union shipping could go safely from St. Louis down to the Gulf of Mexico. Abraham Lincoln said of the pivotal moment, "The Father of Waters
again goes unvexed to the sea."

Lincoln's eloquent words also had meaning in terms of Union strategy, a concept that had been largely lacking in these twenty-six months
of war. Insofar as there had ever been a vision of what the entire Union
military and naval effort should be, in May of 1861 the aged and soon-to-
retire Union general in chief Winfield Scott had devised what came to
be called the Anaconda Plan. The North was to be the great snake wrap-
ping itself around the South, and the South was to be effectively stran-
gled by a combination of blockading its seaports and a careful buildup of
Northern military strength aimed at establishing control of the Mississippi River. The concept had been, in Scott's words, to "envelop the insurgent states and bring them to terms with less bloodshed than by any
other plan," but his strategy had been rejected as being too slow, and
there was little to indicate that it would have brought the Confederacy to
a peace table. With the fall of Vicksburg, Scott's military and naval goals
had to some extent been reached, but in their Western campaigns it had
become ever clearer to both Grant and Sherman that Southern tenacity
could be overcome only by penetrating the South, rather than encircling
it. The Union was still fighting the war on an ad hoc basis, attacking as
opportunity presented itself and defending when attacked.

Whatever the Union lacked in an overarching concept of how to win
the war, Vicksburg was a great victory, and Lincoln fully appreciated the
remarkable accomplishment. He promoted Ulysses S. Grant from major
general of Volunteers to major general in the Regular Army—the highest
rank he could then bestow—and, on Grant's enthusiastic recommendation, promoted Sherman from major general of Volunteers to the higher
permanent rank of brigadier general in the Regular Army. Of Grant, on
the day after Vicksburg surrendered, Lincoln said, "Grant is my man,
and I am his, for the rest of the war."

Despite all this praise for and promotion of Grant, Lincoln was not
finished. "My Dear General," he soon wrote him, "I do not remember
that you and I ever met personally. I write this now as a grateful acknowl-
edgment for the almost inestimable service you have done the country."

Lincoln then set forth the concerns and fears he had felt about some of Grant's movements during the long Vicksburg campaign, and closed with this:

> I now wish to make the personal acknowledgment that you were right, and I was wrong.

Yours very truly,
A. Lincoln

8

PAIN AND PLEASURE ON THE LONG ROAD
TO CHATTANOOGA AND MISSIONARY RIDGE

With Vicksburg's fall, Grant began planning the overall exploitation of the position in which the campaign had placed the forces under his command, while Sherman headed east to find and attack Joseph E. Johnston. His men found themselves marching day after day through the blazing heat of a Mississippi summer. Private Leander Stillwell of the Sixty-first Illinois described their encounters with thunderstorms: "The dirt road would soon be worked into a loblolly of sticky, yellow mud. Thereupon we would take off our shoes and socks, tie them to the barrels of our muskets . . . and roll up our breeches. Splashing, the men would swing along, singing 'John Brown's Body' or whatever else came handy."

Pushing ahead swiftly, Sherman came once again to the Mississippi state capital of Jackson, where Johnston had paused in his retreat to consider whether to make a stand there with his thirty thousand men. By this time, Sherman's exhausted soldiers had little heart for singing "John Brown's Body" or anything else. On July 12, he tried to storm the city in a frontal assault. Although Sherman had twice the number of soldiers Johnston had, his troops were thrown back with losses he did not wish to repeat. He decided to put Jackson under siege, lobbing shells at the enemy every few minutes, but before he could encircle the city, Johnston slipped his army away on the night of July 16.

The following day brought Sherman and his men to a breaking point. As the siege of Vicksburg had come to a close, both Grant and Sherman had thought beyond the city's fall: the plan was not only for Grant to capture Vicksburg but also for Sherman to complete months of campaigning by bringing Johnston to battle in the area east of the captured city. He was to find Johnston's army and destroy it if possible. Sherman had found it and tried to encircle it, but Johnston was gone, again. Learning that Johnston had evacuated his men from Jackson during the night, Grant wired Sherman, "If Johnston is pursued, would it not have the effect to make him abandon much of his [supply wagon] trains and many of his men to desert?" Aware of the conditions under which Sherman's men had been operating, Grant added, "I do not favor marching our men much but if the Cavalry can do anything they might do it."

Sherman and his men had come to the end of their strength. He wired Grant that "the weather is too hot for a vigorous pursuit," and in another telegram added that he would destroy enemy equipment captured in and around Jackson, but "I do not pursue because of the intense heat, dust & fatigue of the men." Grant replied from Vicksburg, "Continue the pursuit as long as you have reasonable hopes of favorable results, but do not wear your men out. When you stop the pursuit return by easy marches to the vicinity of this place."

Trying to explain that his army was in a state of near collapse, Sherman sent a telegram back at nine that evening, saying in part, "All of the Division Brigade & Regiments are so reduced and so many officers of rank sick & wounded determined on furloughs . . . Every officer & man is an applicant for furlough." Half an hour later he sent yet another telegram, written without punctuation and saying that the force with him under Major General Edward Ord "is very much out of order & mine reduced by sickness Casualties & a desire for rest Genl W. S. Smith is really quite ill & says he must go home Cols. Giles, Smith, Tupper, Judy & others are urging their claims to furloughs & I repeat that all the army is clamorous for rest The constant stretch of mind for the past two months begins to tell on us all"

Hearing nothing more from Grant, Sherman closed the exchange with a telegram indicating that, after his men destroyed anything that was left of use to the Confederates in Jackson, he was bringing his spent army back toward Vicksburg. "Our march back, will be slow and easy, regulated by [camping where there is] water."

With Johnston clearly beyond pursuit and posing no threat—he needed to rest his own men—Grant and Sherman settled down for a respite for themselves and their soldiers. Julia Grant and their four children came to be with him in what she described as "a large, white, colonial house" in Vicksburg that he was using for headquarters. Ellen Sherman brought their four oldest children to the vast camp Sherman's divisions constructed beside the Big Black River, thirty miles east of Vicksburg. Writing to his stepfather Thomas Ewing, Sherman spoke of the encampment: "It combines comfort, retirement, safety and beauty . . . I have no apprehensions on the Score of health and the present condition of my command satisfies me on this score." Headquarters was in a grove of large oaks. Two big hospital tents served as quarters for Sherman, Ellen, and their daughters, Minnie, now thirteen, and Lizzie, ten. Nine-year-old Willy and six-year-old Tommy stayed with their uncle Charley, now Sherman's inspector general, in one of the regular military headquarters tents.

It was a happy time. Soon after Julia Grant arrived, she and Grant drove out to call on Sherman and Ellen. Lest Grant take himself too seriously after what he had achieved in capturing Vicksburg, Julia began calling him "Victor" when they were among close friends. She enjoyed Sherman's witty conversation and appreciated his loyalty to Grant. All the Shermans occasionally went into Vicksburg and visited with the Grants and their children; Sherman took his family on a tour of the recently surrendered fortifications and let his children pick up battlefield souvenirs. Out at the Big Black River encampment, the atmosphere was often that of an outing under the trees. In the evenings, a black man known as "Old Shady" sang songs for the Shermans and their guests, and military bands frequently gave concerts. A battalion of the United States Thirteenth Regular Infantry Regiment—the regiment that Sherman was assigned to command at the beginning of the war but that he never led as a colonel because of his duties inspecting the defenses of Washington—treated Willy and Tommy as their own. Tommy had his corporal's uniform from an earlier visit, and a regimental tailor now made Willy a uniform with sergeant's chevrons. The boys were happy in the midst of camp life. Willy, his father's favorite child and a boy who showed real enthusiasm for the military, frequently rode on a pony to accompany his father on inspections and reviews.

During this quiet time, Grant and Sherman each received a letter from General Halleck in Washington, asking them for their views on what forms of civil government should be set up in the areas of the South now firmly under Union control. Halleck added, of the answers he was soliciting, "I may wish to use them with the President."

Although the question was framed in terms of the immediate situation, it opened the subject of how the entire South should be dealt with in the event of a final Union victory. While Grant and Sherman had been making their great contributions toward achieving such a victory, Lincoln had been trying to balance and control the political progress of the war. In Washington, he had his continuing differences with the Radical Republicans, who were adamant in their efforts not only to free every slave swiftly, but looked forward to giving these freedmen the vote as soon as possible in a conquered South that was to be governed under a strict federal rule that would rearrange its entire society. For the Radicals, the question was not whether the freed slaves should be given the vote, but whether white Southern men who had fought against the federal government should not be placed on a form of probation before they were allowed to reenter the political process. Lincoln, while firmly committed to a vigorous prosecution of the military effort and to the eradication of slavery, had as his priority the return of the rebellious states to the Union and took a more measured and conciliatory approach to reaching that goal.

It was a time in the war when much was being tried. In June, the forty-eight counties of western Virginia had been admitted to the Union as the new state of West Virginia. Earlier in the year, the experimental government set up in areas of Louisiana under federal control resulted in two congressmen from that state being seated on the floor of the House of Representatives in Washington, but they were later disqualified. On June 30, the American Freedmen's Inquiry Commission created by the War Department had issued its report titled "A Social Reconstruction of the Southern States." Its three members, all prewar abolitionists, had toured the parts of the South occupied by the Union Army and recommended the creation of a Bureau of Emancipation to safeguard the interests of the slaves, an idea that eventuated in the later Freedmen's Bureau. In addition, the commission called for complete equality for the freed blacks: one member recommended that the lands of Southern planters should be confiscated and redistributed among former slaves—an idea popular among many Radicals.

Answering Halleck's request for ideas on what measures should be instituted in occupied areas, Grant took a conciliatory line. Although the man famously linked with "unconditional surrender" believed that the Confederate Army must be destroyed, he said of the white population now under Union control in Mississippi, Louisiana, and Arkansas, "The people of these states are beginning to see how much they need the protection of Federal laws and institutions. They have experienced the misfortune of being without them." In essence, Grant believed that the white civilian population could be brought back into the Union as full citizens; as for the men of the rebel armies, they must indeed be defeated, but "I think we should do it with terms held out that by accepting they could receive the protection of our laws." As Grant saw it, if these soldiers surrendered and swore allegiance to the United States of America, they too should regain their status as citizens.

Sherman took a harsher line, although he sometimes remembered his happy prewar times in the South and, the past spring, had even written Ellen a letter in which he conjured up the image of his own army being "Rude Barbarians" invading from the north. He was at the moment trying with mixed success to keep his own troops from looting and was distributing food to civilians in the areas under his control, but he kept thinking in terms of a hard policy. Knowing that savage fighting lay ahead, he had little patience with what he had increasingly seen of the hostile attitude of all Southerners, both soldiers and civilians. Ten weeks before, he had written to Ellen, "I doubt if History affords a parallel of the deep & bitter enmity of the women of the South. No one who sees them & hears them but must feel the intensity of their hate."

Now, in a twenty-seven-hundred-word reply to Halleck, Sherman carefully considered many of the problems of dealing with the conquered portions of the Confederacy. As for restoring civil rights to the people who had seceded from the Union, he saw all of those individuals as traitors and said that to give them "a Civil Government now . . . would be simply ridiculous." He added, "I would not coax them, or even meet them halfway, but make them so sick of war that generations would pass" before they thought of taking up arms as a solution to a political problem.

Apart from the questions put to him by Halleck, Sherman had begun to realize that he, whose ideas were solicited by Halleck with the thought that "I may wish to use them with the President," was becoming a national figure himself. With all his fondness for Grant and his occasional paeans of praise for Grant's achievements, Sherman still had some reservations about the man with whom he had, in every sense, come so far. Even after Vicksburg, Sherman seemed not to understand that Grant had intuitive military gifts that simply exceeded his own great abilities. Sherman was better read, a frequently brilliant conversationalist, brave, imaginative, energetic, ambitious, a man who Grant said "boned" [studied hard in planning] his campaigns—how could one have and be and do more than that? After Shiloh, he had written Ellen, of Grant, "He is not a brilliant man . . . but he is a good & brave soldier tried for years, is sober, very industrious, and as kind as a child." More than a year later, writing to Ellen the day after Vicksburg fell, he said that "we have in Grant not a great man or a Hero—but a good, plain, sensible, kindhearted fellow." Two paragraphs later, he tried to do Grant justice, but it was hard for him: "I am somewhat blind to what occurs near me, but have a clear perception of things & events remote. Grant possesses the happy medium and it is for this reason I admire him. I have a much quicker perception of things, but he balances the present & remote so evenly that results follow in a normal course."

The man who later said of Grant, "To me he is a mystery," was demonstrating that this remained true, but he sounded happily confident when he spoke of their demonstrated ability to work together. Looking back on a planning session for the Vicksburg campaign that he and Grant had held the year before, in this same letter he told Ellen, "As we sat in Oxford [Mississippi] in November we saw in the future what we now realize and like the architect who sees the beautiful vision of his Brain, we feel an intense satisfaction at the realization of our military plans." He did not mention that on several occasions, questioning Grant's intuitions, he had wanted to change those blueprints, but their partnership was working. Grant and Sherman were developing an ever-greater respect for each other's views and often listened patiently to each other, but these two West Pointers understood that, once Grant reached a decision, discussion ceased and vigorous action began.

The idyll for the Grant and Sherman families, the Grants in Vicksburg and the Shermans at the encampment on the Big Black River, and the needed rest for the troops themselves came to a sudden end. On September 18, Braxton Bragg threw sixty-two thousand Confederate soldiers at badly positioned Union forces in the mountainous Georgia countryside eleven miles south of Chattanooga, Tennessee. The battle, centering on Chickamauga Creek, went on for three days. At its close, the total casualties suffered on both sides came to thirty-four thousand; among the

Confederates killed was Lincoln's brother-in-law Brigadier General Ben Hardin Helm, a Kentuckian who had married Mary Todd Lincoln's half-sister Emilie Todd. (Lincoln's family was torn apart by the war; in addition to the loss of Helm, three of Mrs. Lincoln's half brothers were killed fighting for the South.)

At Chickamauga, the Union commander, William Rosecrans, was saved from disaster only through the heroic stand made by Major General George Thomas, a Virginian who had chosen to fight for the Union. Because of the skillful rear-guard action under Thomas, who became known as "the Rock of Chickamauga," the demoralized Union forces were able to retreat north to Chattanooga. In their flight, however, the Union regiments reached the city itself but failed to secure the arc of towering ridges just outside the city that hemmed it in from three sides. Bragg's men, advancing behind them, soon looked down on the city from Raccoon Mountain, to the city's west, Lookout Mountain on the south side, and Missionary Ridge to the east. Chattanooga was a vital communications hub, the principal southern rail center, an X that connected lines running southwest-northeast and northwest-southeast. If Chattanooga, only recently taken by Union forces, were recaptured by the Confederate Army, it would be both a great strategic loss for the Union and a rejuvenation for Southern morale after the defeats at Gettysburg and Vicksburg.

In Washington, it became clear that this new flaming area of war needed both troop reinforcements and some new commanders, and needed them swiftly. (Lincoln said that Rosecrans, whom he would soon remove from command, was "stunned and confused, like a duck hit in the head.") Halleck, already sending a reinforcement of eighteen thousand men south from Meade's Army of the Potomac under Joseph Hooker, ordered Grant to send another twenty thousand from his Army of the Tennessee and to go to Chattanooga himself.

For Grant's forces, the movement of so many men, horses, artillery pieces, and supply trains was going to be exceptionally difficult. As the crow flies, Vicksburg is 340 miles southwest of Chattanooga. But Grant's and Sherman's divisions of troops would first go north by riverboat for 220 miles up the Mississippi to Memphis. Then, to reach Chattanooga, they would make their way east through 240 miles of country subject to Confederate raids. (One estimate was that the actual distance, counting river bends and winding roads, came to 600 miles.) Different units would have to use combinations of railways, some of them torn up by the enemy, roads that could deteriorate in bad weather, bridges the enemy would try to destroy, and riverboats steaming slowly on the meandering Tennessee River. Grant, in bed at Vicksburg with a severe leg injury sustained in a fall from a horse during a brief trip to New Orleans, instructed Sherman to take five divisions, which would comprise the required twenty thousand men, and organize them for the movement to Chattanooga. Grant would start for Chattanooga himself as soon as he was able and would probably arrive there ahead of Sherman.

On September 27, Sherman shifted his headquarters to the steamboat *Atlantic*, loaded with troops, including those of the Thirteenth Infantry, ready to head north up the Mississippi. His family was with him. The plan was for Sherman, his staff, and the troops aboard to disembark at Memphis and prepare for the final part of the movement to Chattanooga. Ellen and the children were to go on to Cairo, Illinois, and then travel by train to her family's house in Lancaster, Ohio. Sherman's son Willy, wearing his sergeant's uniform and carrying a shotgun, came aboard, still thinking of himself as a soldier bound for high adventures but complaining of diarrhea.

The ship cast off; as they went on upstream, leaving Vicksburg behind, Sherman stood at the rail, pointing out to Ellen and the children the places where his men had camped and fought. Glancing at Willy, he saw that his son's face was pale and that he was feverish. Ellen hurried Willy to a bunk below. The word was passed that a doctor was needed. The regimental surgeon of the Fifty-fifth Illinois examined Willy, found symptoms of typhoid fever with possible complications of dysentery, and told Sherman that his son's life was in danger. The important thing was to get to Memphis as soon as possible so that Willy could be treated by the physicians there, but the *Atlantic* was a slow riverboat, making its way upstream at the season when the water was low. For a week the ship moved as fast as it could, while Sherman, Ellen, and the doctor remained constantly at their suffering son's bedside.

At ten-thirty on the night of October 2, Willy was carried ashore at Memphis. Every soldier in the battalion of the Thirteenth Infantry wanted to help the nine-year-old boy who was their little mascot, and none could. Sherman summoned two more doctors, who hurried to a room at the Gayoso House and examined their patient as he lay pale in bed. The following morning, Ellen Sherman called in Father J. C. Carrier, a French priest from the University of Notre Dame who was serving as a chaplain for troops who were Catholics. When he visited Willy and they were alone together, "Willy then told me in very few words," the priest recalled, "that he was willing to die if it was the will of God *but that it pained* him to leave his father & mother." Trying to reassure him, the priest "told him it was not certain he would die." Willy seemed unconvinced, and Father Carrier finally promised him that "If God wishes to call you to him—now—do not grieve for he will carry you to heaven & *there* you will meet your good Mother & Father again." Ellen entered and began crying; Willy reached up and patted his mother's face.

At five o'clock on the afternoon of October 3, eighteen hours after the Shermans reached Memphis, Willy died. Sherman said, "Mrs. Sherman, Minnie, Lizzie, and Tom were with him at the time, and we all, helpless and overwhelmed, saw him die." At noon the following day, the battalion of the Thirteenth Infantry, marching to the beat of muffled drums and carrying their rifles reversed in a military funeral march, escorted Willy's body, in a steel casket, to the waterfront. There the *Grey Eagle* had steam up, ready to depart for Cairo, from where the Shermans would go on to Lancaster. Sherman went aboard with Ellen, Minnie, Lizzie, and Tom, said good-bye to them, and returned to his headquarters at the Gayoso. That night he wrote Grant that "this is the only death I have ever had in my family, and falling as it has so suddenly and unexpectedly on the one I most prized on earth has affected me more than any other misfortune could. I can hardly compose myself enough for work but must & will do so at once." He then proceeded to add a report of approximately 750 words, telling Grant in Vicksburg what the situation was at Memphis and his plans for readying his forces for the movement east to Chattanooga. (Three days later, Grant had one of his generals forward to Sherman, who was still in Memphis, what Grant referred to as a "private letter." The contents are unknown.)

Having momentarily discharged his military responsibilities with his report to Grant, Sherman gave way to his emotions in a letter to Captain C. C. Smith, commander of the battalion of the Thirteenth Infantry, which had made Willy an honorary sergeant and had furnished the troops that gave him full military honors as his body left Memphis earlier in the day. Dated "October 4, Midnight," it began with a salutation not usually found in communications from major generals to captains.

My Dear Friend:

I cannot sleep tonight till I record an expression of the deep feelings of my heart to you, and to the Officers and Soldiers of the Battalion, for their kind behaviour to my poor child. I realize that you all feel for my family the attachment of kindred; and I assure you of full reciprocity. Consistent with a sense of duty to my profession and my office, I could not leave my post, and sent for my family to come to me in that fatal climate, in that sickly period, and behold the result! The child who bore my name . . . now floats a mere corpse, seeking a grave in a distant land, with a weeping mother, brother, and sisters clustered about him . . .

But, my poor WILLY was, or thought he was, a Sergeant of the 13th. I have seen his eyes brighten and his heart beat as he beheld the Battalion under arms . . . Child as he was, he had the enthusiasm, the pure love of truth, honor, and love of country, which should animate all soldiers. God only knows why he should die thus young . . .

Please convey to the Battalion my heartfelt thanks, and assure each and all, that if in after years they call on me and mine, and mention that they were of the 13th Regulars, when poor WILLY was a Sergeant, they will have a key to the affections of my family that will open all it has, that we will share with them our last blanket, our last crust. YOUR FRIEND,

W. T. SHERMAN
MAJOR GENERAL

So many of Captain Smith's men wanted copies of the letter that he had it printed and gave each man in the battalion a copy.

Two mornings later, in a letter to "Dearest Ellen" dated as being written at seven a.m., Sherman began:

I have got up early this morning to Steal a short period in which to write you but I can hardly trust myself. Sleeping—waking—everywhere I see Poor Little Willy . . . Why oh Why should that child be taken from us? . . . I will always deplore my want of judgment in taking my family to so fatal a climate at so critical [a] period of the year . . . If human sympathy could avail us aught, I Know and feel we have it—I see it in every eye and in every act— Poor Malmbury, an old scarred Soldier, whom the world would Style unfeeling, wept like a babe as he came to See me yesterday, and not a word was spoken of Poor Willy . . .

I follow you in my mind and almost estimated to the hour when all Lancaster would be shrouded in gloom to think that Willy Sherman was coming back a corpse.

Four days later, in his third letter to Ellen since they parted, he continued his lament and self-recrimination. "The moment I begin to think of you & the children, Poor Willy appears before me as plain as life. I can see him now, stumbling over the Sand hills on Harrison Street [in] San Francisco . . . running to meet me with open arms at Black River & last, moaning in death in this Hotel." Of their children, he said, "Why should I ever have taken them to that dread Climate? It nearly kills me to think of it. Why was I not killed at Vicksburg and left Willy to grow up to care for you?"

Ellen was equally distraught and unable to comfort her husband. "My heart is now in heaven," she wrote him, "and the world is dark and dreary." Everything threatened and frightened her. "Since we lost our dear Willy, I feel that evils of all sorts are likely to come upon us." More earnest a Catholic than ever, she begged Sherman to embrace the religion in which he had been baptized but had never believed in or practiced, so "that you will die in the faith that sanctified our holy one whom we have just given up to God." Sherman made no known response to that, but he soon wrote Ellen of Willy, "He knew & felt every moment of his life our deep earnest love for him . . . God knows and he knows that either of us and hundreds of others would have died to save him." To his daughter Lizzie he wrote, "We must all now love each other the more that Willy watches us from Heaven," and told her always to appreciate "the Soldiers who used to call Willy their brother. I do believe Soldiers have stronger feelings than other men, and I Know that every one of those Regulars would have died, if they could have saved Willy." Usually he signed his letters to his children simply with W. T. Sherman, but in this one he added above that, "Yr. Loving Father."

As Sherman remained in Memphis, grieving for his son as he prepared his forces for the long and difficult move east to Chattanooga, Grant, who said in a letter to another general "I am very glad to say that I have so far recovered from my injuries as to be able to move about on crutches," started his own painful journey from Vicksburg to the besieged but not entirely surrounded city. Even from the outset, his route was a roundabout one. On October 14, he passed Memphis by boat, going on up the river to Cairo, and on October 16, reaching Indianapolis by train, found no less a person coming aboard than Secretary of War Stanton. On their ride to Louisville, Stanton handed him orders that named him commander of all Union forces between the Appalachian Mountains and the Mississippi River. In this reorganization, directed specifically by Lincoln, Grant would have three subordinates. Sherman would take over Grant's position as commander of the Army of the Tennessee. Stanton told Grant that he could replace the defeated Rosecrans as commander of the battered Army of the Cumberland that was now at beleaguered Chattanooga, and Grant decided to give that command to George Thomas, "the Rock of Chickamauga."

The third force, the Army of the Ohio, would continue under the command of Ambrose Burnside, who had succeeded McClellan as commander of the Army of the Potomac, only to be replaced by Lincoln after he failed abysmally when he opposed Lee at Fredericksburg. Burnside was not part of the crisis at Chattanooga; in command at Knoxville, eighty-five miles northeast of Chattanooga, he had a crisis of his own. Facing strong Confederate forces in eastern Tennessee, Burnside was begging for supplies and men; Grant, now responsible for that area as well, felt a great responsibility to save Chattanooga quickly, if it could be done at all, so that he could release forces to come to Burnside's aid. There were questions too about Joseph Hooker, the Union general bringing the twenty thousand reinforcements from the Eastern theater by a circuitous route. Called in to replace Burnside as commander of the Army of the Potomac, Hooker had been soundly outgeneraled by Lee at Chancellorsville. Lincoln had in effect demoted him, giving command of the Army of the Potomac to George Meade, the victor at Gettysburg. "Fighting Joe" Hooker—a nickname he never liked—was in effect on a form of high-level probation.

Grant's responsibilities had just been greatly multiplied, but he had no time to dwell on his rise in the Union Army hierarchy. Staying at the Galt House in Louisville, where he heard reports that the federal forces might be giving up Chattanooga at any time, on October 19 Grant fired off a telegram to General Thomas: "Hold Chattanooga at all hazards. I will be there as soon as possible." The next morning he was assisted onto a train for Nashville, continuing on toward Chattanooga. A Union soldier who looked into the window of the train as it passed through Murfreesboro wrote home that Grant "was seated entirely alone on the side of the car next to me. He had on an old blue overcoat, and wore a common white wool [cap] drawn down over his eyes, and looked so much like a private soldier, that but for the resemblance to the photographs . . . it would have been impossible to have recognized him." At Bridgeport, Alabama, the railroad line had been torn up by Confederate raiders; the only way left open to Chattanooga was to go by foot or on horseback through mountain passes. Grant was placed on a horse, with his crutches strapped to the saddle. In a letter to Julia, he said of the next two days that he endured "a horse-back ride of fifty miles through the rain over the worst roads I ever saw." At times, Grant had to be lifted off his horse and carried across washed-out places where horses might slip and fall. On the second day, his horse slipped coming down a mountain while he was in the saddle, further damaging his leg. In pain, on the evening of October 23, having gotten through the remaining open land route, Grant arrived at George Thomas's headquarters, a one-story frame house in the middle of Chattanooga, and had to be lifted off his horse and helped in out of the rain.

Captain Horace Porter, a twenty-six-year-old West Point graduate, was serving as the ordnance officer on Thomas's staff. He described his first glimpse of Grant: "In an arm-chair facing the fireplace was a general officer, slight in figure and of medium stature, whose face bore an expression of weariness. He was carelessly dressed, and his uniform coat was unbuttoned and thrown back from his chest. He held a lighted cigar in his mouth, and sat in a stooping posture, with his head bent slightly forward. His clothes were wet, and his trousers and top-boots were spattered with mud."

Grant declined Thomas's suggestion, made only after an aide to Thomas quietly mentioned the new commander's bedraggled condition, that he retire to a warm bedroom, change his clothes, and have something to eat. He lit a second cigar and asked for a report on the situation at Chattanooga. Thomas and his chief engineer officer began pointing out the Union and Confederate positions on a large map.

> General Grant sat for a time immovable as a rock and as silent as the sphinx, but listened attentively to all that was said. After a while he straightened himself up in his chair, his features assumed an air of animation, and in a tone of voice which manifested a deep interest in the discussion, he began to fire whole volleys of questions at the officers present. So intelligent were his in-

quiries, and so pertinent his suggestions, that he made a profound impression upon everyone by the quickness of his perception and the knowledge which he had already acquired concerning the army's condition. His questions showed from the outset that his mind was dwelling not only upon the prompt opening of a line of supplies, but upon taking the offensive against the enemy.

The meeting broke up, but Grant detained Porter, asking him questions about the dangerously depleted ammunition supply. Then, at about nine-thirty, when Porter felt certain that Grant would finally eat and go to bed, the new commander began writing telegrams. The first was to Halleck in Washington, telling him that he had arrived. Uncertain of how the orders handed to him by Stanton concerning the reorganization of the Western armies had been distributed, his second sentence read, "Please approve order placing Genl Sherman in command of Dept. & army of the Tennessee with Hd. Qrs. in the field." Porter later wrote of Grant, "He had scarcely begun to exercise the authority conferred upon him by his new command when his mind turned to securing advancement for Sherman." Once again, Grant was using and relying on Sherman as his leading subordinate.

The next day, Grant was taken on an inspection of Union positions. It was a chilling tour. The Union defenders down in the city, which was in a bowl of high hills, numbered forty-five thousand. On the ridges hemming them in on three sides were seventy thousand Confederates. The enemy had cut the principal waterborne supply line that came up the Tennessee River from Bridgeport, Alabama, reducing the defenders' food supply so much that the hungry troops had been subsisting on half rations. Facing south, with the river at his back, Grant had Raccoon Mountain on his right, Lookout Mountain to the front, and Missionary Ridge on his left. Using his field glasses, Grant studied Lookout Mountain, which loomed twelve hundred feet above him. He could see Confederate cannon and artillerymen up there. Those gunners were in perfect position to drop shells anywhere in Chattanooga in support of Southern infantrymen who might be able to swarm down the slopes and engulf the city. Grant saw all this and decided to go on the offensive.

That night, Captain Horace Porter had his second look at Grant. Told to report at headquarters, he found Grant pointing to a chair and saying "bluntly but politely, 'Sit down.'" After Grant asked him several questions concerning the type and placement "of certain heavy guns which I had recently assisted in putting in position," Grant began writing dispatches. When Porter rose to go, Grant said, "Sit still."

My attention was soon attracted to the manner in which he went to work at his correspondence . . . His work was performed swiftly and uninterruptedly . . . He sat with his head bent low over the table, and when he had occasion to step to another table or desk to get a paper he wanted, he would glide rapidly across the room without straightening himself, and return to his seat with his body still bent over at about the same angle at which he had been sitting when he left his chair.

Upon this occasion he tossed the sheets of paper across the table as he finished them, leaving them in the wildest disorder. When he had completed the dispatch[es], he gathered up the scattered sheets, read them over rapidly, and arranged them in their proper order. Turning to me, he said, "Perhaps you would like to read what I am sending."

The captain thanked the general and began to read. He found that Sherman's entire expeditionary force of twenty thousand, en route from the areas of Vicksburg and Memphis by train and road, was being urged to proceed to within "supporting distance" of Chattanooga as quickly as possible. A message to Halleck explained how attacks were going to re-open supply lines. Measures would be taken "for the relief of Burnside in east Tennessee." Most of the hungry and exhausted horses now with the army in Chattanooga were to be sent to quiet areas "to be foraged." Paging through the sheaf of papers, Porter noted that "directions were also given for the taking of vigorous and comprehensive steps in every direction throughout his new and extensive command."

As Grant bade him good night and went off to bed, Porter concluded from what he had seen of Grant while he was writing, and from what he had just read: "His thoughts flowed as freely from his mind as the ink from his pen; he was never at a loss for an expression, and seldom interlined a word or made a material correction." What Porter did not know was that Grant had decided to add him as a future member of his staff.

While Grant came east to take control at Chattanooga and begin planning immediate counterstrokes, Sherman had experienced some perilous moments as he made his own way east toward his commander in their new theater of war. On the morning of Sunday, October 11, Sherman had left Memphis for Corinth on what he described as "a special train, loaded with our orderlies and clerks, the horses of our staff, the battalion of the Thirteenth United States Regulars, and a few officers going forward to join their commands, among them Brigadier-General Hugh Ewing [Ellen's brother]." Some men of Sherman's beloved Thirteenth Infantry were assigned to guard the train by sitting on the roofs of the cars with their muskets beside them. As the train rattled along east of Memphis on this peaceful Sunday morning, these soldiers waved as they passed and left behind them the men of Sherman's Fourth Division who were marching along a road, beginning their long eastward march to the relief of Chattanooga.

At noon, just as the train passed the depot at Collierville, Tennessee, twenty-six miles east of Memphis, a force of enemy raiders that Sherman described as "about three thousand cavalry, with eight pieces of artillery" tried to surround it. Sherman took command of the situation, quickly having the train back up into the station and linking up with the 250 Union soldiers of the Sixty-sixth Indiana who comprised the Collierville garrison. For defensive positions, he had his greatly outnumbered combined force use the train station, a nearby blockhouse, and what he described as "some shallow rifle-trenches near the depot." As the enemy horsemen were about to cut the telegraph line out of Collierville, Sherman sent out a call for help; before the line went dead, he received the words, "I am coming," from Brigadier General John M. Corse, commander of the first brigade of the Fourth Division the train had passed that morning and whose men were still hours away from Collierville.

A blazing battle between five hundred Union soldiers and the three thousand Confederates ensued, in which the South just missed capturing what would have been its most important prisoner of the war. At one point, when a sergeant begged Sherman to stop standing up amid a fusillade of bullets "as though he was standing on parade," Sherman told him to mind his own business. The train's conductor remembered this: "I was somewhat frightened at first, but when I saw such a great man as he so unconcerned amid all the balls flying around him, I did not think it worthwhile for me to be scared." Sherman's matter-of-fact account mentioned none of that.

The enemy closed down on us several times, and got possession of the rear of our train, from which they succeeded in getting five of our horses, among them my favorite mare Dolly; but our men were cool and practiced shots (with great experience acquired at Vicksburg), and drove them back. With their artillery they knocked to pieces our train; but we managed to get possession again, and extinguished the fire . . . The fighting continued all round us for three or four hours, when we observed signs of drawing off, which I attributed to the rightful cause, the rapid approach of Corse's division, having marched the whole distance from Memphis, twenty-six miles, on the double-quick. The next day we repaired damages to the railroad and locomotive, and went on to Corinth.

With Sherman still on the way, at Chattanooga Grant started his surprise counteroffensive about seventy-seven hours after he arrived. The object was to reopen the principal supply line, closed by the enemy. At midnight on October 26, his subordinate commander W. F. Smith started a march down the looping north bank of the Tennessee River with twenty-eight hundred men. Three hours later, in a move that in its way was as bold as Grant's running the Confederate batteries at Vicksburg, Brigadier General William B. Hazen put eighteen hundred men aboard pontoons made for river crossings, and that force glided silently down the Tennessee, passing Smith's troops who were quietly marching along beside the river. At five in the morning, Hazen's soldiers captured the startled Confederate guards at Brown's Ferry on the opposite bank and started using their pontoons to ferry Smith's men across the river as they arrived. Smith's troops, all across by seven a.m., started digging in, and Hazen's soldiers stopped using their pontoons as ferryboats and started lashing them together so that a bridge could be built on top of them. By ten in the morning, a powerful Union position was in place, well behind the Confederate lines, where nothing had existed the day before. The following afternoon, completing the execution of Grant's quickly improvised plan, General Joseph Hooker brought his force up the river from Bridgeport.

Reinforcements began to pour into Chattanooga; the question now was whether this reopened supply line could be kept open. Captain Porter commented on the situation: "As soon as the enemy recovered from his surprise, he woke up to the importance of the achievement; Longstreet was despatched to retrieve, if possible, the lost ground."

For the first time in the war, Grant was facing his old West Point classmate, Julia's cousin who had fought beside him in the Mexican War and was the best man at his wedding. Longstreet waded right in, coming to Wauhatchie (in Lookout Valley, just west of Lookout Mountain) south of Chattanooga at night on October 28 and making a midnight attack on an outnumbered division commanded by Union general John W. Geary. After four hours, Longstreet's men were routed in the darkness by the most unusual charge made during the war. Captain Porter explained what happened.

> During the fight Geary's teamsters became scared, and had deserted their teams, and the mules, stampeded by the sound of battle raging around them, had broken loose from their wagons and run away. Fortunately for their reputation and the safety of the command, they started toward the enemy, and with heads down and tails up, with trace-chains rattling and whiffletrees snapping over the stumps of trees, they rushed pell-mell upon Longstreet's bewildered men. Believing it to be an impetuous charge of cavalry, his line broke and fled.
>
> The quartermaster in charge of the animals, not willing to see such distinguished services go unrewarded, sent in the following communication: "I request that the mules, for their gallantry in action, may have conferred upon them the brevet rank [an honorary promotion] of horses." Brevets in the army were being pretty freely bestowed at the time, and when this recommendation was reported to General Grant he laughed heartily at the suggestion.

Delayed by weather and the need to repair railways and roads in order to keep his men on the move, Sherman arrived at Chattanooga ahead of his troops on November 14, and he and Grant quickly resumed their familiar manner with each other. One of the people who had never seen them together before was Major General Oliver Otis Howard of Maine, who had graduated from Bowdoin College in 1850 and then from West Point in 1854. Howard's right arm had been amputated as a result of a wound received eighteen months before at Seven Pines, Virginia, and earlier in this year he had commanded his XI Corps at both the Union defeat at Chancellorsville and the victory at Gettysburg. Sent to Chattanooga from the Eastern theater of war to serve under Grant in this emergency, he left a vivid description of one of Sherman's entrances into Grant's headquarters. When Sherman came "bounding in after his usual buoyant manner," Grant beamed on seeing him.

"How are you, Sherman?"

"Thank you, as well as can be expected."

Grant offered Sherman a cigar, which Sherman took and managed to light without stopping a flow of words on some subject that had come to his mind. Grant pointed to the best seat in the headquarters, a high-backed rocking chair, indicating that Sherman should sit down.

Sherman demurred. "The chair of honor? Oh, no! That belongs to you, general."

Grant, two years younger than Sherman, came back with, "I don't forget, Sherman, to give proper respect to age."

Sherman surrendered. "Well, then, if you put it on that ground, I must accept."

Howard, accustomed to Grant's businesslike manner with everyone else at headquarters, noted that when Grant talked to Sherman, he was "free, affectionate, and good humored." The friendship was there for all to see.

9

CONFUSION AT CHATTANOOGA

The military situation that developed at Chattanooga in the days from November 23 through November 25 bewildered almost everyone involved. In many battles, commanders lose some measure of their control of the situation, but at Chattanooga, this happened frequently. During the fighting, a number of the Union commanders behaved strangely, and people looking at the same actions on the same terrain said they saw different things. At a crucial point, Grant indecisively delayed a major attack. In the final hours of November 25, the eighteen thousand foot soldiers of Thomas's division disregarded orders, took matters into their own hands, and achieved one of the great successes of the war by advancing to their objective with a nearly fanatical bravery. Miscommunications and misunderstandings occurred among generals on both sides. When it was over, there was reason to think that Grant slanted his report of the Battle of Chattanooga in a way that covered up both Sherman's battlefield failure and his own uncharacteristic hesitation at a critical point. Sherman may have believed his own accounts of what happened in front of Chattanooga, but they were laden with inconsistencies.

There were ample causes for confusion and disharmony. The weather thwarted Grant's plans for Sherman's movements. Union generals from both the Eastern and Western theaters of war were required to work together for the first time, with mixed results. There were rivalries and mistrust: although Thomas had saved the day when Rosecrans failed at Chickamauga, he had liked and admired Rosecrans and was reluctant to become his replacement. Beyond that, Thomas had strongly disagreed with some of Grant's first ideas for movements that Thomas's Army of the Cumberland should make in breaking the siege, plans that Grant himself revised. Hooker was a headstrong, outspoken man, a heavy drinker, and something of a rake. He felt for good reason that Grant's favorite general, Sherman, had a low opinion of his military abilities, and this made him dislike them both.

Friction existed on the Confederate side as well. Despite Braxton Bragg's victory at Chickamauga, Longstreet had soon thereafter fired off a letter to the Confederate secretary of war in which he said of Bragg, "I am convinced that nothing but the hand of God can save us or help us as long as we have our present commander . . . Can't you send us General Lee?" In the Union ranks, Sherman's men, veterans of Shiloh and the Vicksburg campaign, thought of the Easterners as pampered, better-supplied, parade ground soldiers, an opinion that did not sit well with the men who had defeated Lee's Army of Northern Virginia at Gettysburg.

Finally, an eclipse of the moon took place during the first night of battle. A Union major conversant with mythology said that "it was considered a bad omen," not for the Union troops but for the Confederates, because up on Lookout Mountain and Missionary Ridge they were closer to the capricious gods in their heavens. One account of all the discrepancies and unusual behavior reported at Chattanooga simply observed that "the most sensible explanation seemed to be that the eclipse of the moon had made everybody a little crazy."

After the action at Wauhatchie, Grant's Confederate opponent Braxton Bragg had sent Longstreet and his corps off to try to capture Knoxville, then held by the Union general Ambrose Burnside. It is unknown whether this bad idea was Bragg's or was an order from Jefferson Davis, but this sudden removal on November 3 of a force of twenty thousand men and eighty guns that had been facing Grant at Chattanooga, coupled with the arrival of Hooker's eighteen thousand men, had given Grant a superiority in numbers. Sherman, who had arrived in advance of his divisions on November 14, and whose force could have joined him sooner if he had left his wagon trains behind in the last days of his march, found that his entire column was mired in autumn storms that stopped him from getting into position. (General Thomas was always to believe, with some evidence, that Grant delayed because he planned for Sherman to take the major role throughout and get the credit for it, and Sherman wrote Grant, "I need not express how I felt, that my troops should cause delay." As for Grant, he tried to protect Sherman by shifting the blame to himself, saying that he should have ordered Sherman to rush on to Chattanooga without his wagons, and failed to do that.)

On November 23, Grant went ahead, giving only a secondary role to Sherman's forces, which were now in position on his left flank. He swiftly seized Orchard Knob, a big hill three-quarters of a mile west of Missionary Ridge, and positioned Hooker's troops for an attack the next day on Lookout Mountain. Grant said that at Chattanooga, unlike other battlefields, the commanders could see everything—a panorama in which one could in this case study the slopes up which the Union troops would have to attack and the positions near and at the top that the Confederates would be defending. On the next day, however, when Grant sent Hooker and his men up Lookout Mountain, recent heavy rains had created so much cloud and fog that the federal troops disappeared from Grant's view in what became known as "the Battle above the Clouds."

Hooker was making a supreme effort to redeem himself for his failure against Lee and Stonewall Jackson at Chancellorsville. Not only did he throw his foot soldiers at the steep rock-strewn slope, but he also had his batteries of field artillery bring their horse-drawn cannon right up behind them. The horses struggled as they hauled the caissons and cannon up the slope through rocks and bushes, with the gunners heaving at the wheels and helping to pull with ropes. One of Sherman's just-arrived officers, watching this from his temporarily quiet position near Missionary Ridge, found it hard to believe that these Easterners from the Army of the Potomac were doing something so aggressive. Speaking of Hooker, he turned to a fellow officer and said, "It isn't possible the fool is taking artillery up there! . . . They'll never get a gun back [intact]. Didn't I tell you they'd better have stayed at home where they were well off—kid gloves and all?"

The cannon finally could not be tugged farther up the twelve-hundred-foot slope. When they were unlimbered and turned uphill to face the enemy, they could not be elevated high enough to fire and still avoid hitting the Union foot soldiers advancing ahead of them, but everyone on that mountain, federal and Confederate, felt the spirit fueling this attack. When the Confederate artillerymen tried to fire down at the advancing Union troops, they could not depress the muzzles of their cannon sufficiently to aim at them, but they nonetheless kept shooting, to shore up the morale of the outnumbered Confederate infantrymen defending the trenches near and at the crest of the ridge. In the meantime, the storm clouds and fog had thickened. An account described how the attackers disappeared: "Up and up they went into the clouds, which were settling down upon the lofty summit, until they were lost from sight, and their comrades watching anxiously in the Chattanooga valley could hear only the booming of cannon and the rattle of musketry far overhead, and catch glimpses of fire flashing from moment to moment through the dark clouds."

Near the top, the Fortieth Ohio, which had come up the slope in the second wave of attackers, passed through the men of the first wave who were lying on the ground, panting from exhaustion and temporarily unable to advance another step. One of the men who had fought his way up at the very front cried out through the mist, "Here come fresh troops to relieve us. Go to it, boys. We've chased them up for you. Pour it into them! Give 'em hell!" The Fortieth Ohio charged right on over the crest; as they did, their commanding officer fell, shot through the heart. Beside him, the color-bearer carrying the regimental battle flag was killed.

At two in the afternoon, the clouds were thicker than ever. With his men's ammunition nearly gone and no targets left to shoot at that they could see, Hooker ordered his men to cease fire. No one on the top of Lookout Mountain was certain of what the situation was, nor was anyone else. Grant, sending a message to General Thomas about the overall picture, reported that on his left "General Sherman carried Missionary Ridge as far as the [railroad] tunnel with only slight skirmishing," but Grant appeared uncertain of Hooker's fate on Lookout Mountain, which was straight in front of him but wrapped in clouds. All Grant could suggest was to find an alternate route of attack if Hooker sent word that it was "impracticable to carry the top from where he is."

That was all Grant seemed to know about Hooker's situation, and as night fell the rest of the Union soldiers down below Lookout Mountain had no idea of what had happened to their comrades whom they last saw advancing up the slope into the mists. All they knew was that it was silent up there in the dark.

By contrast with this generally held picture, and as an example of the conflicting stories that were to come out of the days at Chattanooga, a seemingly impeccable source offered an entirely different description of the conditions that night. Charles Dana was back with the army. No longer acting as an unofficial spy for Secretary of War Stanton (after

Vicksburg he had written Stanton that yes, Grant drank, but the situation was under control), he had been named assistant secretary of war and, on an inspection trip, had been with the Army of the Cumberland since Chickamauga. He said this of those hours when no one including Grant knew where Hooker and his men were: "A full moon made the battlefield as plain to us in the valley as if it were day, the blaze of their camp fires and the flash of their guns displaying brilliantly their position and the progress of their advance." He did, however, concur that "no report of the result was received that night."

Whatever the midnight weather conditions—and in all accounts the weather did eventually clear—during the night every Confederate soldier left on Lookout Mountain was being quietly marched down the reverse slope in a skillful movement over to Missionary Ridge, where Sherman's men were now ready for full-scale action. At first light the next morning, November 25, thousands of Union soldiers in the valley stared up as the dark mass of Lookout Mountain began to be visible. The dawn sky was clear and the air frosty. As the sun rose, they saw the Stars and Stripes flying at the very top of the mountain, and they began to cheer. A salute from fifty cannon was fired in honor of Hooker's men. At the summit, the exhausted soldiers heard the army's bands far below play "Hail to the Chief" in tribute to them.

Although most of the Confederates had been able to avoid being captured to this point in the battle, a number were taken prisoner, many of them wounded. As one of these men was being herded along to the rear, his group was halted beside a road to make way for several Union generals and their staffs who were crossing a bridge on horseback. The Confederate remembered this: "When General Grant reached the line of ragged, filthy, bloody, starveling, despairing prisoners strung out on each side of the bridge, he lifted his hat and held it over his head until he passed the last man of that living funeral cortege. He was the only officer . . . who recognized us as being on the face of the earth."

After Lookout Mountain, it was Sherman's turn to go into action. Grant's plan apparently was to take Missionary Ridge, to the left of Lookout Mountain as he faced it on Orchard Knob, by having Sherman make the major attack on it from its northeast or left end, with Hooker striking it from its southwest end. One reasonable interpretation of the plan was that Sherman was to break through and keep going along the top of Missionary Ridge, "rolling up" the rebel lines running along the crest by hitting them from the side.

Thomas, who had saved the day at Chickamauga but still wanted to avenge that overall defeat, stood beside Grant on Orchard Knob. In command of eighteen thousand men who also wanted to show what they could do, it appeared that Thomas and his subordinate Gordon Granger were being held back from attacking the center of Missionary Ridge while Grant's favorite, Sherman, and "Fighting Joe" Hooker, who now had the credit for taking Lookout Mountain, were given the chance to collapse the Confederate defense by attacking Missionary Ridge from its flanks. Many of Thomas's officers believed that Sherman's attack was to be the main effort. Sherman, on the other hand, wrote in his official report of the battle that he had received orders from Grant that included the information "that General Thomas would attack *early* in the day." Later in the report, he repeated this, again underlining for emphasis: "I had watched for the attack of General Thomas 'early in the day.' "

Those standing around Grant on Orchard Knob noticed that their usually dead-calm commander seemed nervous. He soon had reason to be. Sherman was marching his divisions down into a gap in Missionary Ridge that, until the day before, Sherman had not known existed. One account had it that the maps furnished to him showed Missionary Ridge as having a continuous crest, whereas, in the skirmishing the day before, he had discovered that, coming at it from the side, he had not reached his objective, Tunnel Hill, but was on another hill, short of that, and still had to deal with a deep ravine that ran between his men and the place they were supposed to be. Stunned by the discovery, Sherman stopped where he was, fortified his side of the ravine, and, having thrown away the momentum Grant so prized in military movements, spent the night there.

Whatever Sherman now expected, this morning his men soon came under a withering fire from several angles. The terrain at the northeast end of Missionary Ridge was such that it was hard for Sherman to send much of his numerically superior force up the narrow slope at one time, and his first attack was decisively thrown back. Then a counterattack made by the outnumbered Confederates captured five hundred of his men and eight of his regimental battle flags. Sherman went on attacking

all morning, losing men and failing to advance, while on Orchard Knob
Grant made no sign that Thomas, who stood a few yards from him,
should throw his eighteen thousand soldiers of the Army of the Cumberland at the long center of the ridge. In his official report, Sherman was to
say that he saw "vast masses" of Confederate reinforcements being sent
to oppose him, reinforcements that would otherwise have remained in
the middle of Missionary Ridge awaiting Thomas's attack. Several Union
generals concurred in Sherman's statement, but that was not the conclusion reached by Colonel James Harrison Wilson, Grant's inspector general. Wilson later reviewed other reports and documents, including a
statement by the Confederate officer commanding the artillery on Missionary Ridge that no reinforcements were sent to oppose Sherman, and
decided that no such movements occurred.

A possible reason for Sherman's belief that his attacks were encountering constantly replenished Confederate forces was that the fighting
abilities of the one division opposing his four divisions may have led him
to overestimate how many enemies his men faced. Its commander was
Major General Patrick Ronayne Cleburne, an Irishman born in Cork on
St. Patrick's Day, who as a young man served in the British Army before
coming to the United States and settling in Arkansas, where he became a
pharmacist and then a lawyer. Cleburne had risen quickly within the
Confederate Army, and the brigade he commanded at Shiloh fought
valiantly in that defeat, losing nearly 40 percent of its men. In subsequent actions on battlefields ranging from Richmond and Perryville in
Kentucky to the Confederate victory at Chickamauga, Cleburne was
wounded three times and earned the confident loyalty of his soldiers.
Promoted to lead the division that was repelling Sherman's attacks on
Missionary Ridge, his regiments, some of which he had led for two years,
were filled with combat veterans from Alabama, Arkansas, Georgia, Mississippi, Tennessee, and Texas. Cleburne's men were fighting with everything they had, and more: in addition to firing their weapons, they
hurled back one of Sherman's attacks by rolling large rocks down the
slope at the advancing Union troops and then threw stones at them.

At noon, still attacking, making no progress, and able to see that
Thomas's forces were not advancing up the center slope of Missionary
Ridge, Sherman had a signalman wave his handheld flags, asking Grant,
"Where is Thomas?" From Orchard Knob, Grant signaled back that

Thomas was starting to move, but in fact Thomas was standing right there beside Grant, just where he was supposed to be and easily accessible for Grant to command, and that was not happening.

Now the officers of Grant's staff began conferring among themselves, some yards away from Grant. Their understanding was that Grant had told Thomas to hold his attack until Sherman turned the enemy's right flank and Hooker turned the left, but both Sherman and Hooker were stopped where they were. Wilson noted that Grant, still standing there silently, looked discouraged. Something had to be done. Grant's chief of staff Rawlins walked up to him and said that surely it was time for Thomas's division to go into action.

Grant turned, went the few steps to Thomas, who was studying the enemy trenches on Missionary Ridge through his binoculars, and said, "Don't you think it's about time to advance against the rifle pits?"

Thomas, who was to say that he was resentful of being held back while Grant gave Sherman the chance to win the day, gave no answer and kept studying the enemy positions through his glasses.

More time passed, with Sherman's men trying to move forward and failing, while Grant and Thomas stood immobile within a few yards of each other. It is certain that both Grant and Thomas, like all the generals in the Union and Confederate armies, knew what had happened at Gettysburg, four months before, when Lee had finally sent Pickett's division and other units up Cemetery Ridge—a slope far less formidable than the rocky, steeper, six-hundred-foot-high face of Missionary Ridge—only to see those able, willing, experienced men slaughtered in a doomed charge that sealed the fate of the battle.

At three in the afternoon, after his fourth major attack was bloodily repulsed, Sherman stopped to rest his men. On Orchard Knob, seeing and hearing the cessation of action, Grant sent Sherman a message by signal flag: "Attack again." At this point, as Sherman remembered it, "I thought 'the old man' was daft, and sent a staff officer [Major L. B. Jenney] to inquire if there was a mistake." Jenney said that Sherman did not send him, but told him, "Go signal Grant. The orders were that I should get as many as possible in front of me and God knows there are enough. They've been reinforcing all day."

Whatever message Major Jenney sent, it stirred Grant into action. Rawlins received it, walked over to Grant, and started badgering him, telling him he must make Thomas attack. Colonel Wilson, also standing

there, said that Grant strode over to Thomas and, "with unusual fire, ordered Thomas to command the attack." Thomas promptly started issuing orders for his regiments to be ready for a signal: six quick cannon
blasts in a row. When they heard the last boom, they were to advance
and take the trenchlike enemy rifle pits at the base of Missionary Ridge.
Grant was later to say that there had been difficulty in passing this order
down to the forward commanders, and that the orders were to take the rifle pits at the bottom of the hill and then stop and reorganize "preparatory to carrying the ridge."

When Thomas's men heard that they were to enter the battle at last,
they were eager to fight: a man of the Sixth Indiana said, "We were crazy
to charge." Thomas's subordinate Brigadier General William B. Hazen
found that every man in his brigade intended to line up and get into the
attack: "All servants, cooks, clerks, found guns in some way." When
the signal of six successive cannon shots started at three-forty p.m., with
the fifth shot everyone started running forward, cheering. A tremendous
fusillade and barrage of enemy rifle and cannon fire poured down Missionary Ridge: a Union soldier said, "A crash like a thousand thunderclaps greeted us." The fire came at them from everywhere: the rifle pits
low on the slope, defensive positions halfway up, and the last line of
trenches, six hundred feet up on the crest. Nothing stopped Thomas's
men. Encountering tree trunks that had either been knocked down by
artillery or felled to slow their advance, the troops jumped, climbed, or
vaulted over these obstacles, shouting and cursing as they rushed ahead.
The enemy soldiers retreated from their rifle pits at the base of Missionary Ridge, firing as they backed up the slope.

As Thomas's men leapt into the abandoned Confederate positions at
the bottom of the slope, their officers began telling them to build up the
back ends of the enemy holes, to protect themselves from the intense enemy fire still pouring down on them. They were to do that, and wait for
further orders. The troops had other ideas; whatever Grant thought that
they were supposed to do in terms of reorganizing "preparatory to carrying the ridge," they were going up right then, and they began running
upward into the enemy fire. For a minute their officers stood on the edge
of the now-empty rifle pits, waving their swords at the backs of their advancing men and shouting orders that they should return; then they too
started running up Missionary Ridge, trying to get in front of their troops.
In a minute the first of the Union soldiers were overtaking the slowest of

the retreating Confederates, and in another few minutes they were at the suddenly evacuated second line of enemy rifle pits, farther up the slope. Alternately shouting and gasping for air, no longer in any semblance of organized formations, Thomas's men kept going.

Watching from Orchard Knob, Grant turned to Thomas, who was still standing in the appropriate place for a commander of one of Grant's armies, and asked sternly, "Thomas, who ordered those men up the ridge?"

Thomas replied, "I don't know. I did not."

Grant turned to Thomas's second in command. "Did you order them up, Granger?"

"No," Granger answered. "They started up without orders. When those fellows get started all hell can't stop them."

Grant muttered that someone would face disciplinary action if the attack failed and went on watching. Regimental battle flags were advancing up Missionary Ridge, moving up through thickets, past boulders, fallen trees, and little suddenly appearing ravines. The enemy was firing, but few Union soldiers stopped to fire back. It had become a race to the top. The man carrying the banner of the Twenty-fourth Wisconsin, shouting the battle cry, "On, Wisconsin!" was Captain Arthur MacArthur, who would one day have a son named Douglas. On another part of the slope, Captain C. E. Briant of the Sixth Indiana had managed to get ahead of his entire company, but as he neared the crest a private named Tom Jackson started to sprint past him. The captain reached out and grabbed Jackson's coattail, yanking him back as he forged ahead, but the private came on again and beat him to the top in the last yards. Looking down the reverse slope, Jackson called out to the winded men of his company who were coming over the top, "My God, come and see 'em run!"

A comrade who walked up beside Jackson recalled, "It was the sight of our lives. Gray clad men rushed wildly down the hill into the woods, tossing away knapsacks, muskets and blankets as they ran." (In the rout, Bragg was nearly captured; four thousand of his scattered troops were eventually taken prisoner.)

Now the higher officers began catching up to the men who had been supposed to reorganize at the bottom of the slope and wait for orders. Brigadier General Thomas J. Wood came over the crest on his horse. After shouting "You'll all be court-martialed!" at a crowd of his men, he laughed delightedly. As General O. O. Howard rode up the slope, he stopped near the top to try to comfort a dying soldier. In answer to his question of where he was hurt, the man replied, "Almost up, Sir." When Howard explained that he meant what part of the man's body had been hit, not where he had been on the slope, the soldier said again, "Oh, I was almost up and but for that"—he finally pointed to his mortal wound—"I'd have reached the top."

Gasping from their efforts, the victors compared notes. The color-bearer of the Thirty-eighth Indiana told his comrades who were still coming over the crest, "A fellow of the Twenty-second Indiana was up here first, but he wouldn't have been if I hadn't had on my overcoat." A captain of the Nineteenth Illinois had come up unscathed but was now examining his overcoat and discovered fourteen bullet holes in it.

Sherman's report of what his men were doing while all this was going on described his own force as having "drawn vast masses of the enemy to our flank" and said that "it was not until night closed in that I knew that the troops in Chattanooga [Thomas's men] had swept across Missionary Ridge and broken the enemy's centre. Of course the victory was won, and pursuit was the next step."

Sherman added that he ordered his reserve "to march at once" and "push forward," but the only officer who successfully exploited the situation was Brigadier General Philip Sheridan, a short, fiery West Pointer who was the son of Irish immigrants. Sheridan had been one of the Union generals brought south with his men to face the emergency at Chattanooga. Now, while Union soldiers of all ranks acted in the spirit of "My God, come and see 'em run!," Sheridan quickly organized a combination of moves to follow the fleeing Confederates; in an effort that did not stop until two in the morning, men of his division captured seventeen hundred prisoners and seventeen artillery pieces. Praised by Grant for his "prompt pursuit," Sheridan had redeemed an earlier indifferent performance at Chickamauga and soon would return to the Northern theater of war, from which he would emerge as the Union's great cavalry leader.

The Battle of Chattanooga, a most important strategic victory for the North, was finally over. The most brilliant part of it had been the impromptu assault on Missionary Ridge, which took the troops exactly fifty minutes to execute. Charles Dana had witnessed this final attack. At four-thirty that afternoon, his first report to Washington began, "Glory to God! The day is decisively ours. Missionary Ridge has just been carried by a magnificent charge of Thomas's troops, and rebels routed." The following day he added, "The storming of the ridge by our troops was one of the greatest miracles in military history. No man who climbs the ascent by any of the roads that wind along its front can believe that eighteen thousand men were moved in tolerably good order up its broken and crumbling face unless it was his fortune to witness the deed . . . Neither Grant nor Thomas intended it."

That was the reality; as soon as the battle finished, the interpretations of what had happened during the great victory began. Hooker recalled that, soon after Missionary Ridge was taken, he heard Grant say, "Damn the battle! I had nothing to do with it." Grant almost immediately sent Sherman a letter that started with, "No doubt you witnessed the handsome manner in which Thomas's troops carried Missionary Ridge this afternoon, and can feel a just pride too in the part taken by the forces under your command in taking first so much of the same range of hills, and then in attracting the attention of so many of the enemy as to make Thomas' part certain of success."

There it was: the beginning of the debate as to whether Sherman's attacks on the flank did in fact draw off large Confederate reinforcements whose absence weakened the enemy center. Sherman wanted to believe not only that this had happened but also that the entire strategy had been to do just that. In a letter to his brother John, he wrote, "The whole philosophy of the Battle was that I should get by a dash the extremity of Missionary Ridge from which the enemy would be forced to drive me," and later commented that "the whole plan succeeded admirably." The overall victory was indeed a military success, but the evidence is that Grant had intended Sherman's effort to be the winning attack that broke through on the northeast flank of Missionary Ridge and went right along its crest, with Thomas playing a secondary role in the center, and that Sherman failed in that assignment.

Like Sherman, Grant believed what he wanted to believe. In an official report of the action, he said, "Discovering that the enemy in his desperation to defeat or resist the progress of Sherman was weakening his center on Missionary Ridge, determined me to order the advance at once. Thomas was accordingly directed to move forward his troops, constituting our center." The "at once" is difficult to comprehend. Grant, a man of proven military intuition who said that at Chattanooga the commanders could see every part of the battlefield perfectly, had been watching Sherman unsuccessfully attack the enemy's right flank all day. Why it took Grant until after three in the afternoon to discover that the enemy was sending reinforcements to face Sherman that were "weakening his center on Missionary Ridge"—something that Colonel Wilson of Grant's own staff said was not the case—is either a mystery, or his statement was simply a convenient way of covering for Sherman and of putting the best face on a day when it was other men who won a crucial Union victory.

Reverting to his usual form, Grant wanted to keep up the momentum gained by scattering Bragg's forces on Missionary Ridge and ordered Sherman and Thomas to pursue the fleeing Confederates, who were heading for Atlanta, as quickly as they could go. Once again, however, as happened after Shiloh and Vicksburg, the Union troops were exhausted, and some of the Confederate forces eluded the follow-up movements that would have destroyed them completely.

Meanwhile, Grant had remained constantly aware of Burnside's threatened situation at Knoxville. On November 27, two days after the victory at Missionary Ridge, Burnside wrote Grant that he had no more than a few days' supplies left and might have to surrender by December 3. Grant turned to Sherman, whose men were still spent from their part in the battle, and ordered Sherman to organize and lead an eighty-five-mile forced march to aid Burnside. (Sherman was to say that he did not want his men to have to make this grueling march, and reluctantly started them off for Knoxville.)

After six days of pushing his men along through cold weather on frozen roads that tore up the soles of their boots, Sherman rode into Knoxville at the head of his troops. The first thing he saw was a pen "holding a fine lot of cattle, which did not look much like starvation." Burnside (whose way of wearing his facial hair gave rise to the term "sideburns") and his officers were "domiciled in a large, fine mansion, looking very comfortable." After having a turkey dinner with them, served at a table complete with linen and silver, Sherman observed to Burnside that this did not look like the headquarters of a starving army on the verge of surrender. Burnside admitted that he had exaggerated his plight; in the meantime, on November 29, his troops had thrown back decisively an attack made by Longstreet, who gave up the effort to take

Knoxville on December 3—the day that Burnside had told Grant he might have to surrender Knoxville—and withdrew his badly beaten forces far into the hills to the north to reorganize. The siege that Sherman's men had made a suffering march to help lift no longer existed. Still, Burnside said, he felt better about the overall situation, now that Sherman's reinforcements had arrived.

By now, Sherman's men were in the condition in which he had expected to find Burnside's troops; he described his soldiers as suffering in the cold with "bleeding feet wrapped in old clothes or portions of blankets that could ill be spared from shivering shoulders." Sherman set about making them comfortable, giving them rest and getting them resupplied and reequipped. Then, while his troops were marched back to the Chattanooga area, Sherman traveled west to Nashville, where Grant, who had been joined there by Julia, was conferring with some generals of his recently enlarged command.

There in the capital of Tennessee, Grant took Sherman and several other generals to pay a call on Andrew Johnson, who was to become far more important in their lives than they—or anyone—could then imagine. The sixty-three-year-old Johnson had been the one senator from the South who did not resign from the United States Senate at the time of secession—an act of loyalty to the Union deeply appreciated by Lincoln, who subsequently appointed the Tennessean a Union brigadier general and named him to the position he now held, that of the state's military governor.

The man who recorded the details of this meeting and the rest of the day was Brigadier General Grenville Dodge, a former civil engineer, businessman, and lobbyist. In addition to commanding troops, Dodge did exceptional service for Grant in constructing and repairing bridges and railroad tracks, and quietly ran the largest and most effective network of spies, including notable women spies, that either side possessed during the war.

As Grant led his handful of generals into Andrew Johnson's handsome, well-furnished new house, he became aware of the contrast in appearance between Johnson, sitting there in comfort, and that of his officers, some of whom had come straight from the rough conditions of living in the field. (Dodge called them "a hard-looking crowd.") When Grant apologized for the way they all looked, Johnson responded by studying them with what Dodge termed "a very quizzical eye." Then the governor began to give these soldiers who did the fighting an exhortation about the evils of their Confederate enemies, saying that he would show the rebels no mercy. To emphasize a point in his gratuitous tirade, Johnson pounded his fist on a piano so hard that these combat leaders, used to cannon fire, jumped when he did it. Dodge felt himself "rather disgusted" by this self-righteousness, because his experience with Andrew Johnson was that "I hardly ever got my hands on rebel stock or supplies that I did not find Johnson trying to pull them off" for his own benefit.

After they left Johnson's house, Sherman told his colleagues that *Hamlet* was to be performed in a local theater that evening, and they went to see it. Looking down from their seats in the first row of the balcony, they saw many Union soldiers in the audience. The play began, and the actors performed so badly that some of the soldiers began laughing. Sherman turned and said angrily, "Dodge, that is no way to play *Hamlet!*" He went on criticizing the performance in such a loud voice that Dodge warned him that the soldiers would look up at the balcony, recognize Grant and Sherman, and begin cheering them. It would bring the play to a halt. Sherman continued making his disparaging comments, "so indignant," as Dodge put it, "that he could not keep still." This went on until Hamlet's graveyard soliloquy, in which Hamlet picks up the skull of "Poor Yorick." At that point a soldier in the back of the theater called out, "Say, pard, what is it, Yank or Reb?" This produced a complete uproar; amid the confusion, Grant led Sherman and his other generals out of the theater and off to have supper. Sherman said that he wanted to have some oysters, and they ended up in a basement oyster shop. Halfway through their meal, the female proprietor, who could see that they were Union officers but had no idea that one of them commanded the entire region from the Appalachian Mountains to the Mississippi River, came to their table. Apologizing, she explained that the hour of the army-imposed military curfew had arrived and that her restaurant must close. Rather than telling her that she was talking to the de facto law of the land, Grant accepted the situation with good grace. He and his fellow generals stood up and left.

While Sherman was still in Nashville, Grant gave him permission to go home to Lancaster for a week's leave at Christmas. Before he left, Sherman had a troubled conversation with Grant about the rumors he had heard of various officers criticizing his leadership at Chattanooga. Grant remained completely supportive of Sherman and discussed plans for campaigns that they would undertake in the coming new year, with Sherman playing his usual important role.

Nonetheless, it was true that Sherman's reputation had suffered, both among some of the Union generals and by newspaper accounts that Grant characterized as "being calculated to do injustice." Thomas, determined to have full recognition of his men's brilliant action at Missionary Ridge, was not the only one with an axe to grind. Hooker, pleased by his own success at Lookout Mountain, wrote his friend Secretary of the Treasury Salmon P. Chase, a man close to Lincoln and also a friend of Sherman's father-in-law, that Sherman's repulses at Missionary Ridge could "only be considered in the light of a disaster . . . Sherman is an active, energetic officer, but in my judgment is as infirm as Burnside. He will never be successful. Please remember what I tell you." (On the other hand, Sherman was a few weeks away from receiving a joint resolution of Congress, thanking him and his men "for their gallantry and heroism in the battle of Chattanooga, which contributed in a great degree to the success of our arms in that glorious victory.")

As 1863 came to an end, with the forces under Grant and Sherman temporarily at rest, Grant was having to face the fact that his increasing military fame now had political dimensions. Barnabas Burns, the chairman of the wing of the Democratic Party in Ohio that favored strongly prosecuting the war effort, wrote Grant asking if he would "permit your name to be used" as a candidate for president of the United States at the Democratic National Convention in the coming May of 1864. Grant replied:

> The question astonishes me. I do not know of anything I have ever done or said that would indicate that I could be a candidate for any office . . .
>
> Nothing likely to happen would pain me so much as to see my name used in connection with a political office. I am not a candidate for any office nor for favors from any party . . .
>
> I . . . above all things wish to be spared the pain of seeing my name mixed with politics . . . Wherever, and by whatever party, you hear my name mentioned in connection with the candidacy for any office, say that you know from me direct that I am "not in the field," and cannot allow my name to be used before any convention.

Sherman, arriving home in Ohio on December 25 to join his family for a heartbreaking Christmas without their beloved Willy, soon realized the full extent of Grant's popularity with the Northern public. In Washington, the Senate and House passed a joint resolution praising Grant and his forces for their victories at Vicksburg and Chattanooga, and instructed that a gold medal be made honoring Grant, to be given him "in the name of the people of the United States." *The New York Herald*, which supported the Democratic Party, said that Grant (who had at this time no meaningful party affiliation) should run against Lincoln in the coming year's presidential election and expressed the belief that he would win. (Lincoln would in fact sound out Grant's congressman Elihu Washburne concerning what political ambitions Grant might have. Washburne turned to J. Russell Jones, a friend of Grant's from Galena days who had recently received a letter from Grant saying, "Nothing could induce me to become a presidential candidate, particularly so long as there is a possibility of having Mr. Lincoln re-elected." When Jones went to Washington at Lincoln's request and handed him that letter, a relieved Lincoln placed his influence behind a movement for further promotion for Grant.)

Thinking about the fame that now surrounded his friend—the man the press had so often defamed earlier in the war—on December 29, Sherman wrote a letter to Grant.

> You occupy a position of more power than Halleck or the President. There are similar instances in European history, but none in ours . . . Your reputation as a general is now far above that of any man living, and partisans will maneuver for your influence; but if you can escape them, as you have hitherto done, you will be more powerful for good than it is possible to measure . . . Preserve a plain military character, and let others maneuver as they will. You will beat them not only in fame, but in doing good in the closing scenes of this war, when somebody must heal and mend the breaches made by war.

Although at earlier moments in the conflict both Grant and Sherman had thought the South might soon collapse, Sherman, despite his reference to "the closing scenes of this war," was not thinking in terms of imminent victory. Before coming home for this bleak family Christmas of 1863, he had written to Ellen that "the next year is going to be the hardest of the war." At the moment, he did not foresee that his policies and actions were in some ways to be the harshest thing in that hardest year, but he was ready to do whatever he thought must be done.

More than ever, Sherman believed in what he said in a retort he finally made to a Southern lady at a dinner party in Nashville who "pecked and pounded away" at him about his troops stealing food as they marched through the countryside: "Madam, my soldiers have to subsist themselves . . . War is cruelty. There is no use trying to reform it; the crueler it is, the sooner it will be over." If ever Sherman had hoped to see political arrangements that might shorten the war, he hoped for them no longer. Writing to Ellen's brother Philemon, he said that he understood his brother-in-law's pleasure in the results of the Ohio gubernatorial election, a popular endorsement of Lincoln's war policy, but added, "The only vote that now tells is the cannon & the musket."

For Grant, December of 1863 had its lighter moments. Brigadier General Isaac F. Quinby, a West Point classmate on recruiting duty in Rochester, New York, wrote Grant that his wife wanted a lock of Grant's hair, which would be auctioned off to the highest bidder at a bazaar being held to raise funds for a wartime charity appeal. This was Grant's reply to his friend's wife.

MY DEAR MADAM,

The letter of my old friend and classmate, your husband, requesting a lock of my hair, if the article is not growing scarse [sic], from age, I presume he means, to be put in an ornament, (by the most delicate of hands no doubt) and sold at the Bazaar for the benefit of disabled soldiers and their families, is just received. I am glad to say that the stock is yet as abundant as ever though time, or other cause, is beginning to intersperse here and there a reminder that Winters have passed.

The object for which this little request is made is so praiseworthy that I can not refuse it even though I do, by granting it[,] expose to the ladies of Rochester that I am no longer a boy. Hoping that the citizens of your city may spend a happy week commensing [sic] to-morrow, and that their Fair may remunerate most abundantly, I remain,

> Very truly your friend,
> U. S. Grant
> Maj. Gen. U.S.A.

In the last two days of 1863, it was back to duty. On December 30, Sherman wrote a letter to his brother John in which he said of Grant, "With him I am as a second self. We are personal and official friends."

He added that he was leaving Ohio to return to Memphis to take up his duties as commander of the Army of the Tennessee, the post to which he had succeeded when Grant was given overall command of that army and two others before the Battle of Chattanooga. Grant wrote Halleck on December 31 that he had just arrived at Knoxville and "will go to the front [this] evening or in the morning . . . Longstreet is at Morristown [Tennessee]."

The year 1864, which Sherman had told Ellen would be "the hardest of the war," began badly for the Grants. In the third week of January, their son Fred suddenly fell ill in St. Louis with what Julia said was "camp dysentery and typhoid fever," the combination of diseases that carried off Willy Sherman. Julia rushed to St. Louis from Nashville, where she had been with Grant, and found that Fred was already beginning to recover.

Leaving his military duties, Grant hurried to see Fred soon after Julia arrived, and the relieved Grants were briefly reunited in St. Louis, near her family's farm where they had met. (From St. Louis, Grant wrote Sherman that "I come here to see my oldest boy who has been dangerously ill of Typhoid Pneumonia. He is now regarded by his physician as Out of danger.") The crisis involving Fred had passed, but Julia received news concerning a less serious medical matter. She was now thirty-seven. All her life she had been conscious of her strabismus, the condition that made one of her eyes go out of focus and squint. When she was younger, an eye specialist in St. Louis had told her several times that he could perform a simple operation that would correct the condition. Julia said of that, "I had never had the courage to consent, but now that my husband had become so famous I thought it behooved me to look as well as possible." Now, with Fred recovering, she had time to consult with the specialist, and he told her that it was, as she wrote of it, "too late, too late." Unhappy, she shared this with Grant.

> I told the General and expressed my regret.
> He replied: "What in the world put such a thought in your head, Julia?" I said: "Why, you are getting to be such a great man and I am such a plain little wife. I thought if my eyes were as others are I might not be so very, very plain, Ulys; who knows?" He drew me to him and said: "Did I not see you and fall in love with you with these same eyes? I like them just as they are, and now, remember, you are not to interfere with them. They are mine, and

let me tell you, Mrs. Grant, you had not better make any experiments, as I might not like you half so well with any other eyes."

Returning quickly to his headquarters at Nashville, Grant plunged back into the problems awaiting him. He referred to the immediate military situation in a typically direct letter to General Thomas: "Longstreet has also been reenforced by troops from the East. This makes it evident the enemy intend to secure East Tennessee if they can, and I intend to drive them out or get whipped this month."

While making plans that resulted in his forces successfully keeping Longstreet away from his objectives, Grant also gave his attention to the many and varied other matters that inevitably came to his desk. He wrote Halleck of the results of an investigation he had ordered as a result of his suspicions "that there was much useless extravagance [sic] in the Quartermaster's Dept." His conclusion: "The result has been already to find that Govt. is being constantly defrauded by those whos [sic] duty it is to protect and guard the public interest. The guilty parties will be relieved and brought to trial." In addition, his headquarters in Nashville was being besieged by the wives of Confederate soldiers who wanted to go farther south to see their husbands. Deciding that these requests should no longer be dealt with on an individual basis, Grant informed General Thomas of his solution to the problem. "As it is rather desirable that all such should be where their affections are set, I propose giving notice through the papers setting a day when all who wish will be permitted to go[,] and fix the point where they will be allowed to pass through our lines. Let me know where and when they should be allowed to go."

Grant had not forgotten the most southerly area of his widespread command and wanted to continue putting pressure on the overall Confederate military effort. Four days after Longstreet broke off his unsuccessful attacks on Knoxville, Grant had written Halleck that he would like to try a previously considered movement to capture the port of Mobile, Alabama. As Grant saw it, whether Mobile fell or not, this would open the prospect of a campaign that would move his forces east from the Mississippi River into Alabama, with a further thrust that might take a Union offensive on into Georgia. This would not only "secure the entire states of Alabama & Mississippi," but also, as Grant saw it, force Robert E. Lee to give up his positions in Virginia, in order to save the Deep South.

This was imaginative strategic thinking, but it received little support in Washington, where there was doubt that anything could pull Lee out of Virginia. Both Halleck and Lincoln felt that a higher priority should be given to following up against Longstreet in east Tennessee—an idea with which Grant did not disagree, but he felt it could not be implemented during winter weather in the mountains north of Knoxville.

Still wanting to keep the ball moving somewhere, Grant approved Sherman's idea of making a massive raid on the Confederate railroad center at Meridian, Mississippi, a hundred miles east of Vicksburg. Here, again, Grant was demonstrating his complete confidence in Sherman, and Sherman responded with a successful performance that could have gone badly wrong. He left Vicksburg on February 3 with twenty thousand men, moving under orders that showed that he had learned from Grant's move against the city of Jackson ten months before. Prefiguring larger moves that he would make, Sherman's orders stressed the need to move swiftly and to carry only essential equipment. Within days, just the news of Sherman's advance, which was blasting aside every enemy in its way, convinced Confederate general Leonidas Polk to give up Meridian without defending it. Reaching Meridian, Sherman's men spent nearly a week destroying everything in the area: 115 miles of railroad track, sixty-one bridges, and twenty-one locomotives, in addition to arsenals, warehouses, and workshops. They returned to Vicksburg, herding along five thousand slaves they had freed, along with another, newer category of refugee—a thousand white Southerners who wanted to place themselves under Union control. The destruction of the Confederate warmaking capacity was the greatly successful side of the military ledger, but on his way to Jackson, Sherman was nearly captured. Another part of the effort, in which seven thousand horsemen under Grant's cavalry chief General William Sooy Smith were to defeat the four thousand cavalrymen led by the mercurial Nathan Bedford Forrest, failed when Smith did not coordinate successfully with Sherman, and Forrest's lesser numbers outmaneuvered and occasionally routed the federal troopers. (On the day he got back to Vicksburg, Sherman wrote Ellen, "Somehow our cavalry is not good. The Secech with poor mean horses make 40 & 50 miles a day, whereas our fat & costly horses won[']t average 10. In every march I have ever made our Infantry beats the Cavalry & I am ashamed of them.")

Grant and Sherman were to be remembered for their dramatic campaigns, but both men valued the clandestine side of warfare represented by military intelligence. As he was leaving Vicksburg aboard the naval gunboat *Silver Cloud* to return to his headquarters in Memphis, Sherman wrote Grant concerning Confederate troop strength remaining in Mississippi; as for the information he did not yet have, Sherman said, "I have one of my best Memphis female spies out, who will be back in time to let me know all we want." The Meridian campaign had given Sherman added confidence: a few days after he sent Grant his letter about the female spy, he telegraphed Grant from Memphis, regarding another thrust he intended to make: "Enemy is scattered all over Mississippi and I think the movement indicated will clean them out."

During the time that Sherman was conducting what became known as the Meridian Campaign, Grant wrote a brief letter to Julia. In it he said, "It now looks as if the Lieut. Generalcy bill was [*sic*] going to become a law. If it does and is given to me, it will help my finances so much that I will be able to be much more generous in my expenditures."

This was Ulysses S. Grant's way of telling his wife that Congress was going to name him as the first lieutenant general since George Washington received that rank. The promotion would automatically make him general in chief, placing him above Halleck, who under General Orders 98 ceased to hold that position and was "assigned to duty in Washington as Chief of Staff of the Army, under the direction of the Secretary of War [Stanton] and the lieutenant general commanding [Grant]." Ulysses S. Grant would command all the Union armies. He could therefore stay in the West or go east, but he decided to leave his vast Western theater of war, which he intended to turn over to Sherman, and base himself in or near Washington, ready to face Robert E. Lee's Army of Northern Virginia.

There were things that Grant already knew he wanted to do. The man who had grown up with horses was to say that, until now, the campaigns of the armies of the Union had reminded him of draft horses that were pulling the same wagon, but doing it in an awkward and inefficient way. To this point in the war, despite individual Union victories such as Shiloh, Vicksburg, Chattanooga, and Gettysburg, the Union Army was divided into nineteen geographical military departments, with the Army of the Potomac an entity unto itself. Generals in all those sectors had been acting, when they did, on their own initiative, often without consultation or coordination with their peers in adjoining departments. This had resulted in sporadic, uncoordinated attacks and campaigns: the South, often given time to recover after a limited Union offensive ground to a halt in one area, was able to move its troops considerable distances and consolidate its forces to counter a new Union threat.

Grant intended to impose a cohesive Union strategy. He was going to be one of the two major figures in implementing that—the other was Sherman. As usual, any officer talking to Grant about a military matter was left in no doubt as to what he wanted and was left with great latitude in accomplishing the objective. Soon, Grant and Sherman would be conferring face-to-face, and Sherman would always remember the over-riding concept: "He was to go for Lee, and I was to go for Joe Johnston."

There was no date on the letter in which Grant told Julia the news of his promotion—Sherman would be informed of it as soon as the promotion became official, and a warm exchange of letters between the two men would ensue—but Grant's letter to Julia was written about February 10, 1864. On that date three years before, Ulysses S. Grant was sitting in his father's leather goods store in Galena, Illinois, bored and doing poorly at the job his father had created for him—a retired army captain who had resigned from the service rather than face a court-martial on charges of drinking while on duty. Now he commanded a continually growing army of seven hundred thousand men—seven hundred times the size of the Twenty-first Illinois, the regiment he began to lead thirty-one months before—in the struggle that would decide whether the United States would be two nations or one.

10

GRANT AND SHERMAN BEGIN
TO DEVELOP THE WINNING STRATEGY

On March 2, 1864, Grant learned of Sherman's success in the Meridian Campaign—the march through Mississippi that demonstrated Sherman's ability to operate independently deep in enemy territory, far from Grant's headquarters in Nashville, destroying more of the Confederate capacity to make war. Despite the failure of Sooy Smith and his cavalry to carry out their role in the campaign, Sherman's execution of the large-scale raid, going to and from Meridian, fully justified Grant's belief in him and foreshadowed the far greater movements that Sherman would soon be making. Grant said of that moment in March, "I was ordered to Washington on the 3rd to receive my commission."

Grant's promotion to lieutenant general and commander of all the Union armies was now official. On the same day, he wrote Sherman before he was to leave Nashville for Washington the following morning. In his letter, which he marked "Private" and began with, "Dear Sherman," he included the name of the gifted and enterprising thirty-five-year-old Major General James B. McPherson, of whom they were both fond, and said:

I want to express my thanks to you and McPherson as *the men* to whom, above all others, I feel indebted for whatever I have had of success. How far your advice and suggestions have been of assistance you know. How far your execution of whatever has been given you to do entitles you to the reward I am receiving you cannot know as well as me . . .

Your friend
U.S. Grant
Maj. Gen.

This produced an effusive response from Sherman, writing from Memphis. Marked "(Private and Confidential)," it said in part:

> You are now Washington's legitimate successor, and occupy a position of almost dangerous elevation, but if you can continue as heretofore to be yourself, simple, honest, and unpretending, you will enjoy through life the respect and love of friends, and the homage of millions of human beings that will award to you a large share in securing to them and their descendants a Government of Law and Stability . . . You do General McPherson and myself too much honor . . . The chief characteristic in your nature is the simple faith in success, which I can liken to nothing else than the faith a Christian has in a Savior.

Looking back on their campaigns together, Sherman now expressed his feeling for Grant: "I knew wherever I was that you thought of me, and if I got in a tight place you would come if alive." He continued, with equal candor, "My only points of doubt were in your knowledge of Grand Strategy and of Books of Science and History. But I confess your common sense seems to have supplied all these."

Closing this warm statement of appreciation and praise, he said, "We have done much, but still much remains to be done." Then Sherman, a man of the West speaking to another man of the West, urged Grant to leave Halleck in Washington, where Halleck knew how "to stand the buffets of Intrigue and Policy." Sherman wanted Grant to run the whole war from the Western theater, and in expressing this he showed his willingness to give up his chance to be the man clearly in command in the West. "Come out West, take to yourself the whole Mississippi Valley. Let us make it dead sure, and I tell you the Atlantic slope and Pacific shores will follow its destiny as surely as the limbs of a tree live or die with the main trunk . . . From the West when our task is done, we will make short work of Charleston, and Richmond, and the . . . coast of the Atlantic."

That was not to be. Grant, accompanied by his son Fred, now thirteen, arrived in Washington on March 8. A welcoming committee met the wrong train, so Grant, dressed in a nondescript linen duster that concealed the general's stars on his uniform, made his own way to the Willard Hotel with Fred and asked the desk clerk for a room. Unimpressed by the appearance of this rumpled traveler, the clerk handed him a key to a small room on the top floor and asked him to register.

When he saw the signature, "U. S. Grant and son, Galena, Illinois," he took back the key, and Grant and Fred were escorted to the best suite in the hotel. After dinner at the hotel, during which everyone in the dining room rose and gave "three cheers for Lieutenant General Grant," he found a note from the White House: President Lincoln was holding his weekly evening reception and would like General Grant to join him. Grant put Fred to bed and was soon shaking hands with the six-foot-four Lincoln, who looked down at his five-eight choice to lead the armies of the Union and said, "Why, here is General Grant! Well, this is a great pleasure, I assure you." Lincoln then introduced Grant to Secretary of State William H. Seward. It was Seward who presented Grant to Lincoln's wife, the mentally erratic and unpredictable Mary Todd Lincoln. On this occasion Mrs. Lincoln was calm and friendly, and began making social conversation with Grant.

The hundreds of guests at first tried to restrain themselves from walking over to get a close look at the man in whom the Union now reposed its hopes, but soon Grant found himself surrounded by a crowd of well-wishers; one guest said that Grant "blushed like a schoolgirl" as he tried to shake the scores of outstretched hands. When the room began to rock with cheers of "Grant! Grant! Grant!" he was persuaded to stand on a sofa so more people could see him, which produced louder cheers. A journalist who was present wrote: "It was the only real mob I ever saw in the White House . . . For once the President of the United States was not the chief figure . . . The little, scared-looking figure who stood on the crimson-covered sofa was the idol of the hour."

The next morning, Grant was back in the White House, where Lincoln presented him with his commission as lieutenant general. After the short ceremony, the two men went upstairs to talk. They had a rapid and complete meeting of minds. As Grant remembered it, Lincoln told him that, in military matters, "all he wanted or ever wanted was someone who would take the responsibility and act, and call on him for all of the assistance needed, pledging himself to use all the power of the government in rendering such assistance." Grant's response: "Assuring him that I would do the best I could with the means at hand, and avoid annoying him or the War Department, our first interview ended."

A British war correspondent who saw Grant during his initial visit to Washington underscored the same qualities that Lincoln liked so much in this general. "I never met a man with so much simplicity, shyness, and decision . . . He is a soldier to the core, a genuine commoner, commander of a democratic army from a democratic people. From what I learn of him, he is no more afraid to take responsibility of a million men than of a single company."

After further conferences with Lincoln and Stanton, and an inspection trip to see General George Meade at the headquarters of the Army of the Potomac sixty miles southwest of Washington, Grant got on a train to return for a short time to Nashville and close up his headquarters there. He knew what he wanted to do, and now he had the authority to do it. In line with a suggestion from Sherman that he stay out of Washington with all its intrigues and bureaucracy, Grant would leave Halleck to run the detailed administration of the army from the War Department in Washington, while he set up his headquarters as general in chief near those of Meade on the fighting front in Virginia. From that headquarters in the field, he would plan and oversee the overall campaigns of the Union military effort in the Eastern and Western theaters of war. In addition, he would become the de facto commander of the Army of the Potomac, fighting Robert E. Lee and Lee's Army of Northern Virginia in the broad battlefront area between Washington and Richmond, but he would issue those orders through Meade. (Grant's worries about Meade's potential resentment of being superseded vanished at their first meeting. Meade, the greatly famous victor of Gettysburg, immediately told Grant that he would understand if Grant wished to replace him with Sherman or any of the other generals who had served him well in the West. The important thing was to get on with the job, and he pledged, as Grant admiringly remembered it, to "serve to the best of his ability wherever placed." Grant "assured him that I had no thought of substituting anyone for him.")

Grant also had a number of ideas about promotions, demotions, and transfers. Although the Navy Department controlled the assignments of Admiral Porter, with whom Grant and Sherman had worked so well during the Vicksburg campaign, Porter would eventually move from the Mississippi theater to take command of what was known as the Northern Blockade Squadron, on the Atlantic coast. Addressing another aspect of the Union's military posture and practices, Grant decided to end the virtual autonomy of the army departments such as those controlling supplies and commissary matters, and the legal department run by the adjutant general. Nearly half the soldiers in the Union Army were serving in various assignments well behind the fighting fronts, and Grant determined to reduce those positions "to the lowest number of men necessary for the duty to be performed."

On March 17, thirteen days after receiving the official notification of his promotion, Grant arrived back in Nashville. At Grant's request, Sherman had come from Memphis to meet him, bringing four other generals, including Grenville Dodge, who had done such remarkable work in building and repairing railway lines and bridges when construction rather than destruction was needed, and in running the Union spy network in the Western theater—the largest one operated by either side during the war. For two days these men conferred, as Grant handed over to Sherman the daily conduct of the war in the West and the Deep South. Adam Badeau, a journalist who had joined Grant's staff as military secretary, now for the first time saw Grant and Sherman in the same room.

> Sherman was tall, angular, and spare, as if his superabundant energy had consumed his flesh. His words were distinct, his ideas clear and rapid, coming, indeed, almost too fast for utterance, in brilliant, dramatic form . . .
>
> Grant was calmer in manner a hundred-fold. The habitual expression on his face was so quiet as to be almost incomprehensible . . . In utterance he was slow and sometimes embarrassed, but his words were well-chosen, never leaving the remotest doubt of what he intended to convey . . . Not a sign about him suggested rank or reputation or power . . . [but] in battle, the sphinx awoke.

In a hurry to return to Washington, Grant had Sherman and General Dodge accompany him on the train to Cincinnati, with Sherman and Grant smoking cigars as they discussed the campaigns to come. In Cincinnati, Sherman had a brief, bittersweet reunion with Ellen. Her mother, who had raised Sherman from the time he came to live in the Ewings' house at the age of nine, had died, which brought sorrow to him as well as her. Ellen was pregnant again; they had by letter discussed the idea of naming the baby Willy if it were a boy but decided that it would be too painful a reminder of their son who died the previous summer. Writing Ellen, Sherman had spoken of their feelings in these words: "On reflection I agree with you that his name must remain sacred to us forever[.] He must remain to our memories as though living, and his name must not be taken by any one. Though dead he is still our Willy and we can love him as God only knows we loved him."

Grant rented a room in a Cincinnati hotel, and for two days he and Sherman pored over maps, as Dodge kept track of all the paperwork involved in their deliberations. The grand strategy was, as Sherman would famously say, "He was to go for Lee, and I was to go for Joe Johnston." In the Confederate military hierarchy, Lee and Johnston were at this time de facto equals under the civilian direction of Jefferson Davis, with Lee leading his Army of Northern Virginia against the Army of the Potomac on the South's northern front, while Johnston was reorganizing the South's Army of Tennessee at Atlanta and intending to begin offensive operations. If both men and their armies could be defeated in their separate theaters of war, the South's ability to fight on would be virtually at an end.

While Grant gave Sherman much latitude in how he was to "go for Joe Johnston," he stressed certain points. Lee and Johnston had the advantage of operating at relatively short distances from their bases in the interior of the South, while Grant would be attacking from the north and Sherman would have to start his movements from a point 530 miles southwest of Grant and the Army of the Potomac. The danger was that Lee, who had shown immense skill in moving his forces from one critical point to another by railroad, might be able to send reinforcements to Johnston when Johnston needed them, and that Johnston could similarly send large and swift support to Lee. To forestall that, Grant's and Sherman's armies had to act in close cooperation, keeping constant pressure on their respective fronts so that there were no quiet moments when either Lee or Johnston could spare troops to send to the other.

Even after their lengthy session in Cincinnati, when Grant got to Washington he reinforced his priorities in a letter to Sherman in which he said, "You I propose to move against Johnston's army, to break it up, and to get into the interior of the enemy's country as far as you can, inflicting all the damage you can against their war resources." Sherman responded by saying, "Like yourself you take the biggest load and from me you shall have thorough and hearty cooperation." To reassure Grant that he really understood what was wanted of him, he added, "I will not let side issues draw me off from your main plan in which I am to Knock Joe

Johnston, and do as much damage to the resources of the Enemy as possible." (Sherman was to characterize all this as being a policy of "Enlightened War.")

Using this strategy, Grant hoped to close out the war in 1864. To strengthen this military policy and objective, he intended to issue orders that would bring back to the main Eastern and Western Union armies the smaller forces then operating in such places as Florida and Arkansas. If he and Sherman could accomplish what he hoped—smash and wear down the main Confederate armies in a two-pronged, coordinated effort—the enemy's out-of-the-way outposts would wither on the vine.

As Sherman prepared to go on the offensive, Grant returned to Washington, took up his headquarters near Meade's Army of the Potomac, and reiterated the pertinent part of this philosophy to General Meade. On April 9, he told Meade, "Lee's army will be your objective point. Wherever Lee goes, there you will go also." Although the capture of Richmond, the Confederate capital, remained an inevitable goal, the emphasis was going to be less on control of territory and more on destroying the Confederate armies and the South's means of waging war. (In this latter area, that of drying up the South's source of men and of supplies ranging from weapons to food, both Grant and Sherman had increasingly come to realize that the South was indeed a nation in arms and that the common European practice of having standing armies engage each other in set-piece battles to determine the outcome of a war was not enough to win this struggle. When Grant had told Sherman right after Vicksburg to set out after Joseph E. Johnston "and inflict on the enemy all the punishment you can," he had already demonstrated at Jackson that he regarded all kinds of supplies as legitimate military targets, and Sherman's burning of Randolph, Tennessee, and his Meridian Campaign had shown that he too was ready to lay waste anything and anyplace that could sustain the enemy's ability and will to resist. Both men were ready to engage in what became known as total war.)

Bold as Grant was, he did not at that moment realize that Sherman was thinking even more boldly than he, in terms of getting "into the interior of the enemy's country as far as you can." By the end of April, Sherman had assembled ninety-eight thousand men in Chattanooga, ready to march southeast in an attempt to take Atlanta, a hundred miles away. This was not going to be a massive raid like the Meridian Campaign, in

Ulysses S. Grant. One of the finest horsemen ever to graduate from West Point, Grant was the most aggressive and resolute general in the Union Army. (Courtesy of the Abraham Lincoln Presidential Library)

Grant's wife, Julia. Highly intelligent and charming, always believing in her husband's destiny despite his prewar failures, she and Grant lived one of the great American love stories. (Courtesy of the Library of Congress)

St. Louis, Dec 23rd 1857

I this day consign to J. S. FRELIGH, at my own risk from loss or damage by thieves or fire, to sell on commission, price not limited, 1 Gold Hunting Detached Lever & Gold chain on which said Freligh has advanced Twenty two Dollars. And I hereby fully authorize and empower said Freligh to sell at public or private sale the above mentioned property to pay said advance—if the same is not paid to said Freligh, or these conditions renewed by paying charges, on or before Jan 23/58

U. S. Grant

This receipt shows that on December 23, 1857, at a time when he was down and out, Grant pawned his gold watch for twenty-two dollars so that he could buy Christmas presents for his family. Eleven years later, he was elected President of the United States. (Courtesy of the Abraham Lincoln Presidential Library)

During the war, numerous photographs were made of Grant by himself and with his higher-ranking officers, but this is the only one showing him against a background of his troops in the field. (Courtesy of the Library of Congress)

William Tecumseh Sherman. He said of his immensely successful military partnership with Grant, during which they constantly supported each other's efforts, "We were as brothers, I the older man in years, he the higher in rank." (Courtesy of the Library of Congress)

Sherman's wife, Ellen Ewing Sherman. Their marriage was a difficult one, but it was impossible to imagine either of them being married to anyone else. (Smithsonian American Art Museum, Washington, D.C. / Art Resource, N.Y.)

Sherman's devoted younger brother John. Already a United States senator when the Civil War began, John Sherman served in the Senate for thirty-two years, and is best known as the author of the legislation known as the Sherman Anti-Trust Act. (University of Notre Dame Archives)

General Henry Halleck, sometimes wise and sometimes duplicitous. At various times he commanded both Grant and Sherman. (Courtesy of the Library of Congress)

Admiral David Dixon Porter of the United States Navy, who worked effectively with Grant during amphibious operations such as those on the Mississippi that led to the great Union victory at Vicksburg. (National Archives at College Park)

One of the Navy's "mud turtles," flat-bottomed gunboats that furnished vital support for Grant and Sherman's campaigns along the rivers of the South. (Collection of the New-York Historical Society)

Confederate general James Longstreet. Grant's West Point classmate, he was a cousin of Grant's wife, Julia, and best man at their wedding. (Courtesy of the Library of Congress)

General Joseph E. Johnston. Robert E. Lee's West Point classmate and friend, Johnston was the Confederacy's master of defensive and evasive maneuvers. (Courtesy of the Library of Congress)

During Grant's attack on Lookout Mountain at Chattanooga, the weather deteriorated so rapidly that the upper part of the mountain disappeared from view. The ensuing fight became known as "The Battle Above the Clouds." (U.S. Army Center of Military History, Army Art Collection, Fort McNair, Washington, D.C.)

Johnston's decisive repulse of Sherman's attacks on Kennesaw Mountain in Georgia underscored the fact that Sherman was better at executing such sweeping moves as his March to the Sea than at fighting pitched battles like this. (Rights owned by the University of Mississippi Press)

SANTA CLAUS SHERMAN PUTTING SAVANNAH INTO UNCLE SAM'S STOCKING.

Sherman's bold and brutal March to the Sea, moving from Atlanta to the coast, resulted in the capture of Savannah on December 21, 1864. Sherman wired Lincoln, "I beg to present you as a Christmas gift the city of Savannah." (Courtesy of the Library of Congress)

A representation of the historic meeting at City Point, Virginia, on March 28, 1865, as Lincoln met with Grant and Sherman to discuss the closing phases of the war. LEFT TO RIGHT: Sherman, Grant, Lincoln, Admiral Porter. (Courtesy of the Library of Congress)

General Robert E. Lee. This photograph was taken by Mathew Brady a few days after Lee's return to his house in Richmond, Virginia, after surrendering his Army of Northern Virginia to Grant at Appomattox Court House. (Courtesy of the Library of Congress)

Lincoln's able, irascible, dictatorial secretary of war, Edwin M. Stanton. (Courtesy of the Library of Congress)

A contemporary drawing of the two-day Grand Review, held in Washington in May of 1865. It was the Union's farewell to its victorious army. With Grant and Sherman on the reviewing stand, 135,000 men, most of them soon to be demobilized, marched past wildly cheering crowds. (Rights owned by the University of Mississippi Press)

which he left Memphis, marched a hundred miles, struck, and then returned to Memphis. This was going to be straight-ahead fighting, with no intention of turning back.

When Grant understood what Sherman wanted to do, he gave his approval for the campaign to march into the heart of the South, headed for Atlanta, which was a vital center for manufacturing and the storage of supplies, as well as being a major railroad hub. Whether Sherman hoped even then to extend this immense movement another 225 miles from Atlanta to Savannah, to complete his epic March to the Sea, is not clear. What is abundantly clear is that, in good part due to his association with Grant, the Sherman of 1864 bore no resemblance to the man who in 1861 had begged Lincoln not to make him the departmental commander in Kentucky but to keep him always as a second in command, serving directly under a superior officer. Even to move his headquarters down to Chattanooga from Nashville meant that Sherman had to leave his most secure, heavily supplied base 110 miles to his rear, connected to his forward headquarters by a single-gauge railroad track that could be struck by Confederate cavalry raids, but he was not looking to his rear. Sherman needed thirteen hundred tons of supplies a day for his hundred thousand men and hoped that the railroad would bring forward as much of that as possible, but he also intended to live off the land, in the heart of the Confederacy, and travel light as he went. (Sherman underscored his intention to carry a minimum of supplies and equipment when he wrote, "My entire headquarters transportation is one wagon for myself, aides, officers, orderlies, and clerks.")

As spring came to Washington, Grant first had a different kind of battle to fight. In his effort to add to the combat strength of the Union Army, he was sending to the front thousands of soldiers who had been assigned to garrison duty or to guard supply lines. Relying on the traditional civilian control of the military, Stanton told Grant that he was pulling too many men out of the defenses of Washington and ordered it stopped. When Grant politely replied, "I think I rank you in this matter, Mister Secretary," Stanton answered, "We shall have to see Mister Lincoln about that," and they walked from the War Department building over to the White House, which was next door.

While Lincoln listened, with Grant remaining silent, Stanton set forth the matter as he saw it. When he finished, Lincoln smiled and said,

"You and I, Mr. Stanton, have been trying to boss this job, and we have not succeeded very well with it. We have sent across the mountains for Mr. Grant, as Mrs. Grant calls him, to relieve us, and I think we had better leave him alone to do as he pleases."

As Grant readied himself to fight Lee, some of the officers of the Confederacy's Army of Northern Virginia made disparaging remarks about him. Now, they said, Grant would find out what real opposition was. Julia Grant's cousin James Longstreet, back with Lee's army in Virginia after seeing how Grant had turned around the situation after Chickamauga by the victory at Chattanooga, warned them not to underestimate his old West Point classmate and friend, whom he had seen in constant fiery action when they fought in the same regiment during the Mexican War. He said to one officer, "We must make up our minds to get into line of battle and stay there, for that man will fight us every day and every hour till the end of this war."

Sherman was later to speak of the way the "magnetic telegraph" enabled him to keep in constant touch with Grant, despite their geographical separation. At this moment, he was sending daily reports to Halleck's office in Washington, and Halleck was relaying what was relevant to Grant's headquarters in the field. By the beginning of May, the only question was whether it would be Grant or Sherman, each now in his new role, who first engaged the enemy. As it happened, Grant was sitting in his tent near Germanna Ford on the Rapidan River in Virginia on the evening of May 4, smoking a cigar and talking with Meade as they prepared for the first day's fighting in what became known as the Battle of the Wilderness, when he received by telegraph from Washington the news that Sherman was advancing from Chattanooga into Georgia.

Sherman's march toward Atlanta began with several days of maneuvering toward Resaca, Georgia, but on his own front in Virginia, Grant ran straight into some of the war's fiercest and most constant fighting. The two opposing leaders, Grant and Lee, were certainly different. Lee was a strikingly handsome, courtly Virginia aristocrat, while Grant was an ordinary-looking man who was once described as having a genius for vanishing into a crowd. Nonetheless, when Adam Badeau said of Grant, "In battle, the sphinx awoke," he was paralleling a comment by a Southerner who observed that General Lee on the battlefield was a different man from General Lee in the drawing room, a remark echoed by Confederate general Henry Heth, who said that he found Lee to be "the most belligerent man in the Confederate Army." There was no doubt that Grant and Lee were the two most aggressive generals to fight in the war; in this first clash it was Grant who crossed the Rapidan in an effort to move around Lee's right flank and get between him and Richmond, and it was Lee who saw what he was trying to do and threw everything he had straight at him.

For two of the bloodiest days of the war, May 5 and 6, the armies of these two determined opponents fought each other in the tangled underbrush, fallen trees, and marshes of Virginia's Wilderness, a rectangular area of sixty-four square miles. Before it was over, Grant had poured into the battle 101,895 foot soldiers and artillerymen, while Lee committed an estimated 61,000. A combination of gunsmoke and the smoke of fires started by battlefield explosions created such thick clouds that some commanders, seeing only murky thickets and losing all sense of direction, had to move their units by looking at pocket compasses. Captain Horace Porter of Grant's staff described the inferno. "At times the wind howled through the tree-tops, mingling its moan with the groans of the dying, and heavy branches were cut off by the fire of artillery, and fell crashing upon the heads of the men, adding a new terror to battle. Forest fires raged; ammunition-trains exploded; the dead were roasted in the conflagration; the wounded, roused by its hot breath, dragged themselves along, with their torn and mangled limbs, in the mad energy of despair, to escape the ravages of the flames, and every bush seemed hung with shreds of blood-stained clothing." In an almost perfect metaphor of war, correspondent Charles A. Page of the *New York Tribune* described watching stretcher-bearers carrying wounded men to the rear and then seeing those stretcher-bearers rush back to the front, using the same stretchers to carry forward boxes of cartridges to maintain the supply of ammunition.

At the end of his first day fighting Lee, Grant threw himself down in despair on the cot in his tent: one of his staff, Captain Charles Francis Adams Jr., a member of the distinguished Massachusetts family that included two presidents, said, "I never saw a man so agitated in my life." Other accounts describe him as being composed, but in any case on the next day Grant was right back in the thick of battle. A Northern soldier described the kind of combat he experienced: "We fought them with bayonet as well as bullet. Up through the trees rolled dense clouds of battle smoke, circling about the pines and mingling with the flowering dogwoods. Each man fought on his own, grimly and desperately." The generals on both sides were right at the front. On the second day, Union General Alexander Hays of Pennsylvania was killed, as was Confederate General Micah Jenkins of South Carolina, and Union General James S. Wadsworth of New York was mortally wounded. In a nearly fatal repetition of what happened to Stonewall Jackson the year before, Confederate soldiers mistakenly fired at Longstreet, seriously wounding him in the throat and shoulder.

There were examples of how Lee, fighting Union forces that in this case outnumbered his men nearly two to one, could inspire the martial feats of the Army of Northern Virginia. Soon after sunrise on this same singularly bloody second day in the Wilderness, Lee found himself almost alone, mounted on his horse Traveller, as veteran Confederate regiments streamed past him, retreating in the face of a powerful federal assault that was about to capture a Confederate artillery battalion. The advancing blue lines were only two hundred yards away. Then, out of the drifting smoke through which Southern regiments were retreating, twenty men in ragged clothes ran forward with their muskets at the ready, entering the field at the end of a long forced march to reach the front.

"Who are you, my boys?" Lee shouted to these scarecrows, as scores more dashed up to form a line of battle.

"Texas boys!" they yelled. In a few more seconds, there were hundreds of them.

"Hurrah for Texas!" Lee stood in his stirrups and waved his hat. "Hurrah for Texas!" He rode to the left of the line, and the Texans realized that he intended to lead the counterattack, right at the blue lines.

"Go back, General Lee!" they shouted. "Go back! We won't go on unless you go back!"

"Texans always move them!" Lee roared, about to spur Traveller right into the enemy. It was only when the combination of a sergeant, a colonel of his staff, and Brigadier General John Gregg of the Texans closed in on him, the sergeant grabbing Traveller's reins and Gregg maneuvering his horse to block Traveller from plunging forward, that Lee was led back from the very front, still waving his hat and cheering on the Texans as they swept forward to save the Confederate artillery positions.

The men of the Army of the Potomac were learning about their new leader, Grant. Even before the horrible struggle in the Wilderness, with Grant ordering troops forward under conditions in which the Army of the Potomac's previous commanders would have stopped them where they were, a veteran soldier from Wisconsin saw him for the first time and said, "He looks as if he meant it." At one point on the morning of the terrible second day, when it appeared as if Lee's forces might overrun Grant's headquarters, an officer came up to Grant, who was standing on a knoll smoking a cigar as he studied the battlefield. "General," the man said, "wouldn't it be prudent to move headquarters to the other side of the Germanna road until the result of the present attack is known?" According to Horace Porter, "The general replied very quietly, between puffs of his cigar, 'It strikes me it would be better to order up some artillery and defend the present location.' "

That evening, Porter saw his chief in yet another revealing situation. At sundown, the fighting had seemed to come to an end. Then, with darkness, a roar of gunfire began, as Lee struck in a surprise attack. As Porter recalled, "Aides came galloping in from the right, laboring under intense excitement, talking wildly, and giving the most exaggerated reports of the engagement." One of the generals of the Army of the Potomac—Porter never did say who it was—appeared out of the night, "speaking rapidly and laboring under considerable excitement." Telling Grant that he knew from experience what Lee was going to do next, he said that Lee was about to put his entire army behind Grant's and cut them off from their communications and supplies. Grant "rose to his feet, took his cigar out of his mouth, turned to the officer, and replied, with a degree of animation he seldom manifested: 'Oh, I am heartily tired of hearing what Lee is going to do. Some of you always think he is about to turn a double somersault, and land in our rear and on both our flanks at the same time. Go back to your command, and try to think what we are going to do ourselves, instead of what Lee is going to do.' "

The following morning, it appeared that Lee's surprise attack had resulted in the capture of eight hundred Union soldiers, but this was a small part of the total picture: the losses on both sides during the two-day battle were frightful. Of the 100,000 troops that Grant had thrown into the Wilderness, 17,666 were either killed, wounded, or missing, while Lee, beginning with 60,000, lost about 11,000 of his gallant, outnumbered men. Except for some minor skirmishing on this third morning, the struggle in the Wilderness was over. Both sides had lost 18 percent of the men they sent into the battle, but it was easier for Grant, who smoked *twenty* cigars during the second day, to replenish his brave ranks. Of itself, the Wilderness was not an area worth fighting for; it was simply an obstacle between the federal troops and Richmond, an objective the Army of the Potomac had been fitfully trying to reach for three years. Grant's army now possessed the smoldering battlefield; Lee had withdrawn into defensive positions some distance away.

In the confused aftermath of the terrible carnage, the men of the Army of the Potomac thought they might well have been beaten: a soldier from Massachusetts recalled: "Most of us thought it was another Chancellorsville." Two correspondents, one from *The New York Herald* and one from the *Tribune*, certainly thought so, and that conviction was reinforced by the fact that Grant refused to let the journalists accompanying his army use the telegraph to send out their stories of the fighting.

For the Union troops, who had seen nearly one in five of their comrades become casualties in just two days, their past experiences with inadequate leadership led them to feel certain that, whatever the outcome of a specific battle, their sacrifices would probably come to naught. They would either be held in place, handing the initiative back to Lee, or they would be pulled back toward Washington—in this instance that would involve marching back across the Rapidan—to recuperate and reorganize. That is what their former commanders had done. McClellan, failing to exploit his advantages in the Peninsular Campaign of March to July of 1862, had been ordered by the then new general in chief Halleck to withdraw and reinforce Major General John Pope in the failed Second Bull Run campaign, also known as Second Manassas, which had ended with withdrawal into the defenses of Washington at the beginning of September; McClellan, once again commanding the Army of the Potomac after Pope's failure, failed to move forward against the withdrawing Lee after the terrible Battle of Antietam with its action at Sharpsburg on September 17, 1862, in which on the war's single bloodiest day Union forces had twelve thousand casualties in all categories, with the Confederates losing nearly fourteen thousand. When this failure to exploit Lee's retreat ended McClellan's career, he was relieved by Ambrose Burnside, whose Fredericksburg Campaign of November–December 1862 ended with the Battle of Fredericksburg on December 13, in which

Burnside lost 12,700 men while Lee lost 5,300. Joseph Hooker then commanded the Army of the Potomac from January 1863, being relieved in late June after Lee's brilliant victory over him at Chancellorsville at the beginnng of May, a defeat that resulted in yet one more withdrawal to the north, this one being across the Rapahannock on May 6.

With the appointment of General George Gordon Meade as its new commander only two days before the Battle of Gettysburg on July 3, 1863, the Army of the Potomac's fortunes had changed, but even after that great victory for the Union, Meade was so slow in trying to follow Lee's retreating Army of Northern Virginia that Lincoln had been greatly disappointed in his failure to exploit his success.

Thus, the frustrations of the long-suffering rank and file of the Army of the Potomac had begun in 1862: in one week of that summer, from June 25 through July 1, during the Seven Days' Battles of the Peninsular Campaign, Lee had beaten McClellan at Oak Grove, Mechanicsville, Gaines's Mill, White Oak Swamp, and Malvern Hill. Now, two years later, after the ghastly fighting in the Wilderness, on the afternoon of May 7 the orders came down to sling their packs and be ready to move out. The troops had no doubt that once again they had fought and bled on the soil of Virginia, only to march away from a battlefield and head back in the direction from which they had so often come. When the first column reached the crossroads where they would turn right to head back over the Rapidan, their officers on horseback turned to the left—south, toward the enemy, toward Lee, toward Richmond. Excited comments went up and down the line. Regiment after regiment turned left; there was no mistake. Grant was not giving up an inch; he was taking them south. "Our spirits rose," a soldier from Pennsylvania said. "We marched free, and men began to sing."

Dusk came; no one stopped. Everything was moving south: the artillery, the cavalry, the engineers who would build bridges, the ambulances, the wagons carrying food and supplies. Around nine in the evening, the word was passed down from the rear, "Give way to the right. Give way to the right." Something was coming down the road, heading south, heading to the very front, and must be let through.

Ulysses S. Grant came down the road on his big bay horse Cincinnati, accompanied by Meade and their staffs. Horace Porter described what happened when the men saw Grant coming.

Wild cheers echoed through the forest. Men swung their hats, tossed up their arms [muskets], and pressed forward to within touch of their chief, clapping their hands, and speaking to him with the familiarity of comrades. Pine-knots and leaves were set on fire, and lighted the scene with their weird flickering glare. The night march had become a triumphal procession.

Regimental bands, going along through the night carrying their instruments with them, brought them out and began to play spirited marches. Thousands of men sang. When Sherman later spoke of that night, he called Grant's decision to move on south the most important act of his life. Grant was going after Lee, whatever the cost, whatever lay ahead. Sherman was not there, but everything he learned confirmed what he knew about his friend's mind, heart, and methods: "Undismayed, with a full comprehension of the work in which he was engaged, feeling as keen a sympathy for his dead and wounded as anyone, and without stopping to count his numbers, he gave his orders calmly, specifically, and absolutely."

Grant was taking the Army of the Potomac to Spotsylvania Court House, nine miles southeast of his position in the Wilderness and forty-five miles northeast of Richmond, and into eleven more days of brutal fighting. Lee maneuvered his troops there just ahead of Grant, and on the morning of May 8 the two armies began blasting away at each other. Referring to the Wilderness and the continuing action at Spotsylvania collectively in a letter that he wrote Halleck at eight-thirty in the morning on May 11, Grant said, "We have now ended the sixth day of very hard fighting." With an optimism he may not have felt, he added, "The result up to this time is much in our favor." It was favorable in some ways—Grant was moving into Lee's territory, meeting the great Southern leader move by move, causing Confederate casualties Lee could ill afford—but he also had to report this to Halleck: "We have lost to this time eleven general officers killed, wounded, and missing, and probably twenty thousand men." (Lee had lost for the time being the severely wounded Longstreet, and his cavalry leader Jeb Stuart had been killed at Yellow Tavern.) In his way of combining the prosaic with the memorable, Grant finished that paragraph of his report with a statement of his commitment: "I am now sending back to Belle Plain all my wagons for a fresh supply of provisions and ammunition, and purpose to fight it out on this line if it takes all summer." Two days after Grant wrote to Halleck, the regimental surgeon of the 121st New York Infantry underscored the intensity of the continuing clash in a letter to his wife: "After eight days of the hardest fighting the world has witnessed, I . . . am still alive . . . The rebels fight like very devils! We have to fairly club them out of their rifle pits. We have taken thousands of prisoners and killed an army; still they fight as hard as ever."

11

SHERMAN SAVES LINCOLN'S
PRESIDENTIAL CAMPAIGN

As Grant pressed on into almost constant bloody action against Lee in northern Virginia during the summer of 1864, Sherman would determine the fate of more than the success or failure of his march into Georgia to reach Atlanta. Without at first realizing it, Sherman and his army had become a force that could not only fulfill Grant's strategy to destroy the Confederate military but also save Lincoln's political future and ensure the prospects for the kind of peace for which the war was being fought.

On June 8, the Republican Convention in Baltimore nominated Abraham Lincoln for a second term as president, with Andrew Johnson to run for vice president, but the North was sick of the war. Later in the summer the Democrats chose as their presidential candidate General George B. McClellan. Relieved of command by Lincoln in September of 1862 because of his delay in pursuing Lee after Antietam, McClellan had gone to his home in New Jersey, still a general but waiting for orders that never came. The Democrats, many of whom were eager to explore ways to end the fighting, saw the well-known and personable McClellan as a candidate who could not be faulted for lack of patriotism and could add legitimacy to the desire for peace. Even among Republicans, a split developed; as a concession to the defeatists among them, the party's executive committee went so far as to ask Lincoln to open communication with Jefferson Davis, to see if some kind of mutually acceptable terms for peace could be reached.

All this moved forward against a background of rising Union casualty figures. From the time Grant moved into the Wilderness on May 5 until he finished three unsuccessful days of throwing his army against Lee's defenses at Petersburg on June 18, he had lost sixty-five thousand men killed, wounded, and missing—nearly 10 percent of the entire Union Army at that time.

Among the terrible days were those of the battle at Cold Harbor, only twelve miles northeast of Richmond, an engagement Grant later admitted he should never have fought. Although Grant, Lincoln, and Halleck realized that the destruction of Lee's army was more important than the capture of Richmond, from June 1 to June 3, Grant made a series of attacks in an effort to break through Lee's defenses that were so near Richmond, capture the Confederate capital, and possibly end the war right there. On June 3, the last day of these attacks, seven thousand Union soldiers were killed or wounded in thirty minutes at Cold Harbor, and the three-day total of Northern casualties came to twelve thousand.

None of this bloodshed had produced a decisive victory—no Shiloh, no Gettysburg, no Vicksburg, no Chattanooga—and, despite Grant's expressed determination to finish the war in 1864, the people of the North could see no end in sight. President Lincoln's highly nervous wife, who had chatted pleasantly with Grant at the White House the night he arrived from the West to be hailed as the man who would save the Union cause, now frequently said of him, "He is a butcher." The Northern public started to distrust newspaper reports of what was happening to their soldiers in Virginia. The first bulletins from Petersburg were of a brilliant capture of the city. The next accounts said that Union troops were not in the city but were taking some of its outer defenses, suffering heavy losses as they did. Next came what proved to be the truth: after being thrown back repeatedly, Grant's army was digging in opposite those outer defenses and starting to besiege Petersburg. In an effort to breach the formidable Confederate earthworks, a regiment of coal miners from Pennsylvania was set to work digging a tunnel under a sector of the enemy trenches, and this was filled with four tons of gunpowder. Grant was skeptical about the scheme, but on the morning of July 30, the explosion was set off, killing 278 of the Confederate defenders and creating a crater 170 feet long, 70 feet wide, and 30 feet deep. Rather than opening an avenue through which Grant's forces could rush through Petersburg's defensive line, this gigantic hole proved to act as a trap—in the crater and the area surrounding it, fifteen thousand Union soldiers milled about ineffectually under enemy artillery and musket fire, and in eight hours four thousand of them were killed or wounded in the disaster that became known as the Battle of the Crater. After this fiasco, which Grant himself termed "a stupendous failure," there began to be calls for him to be relieved of command. One estimate was that he had lost ninety thousand men in a two-month period.

In fact, despite the severe losses, Grant was winning the war. (A month before Grant started into the Wilderness, Sherman wrote his brother that "Grant is as good a Leader as we can find . . . Let him alone.") Lee had fought so hard and well in an effort to throw back Grant toward Washington, precisely to stop him from getting to a strategic position such as the one where he now stood. It was ironic that Grant, who had maneuvered so boldly in all his Western campaigns, was to be thought of only as the bulldog that fought it out toe-to-toe with Lee. Nine days after his costly failure at Cold Harbor, he had taken an enormous gamble: slipping his army away from their entrenchments facing Lee at Cold Harbor in darkness during the night of June 12–13, he had 115,000 of his men moving rapidly to the southeast by the time Lee's picket line saw in the morning that the Union trenches were empty. The next night, going swiftly to the James River, risking destruction of some of his long columns that were vulnerably strung out on the march, he had an entire corps ferried across the river. By eleven o'clock the following night, his engineers had completed throwing across the James the longest pontoon bridge ever assembled, twenty-one hundred feet in length, and division after division poured across.

Only Lee could have recovered so quickly from this complete surprise. He countered rapidly enough to stop Grant, who was moving south along the east bank of the Appomattox River, from taking Petersburg, twenty-five miles south of Richmond. Nonetheless, Grant had gained a tremendous strategic advantage: he was facing and besieging Petersburg, with his headquarters at City Point on the James River, where his army could be supplied both by land and sea, the latter avenue secured by naval forces under Admiral Porter, his colleague from the Vicksburg campaign. Relentless military logic dictated what Grant could eventually accomplish: with superior forces, he could keep extending his lines to the west into Virginia south of Petersburg, forcing Lee to spread his lesser numbers thinner and thinner to avoid being outflanked, until Grant could break through somewhere. The public in the North, and many Northern politicians, saw only the casualty figures and not the strategy, but at seven a.m. on June 15, commenting on Grant's enormously resourceful move away from Cold Harbor less than seventy-two hours after he got it under way, Lincoln had sent him a message that said, "I begin to see it. You will succeed—God bless you all." At just the period when a storm of criticism began to surround Grant, Lee admitted privately that if Grant could get his army in position to besiege Petersburg—something he had now done—Confederate defeat would be "a mere question of time."

Lee's prediction, however, assumed that the North would maintain its will to win. From the war's outset, the stakes had been different for the South and the North. The Confederacy had never dreamed of marching into the more distant Northern cities, nor did it need to do so. From the beginning, the South had hoped to inflict such dramatic defeats upon the North, such bloody losses, that the Northern public would lose heart and cease to support its invasion of an area which, even cut into as it now was in 1864, was larger than any European nation. To win, all that the Confederacy needed was not to lose, and to be left to go its own way, a separate American nation with its institution of slavery intact. For the North to have a fully meaningful victory, federal troops had to destroy the Confederate armies, bring the South to its knees, and impose a peace founded on the concept of one nation, a nation in every part of which slavery was abolished. (Even now, after three years of war, there was no guarantee that a Union military victory would bring an end to slavery. Lincoln's Emancipation Proclamation, issued on January 1, 1863, after his preliminary proclamation in 1862, declared that all slaves in the states "in rebellion" were free, but this left many questions unanswered. Of itself, the document did not free slaves in Northern or border states, and Lincoln's authority to take action was based on his wartime powers rather than on an act of Congress. There was a widespread perception that a Union victory, bringing to an end the authority under which Lincoln acted, could render the freeing of the slaves invalid—a concern that would not be remedied until the Thirteenth Amendment to the Constitution, clearly abolishing slavery throughout the United States, was enacted on December 18, 1865, more than six months after the fighting ended.)

The Union's will to see its overall war aims through appeared to diminish or be undercut every day during the summer of 1864. In a news leak, the Northern public learned that Lincoln had authorized Horace

Greeley, the famous editor of the *New York Tribune* and a Republican who was an advocate of negotiations for peace, to meet with Confederate emissaries on the Canadian side of Niagara Falls. The meeting on July 18 had come to naught—the Southerners proved not to have the authority they claimed to have, and there was no indication that they would have accepted anything but the recognition of the Confederate States of America as a permanent sovereign nation, practicing slavery—but the fact that Lincoln had agreed to such a secret conference caused widespread worry and confusion.

Things got worse. Thurlow Weed, the editor of the *Albany Evening Journal* and the boss of New York State's powerful Republican political machine, came out in print with "Lincoln's re-election is an impossibility." Even Henry J. Raymond, founder of *The New York Times*, who was chairman of the Republican National Executive Committee, wrote Lincoln privately that "the tide is setting strongly against us." Raymond laid it right out for Lincoln: Grant's stalled drive in Virginia was a major and growing election issue. Fewer voters were ready to continue making such bloody sacrifices to ensure the emancipation of slaves, and more of them were ready to accept Southern secession as the price for peace. It was also true that in the coming election, the soldiers of the Union Army would be voting; nineteen Northern states had arranged for their troops to vote by absentee ballot, and the others expected their citizen-soldiers to receive furloughs to come home and vote. Would those hundreds of thousands of military voters endorse Lincoln and a national policy to fight on in a war in which increasing numbers of their comrades were being killed?

Even though Lincoln was the official Republican nominee and the election was scheduled for November 7, during August a number of prominent Republicans began planning to hold a political convention of their own in Cincinnati, to put forward what would in effect be a third-party candidate. On August 23, with the Democratic Convention in Chicago six days away, Lincoln wrote a remarkable memorandum that he intended to use if the Democrats won at the polls in November. He somehow got every member of his cabinet to sign the back side of the paper, without seeing his words, and commit themselves to act on what it said. Then he put it in a desk drawer. It read: "This morning, as for some days past, it seems exceedingly probable that this Administration will not be reelected. Then it will be my duty to so co-operate with the President elect, as to save the Union between the election and the inauguration . . ."

Lincoln wrote this memorandum to guarantee continuity and cooperation in the transition of administrations that would occur if he were defeated. As he did this, word was just reaching Washington that two days earlier the Confederate cavalry leader Nathan Bedford Forrest boldly attacked Memphis, which was supposed to be solidly in Union possession, and held it for the day; two Union generals fled just in time to escape capture. At the same time came news that, in an action south of Petersburg, federal troops had repelled a Confederate counterattack but in doing so lost 4,445 of the 20,000 Union soldiers engaged. Lincoln, Grant, and the Northern cause desperately needed some good news. *The New York World* asked, "Who shall revive the withered hopes that bloomed at the opening of Grant's campaign?"

Sherman, starting toward Atlanta from Chattanooga at the same time in May that Grant began fighting Lee in the Wilderness, found himself in a war of maneuver against his and Grant's old opponent, Joseph E. Johnston. Robert E. Lee's friend and West Point classmate, Johnston had been successfully aggressive at Bull Run in July of 1861; after he was wounded in 1862 at Seven Pines, Virginia, his command of Confederate forces in northern Virginia had been taken over by Lee. When he recovered, Johnston was sent west, where his actions during the Vicksburg campaign could be seen either as having been indecisive, contributing to Vicksburg's fall, or as realistic decisions and clever elusive movements that saved tens of thousands of Confederate soldiers for further service to the Southern cause.

During the time that Grant and Lee locked their armies in close-quarters fighting for eleven days at Spotsylvania, Sherman's forces outflanked Johnston at Resaca, Georgia. Although Grant and Sherman's favorite young general James B. McPherson had failed to push through Snake Creek Gap to Resaca in a move that might have destroyed most of Johnston's army, Sherman kept on outmaneuvering Johnston until the Union forces reached Allatoona Pass, only thirty miles from Atlanta. There Johnston's divisions consolidated, and the Southern defenses stiffened; by the end of May, the running war between Sherman and Johnston had cost each side about nine thousand casualties, with Confederate general John Bell Hood playing an increasingly important part in the unfolding campaign. Hood, a blond, sad-eyed, aggressive six-foot-two Kentuckian and West Pointer who had been wounded in prewar frontier fighting against the Indians, had first gained fame commanding the Texas Brigade. He was then badly wounded in the arm at Gettysburg and subsequently lost his right leg at Chickamauga. Strapped into the saddle to keep him from falling off his horse when he went into action, Hood was a fearless leader who disliked Johnston and thought he was far too cautious a general. (Lee's evaluation of Hood, expressed in a letter to Jefferson Davis, was, "Hood is a bold fighter . . . I am doubtful as to other qualities necessary.") Later in the campaign, when a worried Jefferson Davis sent General Braxton Bragg south from Richmond to talk with Johnston about his continued withdrawals toward Atlanta and to make his own observation of Sherman's march into Georgia, Hood wrote an out-of-channels letter to Bragg, subversively criticizing Johnston for being "so directly opposite" to his own view "that we should force the enemy to give us battle."

After more maneuvering against Johnston and two weeks of rain, on the morning of June 27 Sherman entered into a situation reminiscent of what happened to his forces at Missionary Ridge, although this time he moved forward on a far wider front. It was the worst battlefield mistake he made as an independent commander, and he did it at a moment when the Confederates had eighteen thousand men to oppose the sixteen thousand he threw into this action.

Attacking Johnston's formidable positions on Kennesaw Mountain, near Marietta, Georgia, eighteen miles northwest of Atlanta, Sherman mounted a major assault up the slopes and saw it fail and be thrown back. He ordered a second assault and saw it fail, and sent his forces up a third and final time, to be defeated again. One of the Confederate defenders described the scene: "A solid line of blue came up the hill. My pen is unable to describe the scene of carnage that ensued in the next two hours. Column after column of Federal soldiers were crowded upon that line. No sooner would a regiment mount our works than they were shot down or surrendered. Yet still they came . . . All that was necessary was to load and shoot. In fact, I will ever think that the only reason they did not capture our works was the impossibility of their living men to pass over the bodies of their dead men." In his memoirs, Sherman succinctly summed up the morning's action: "At all points the enemy met us with determined courage and great force . . . By 11:30 the assault was over, and had failed." Sherman lost 1,999 men killed and wounded, while the defenders had only 270 casualties.

That was Sherman's only attempt to win a head-on battle with Joseph E. Johnston. He kept up his war of maneuver, trying to get through to Atlanta by a number of routes; Johnston kept checking him. Despite demands from Jefferson Davis and John Bell Hood that Johnston stand his ground and produce another Kennesaw Mountain victory, there was no terrain quite like the slopes of Kennesaw Mountain near Atlanta, and Johnston managed to fight a remarkable series of delaying actions. A man of Sherman's 104th Illinois, frustrated by seeing the smoke of the factories of Atlanta in the distance day after day, shouted across to an enemy outpost, "Hello, Johnny, how far is it to Atlanta?" and received the answer, "So damn far you'll never get there!"

As Johnston continued his defensive maneuvers, he may also have been thinking in terms of more than military delay. There are questions as to how aware of Lincoln's political crisis he was at that moment, but Johnston later said that his goal was to keep Sherman from taking Atlanta before November and thus help the Democrats "to carry the presidential election . . . [which] would have brought the war to an immediate close."

In the midst of this, on July 17, Jefferson Davis replaced Johnston with John Bell Hood. The idea was to have the offensive-minded Hood drive Sherman away from Atlanta. Johnston took the news calmly, leaving his headquarters for his home in Macon within twenty-four hours. During this period, Hood, having received the command for which he lobbied, went through what seemed to be the charade of asking Jefferson Davis to suspend the order, on the grounds that it was not good to change commanders at just that moment. As for Grant, years later he said to Julia, "My satisfaction at Hood's being placed in command was this . . . [Johnston] was a most careful, brave, wise soldier. But Hood would dash out and fight every time we raised a flag before him, and that was just what we wanted." (Union intelligence about this Confederate change of command was good: Sherman read of it the same day in a copy of an Atlanta newspaper slipped out of the city by one of Grenville Dodge's spies. After conferring with his division commander John M. Schofield, who had known Hood at West Point, Sherman readied his forces in conformance with Schofield's opinion that Hood would attack within forty-eight hours.) On the next day, Hood attacked in what became known as the Battle of Peach Tree Creek, and Sherman's prepared counterattack threw Hood's men back so violently that they retreated to the earthworks that had been prepared for a final defense of Atlanta.

Four days later, Hood attacked the Union corps positioned east of the city and was defeated again. Midway through the battle, however, the commander of the federal troops engaged, Major General James B. McPherson, the man among the younger Union generals for whom both Grant and Sherman had the most admiration and affection, was killed. When McPherson's body was brought to Sherman's headquarters while the battle went on, tears ran down Sherman's face, and he wept as he continued to receive reports and issue orders. That night, speaking to an aide, Sherman said, "I expected something to happen to Grant and me, either the Rebels or the newspapers would kill us both, and I looked to McPherson as the man to follow us and finish the war."

The news had an equally strong effect on Grant, who earlier in the war brought McPherson onto his staff and later recommended him for his series of promotions. A man who saw Grant learn of McPherson's death described his reaction: "His mouth twitched and his eyes shut . . . Then the tears came and one followed the other down his bronzed cheeks as he sat there without a word of comment." Grant wrote McPherson's eighty-seven-year-old grandmother, telling her that "the nation had more to expect from him than almost any one living . . . He formed for some time one of my military family. I knew him well. To know him was but to love him . . . Your bereavement is great, but cannot exceed mine."

During this bloody summer of what Sherman had rightly said would be "the hardest year of the war," Grant and Sherman stayed in close contact, despite the enormous challenges each faced in his day-to-day operations. When in early August Sherman sent Grant a report in which he sounded apologetic for his inability to cut right through to Atlanta, Grant replied with a telegram in cipher that said, "Your progress instead of appearing slow has received the universal commendation of all loyal citizens as well as of the President, War Dept, and all persons you would care for." Sherman answered this on the same day with a telegram that began with, "I was gratified to learn you were satisfied by my progress," and, in language that stressed the concept of their being a team, continued, "Let us give these southern fellows all the fighting they want and when they are tired we can tell them we are just warming up to the work[.] Any sign of a let up on our part is sure to be falsely construed and for that reason I always remind them that the siege of Troy lasted six years and Atlanta is a more valuable town than Troy."

At a time when he was in almost daily battle with Lee, Grant never forgot that Sherman also needed reinforcements. On August 10, he sent a coded telegram to Halleck in Washington, saying in part, "We must try and get ten thousand replacements to Sherman by some means . . . I would like to hear of 1000 a day going." Three days later, Grant wired Halleck, "Is [sic] there any recruits from the Western states going to Sherman?" Referring to troops en route to General John Pope, in the West, he added, "All the troops that Pope can relieve by this increase I want sent to Sherman." During these same days, Grant wired this recommendation to Stanton: "I think it but a just reward for services already rendered that Gen. Sherman be now appointed a Maj. Gen. in the Regular Army." Stanton wired back within hours, saying that the promotion "will be immediately made."

In another development with implications not only for Sherman but also the entire Union cause, Halleck informed Grant that the mustering out of more than sixty regiments whose three-year enlistments had expired might require removing tens of thousands of Grant's troops from the front. They would be needed to replace the newly discharged men in duties such as guarding prison camps and manning key garrisons. There was, however, something more, laden with explosive political problems. The men who were being released would be replaced by conscription, and Halleck explained: "The draft must be enforced, for otherwise the army cannot be kept up. But to enforce it may require the withdrawal of a very considerable number of troops from the field."

Halleck went on to sketch out the possibility of numerous draft riots such as those that had caused a thousand deaths the year before in New York City when, running out of volunteers, the Union had begun conscription: "The people in many parts of the north and west now talk openly and boldly of resisting the draft, and it is believed that the leaders of this 'Peace' branch of the Democratic Party are doing all in their power to bring about this result . . . It is thought the attempt will be made. Are not the appearances such that we ought to take in sail and prepare the ship for a storm?"

This was the biggest challenge to the Union cause off the battlefield that Grant had encountered. As the Union's supreme military commander, he was being asked to consider withdrawing so many men from the front, to enforce the essential induction of more soldiers from civilian life, that he might well have to give up his bitterly won strategic position in front of Petersburg. More than that was at stake: Grant realized that if he had to abandon many of the gains he had made since May, Lee would be able to send major reinforcements south to defend Atlanta. Grant replied to Halleck that "My withdrawel [sic] now from the James River would insure the defeat of Sherman," and he urged Halleck to have Lincoln ask the Northern governors to mobilize their state militias to keep the peace and "deter the discontented from commiting any over act." In this, Grant quickly found his most powerful ally: Lincoln was shown Grant's reply to Halleck, and wired, "I have seen your dispatch expressing your unwillingness to break your hold where you are. Neither am I willing. Hold on with a bulldog grip, and chew and choke as much as possible."

In the midst of the gathering political storm, on September 3 Sherman sent Halleck these electrifying words: "Atlanta is ours, and fairly won." Hood had miscalculated. First he had sent out his cavalry under General Joseph Wheeler in an effort to cut Sherman's supply line to the north, an effort that came to naught; the Confederate troopers had inflicted little damage and had ridden on into east Tennessee, thus depriving Hood of the force on which he depended for knowledge of Sherman's movements. Thus Hood missed the fact that Sherman next boldly marched away from his base of supplies near the Chattahoochie River, which ran on a north-south axis west of the city, and, letting his men feed themselves on the march by roasting the ears of corn that had now ripened along the way, quietly moved almost his entire army from a point west of Atlanta to its more vulnerable southeast side. When spies told Hood that at various places they had seen Sherman's men marching along without supplies, he concluded that Sherman was breaking off the siege, rather than moving to attack, and telegraphed Richmond that this was a "great victory." Hood finally realized that Sherman's forces had much of the railway mileage around Atlanta in their hands and might soon surround it completely, and evacuated his forces—he had approximately forty thousand men in or south of the city—detonating powder magazines and burning warehouses filled with supplies as he withdrew.

Sherman, having climaxed a successful campaign with the capture of the Confederacy's second most important city after Richmond, in effect let Hood go without further bloodshed—"I do not wish to waste lives by an assault," he told his subordinate General O. O. Howard—and Sherman's columns marched in.

Triumphant joy swept the North. Grant ordered every one of his artillery batteries on the Petersburg front to prepare to fire a salute in honor of the victory (using real cannonballs aimed at the enemy), and wired Sherman that the guns would go off "amid great rejoicing." Grant urged Sherman to start another campaign soon: "We want to keep the enemy continually pressed to the end of the war." When Grant had a little more time a few days later, he wrote Sherman that "you have accomplished the most gigantic undertak[ing] of any General in this War and with a skill and ability that will be acknowledged in history as unsurpassed if not unequalled."

It had been a remarkable performance. Sherman had outmaneuvered first Joseph E. Johnston and then John Bell Hood, and inflicted casualties upon their forces across a large area of northern Georgia. In the process he had tied down tens of thousands of Confederate soldiers who could otherwise have been sent north to aid Lee in his pitched battles against Grant, and had captured the South's second most valuable city. Fully as important, his timely march into Atlanta muffled the cries for peace throughout the North and saved the election for Lincoln. The Democratic candidate McClellan promptly repudiated the plank of his party's platform that called the war a failure, and the prominent Republicans who had been planning to hold a rump convention in Cincinnati to produce a candidate other than Lincoln abandoned that idea. Demands that Grant be removed ceased.

Despite Grant's desire to keep driving on and end the war in 1864, both Sherman's forces and Grant's Army of the Potomac needed some rest, as did the commanders themselves. Grant, whose health had been deteriorating and who had been suffering from migraine headaches, was restored by visits from Julia and the children to his headquarters on a bluff at City Point, Virginia, on the west bank of the James River where it is joined by the Appomattox River, nine miles northeast of besieged Petersburg and twenty-three miles southeast of Richmond. One morning, Horace Porter of Grant's staff stepped into Grant's tent and saw Grant playing with his children.

I found him in his shirt-sleeves engaged in a rough-and-tumble wrestling match with the two older boys . . . The lads had just tripped him up, and he was on his knees grappling with the youngsters, and joining in their merry laughter, as if he were a boy again himself. I had several despatches in my hand, and when he saw that I had come on business, he disentangled himself with some difficulty from the young combatants, rose to his feet, brushed the dust off his knees with one hand, and said in a sort of apologetic manner: "Ah, you know my weaknesses—my children and my horses."

Speaking of the relationship between Ulysses and Julia, he said that, in the log cabin in which Grant and his family lived, "They would seek a quiet corner of his quarters of an evening, manifesting the most ardent devotion; and if a staff-officer accidentally came upon them, they would look as bashful as two young lovers spied upon in the scenes of their courtship." On an evening earlier in the year, when Grant had asked Julia to come to City Point and she had been there without their children, a Confederate ironclad broke through on the river and was expected to begin firing at City Point. Although nothing finally came of the threat, no one could have known it then, and Porter observed: "Mrs. Grant, who was one of the most composed present, now drew her chair a little closer to the general, and with her mild voice inquired, 'Ulys, what had I better do?' The general looked at her for a moment, and then replied in a half-serious and half-teasing way, 'Well, the fact is, Julia, you oughtn't to be here.' " (Julia later wrote in mock indignation, "And he had sent for me, mind you!")

As for the Shermans, on June 11, in the middle of the Atlanta campaign, Ellen had given birth to a son, Charles, and had been sick for the next two months. The baby also was not well. After the death of her mother earlier in the year, Ellen had tried to keep house for her ailing father in Lancaster, Ohio, but had recently found the combination of responsibilities too stressful. She decided to move herself and her children to South Bend, Indiana, which she had chosen both because it had Catholic schools into which she was putting the three older children and because of the good medical care it offered her and her baby. Ellen appears to have done this with a minimum of consultation with Sherman, who had always considered her tied to her parents and to Lancaster, but he consented to the move and was perhaps a bit bemused that after all these years she would leave the place and family that he had always found to be such powerful rivals for her affections. In a letter he wrote her just after taking Atlanta, he explained that "there is no chance of my getting north again and therefore you can choose a house utterly regardless of my movements."

The letters that Sherman and Grant sent their wives throughout the war demonstrated the great respect they had for their intelligence. Sherman, tending to be far more verbose although not more profound, went much further than did Grant in discussing with Ellen the war's military and political aspects. All through the summer's campaign he had written her long letters that included clear accounts of his movements. Usually he went into some detail, but at one point he summed up the situation this way: "We have Atlanta close aboard as the Sailors say but it is a hard nut to handle. These fellows fight like Devils & Indians combined, and it calls for all my cunning & Strength." In another letter, asking Ellen to help the children realize that at the moment he could not answer all their letters individually, he referred both to his army and the masses of both black and white refugees he was feeding when he wrote, "They must understand my present family is numbered by hundreds of thousands all of whom look to me to provide for their wants."

At this point, the outlines of Grant and Sherman's grand strategy were in some ways visible, but much remained to be seen. Sherman had indeed penetrated the enemy's heartland, and Grant was bloodily engaging Lee on the Confederacy's northern front in Virginia. Within the scope of the objectives still before them, Grant's task was enormous but clear: defeat Lee's Army of Northern Virginia. When Grant had campaigned in the West, he had the opportunity to exploit those wider areas, which frequently offered him choices of where and how to maneuver his troops. He no longer had those advantages; Lee stood firmly before him in a smaller fixed area.

Sherman certainly had a goal—to tear up the South and defeat Joseph E. Johnston—but he still had the strategic luxury of deciding where to go and when to make his moves. Nonetheless, it was time to make decisions of the greatest importance. Sherman, in concert with Grant, had to decide exactly what to do next with this force of his, which had proved to be such a potent and flexible weapon.

To plan this next phase of the war, Grant sent his aide Horace Porter south, carrying a letter concerning the strategic options and to get Sherman's ideas as to what should come now. Porter had never seen Sherman and wrote his impression of their first meeting at Sherman's headquarters in Atlanta.

> He was just forty-four years of age, and almost at the summit of his military fame. With his large frame, tall, gaunt form, restless hazel eyes, aquiline nose, bronzed face, and crisp beard, he looked the picture of "grim-visaged war" . . . I approached him, introduced myself, and handed him General Grant's letter. He tilted forward in his chair, crumpled the newspaper in his left hand while with his right he shook hands cordially, then pushed a chair forward and asked me to sit down . . .
>
> He exhibited a strong individuality in every movement, and there was a peculiar manner of energy in uttering the crisp words and epigrammmatic phrases which fell from his lips as rapidly as shots from a magazine-gun . . . He said, "I knew Grant would make the fur fly when he started down through Virginia. Wherever he is the enemy will never find any trouble about starting up a fight. He has all the tenacity of a Scotch terrier. That he will accomplish his whole purpose I have never had any doubt."

12

PROFESSIONAL JUDGMENT AND PERSONAL FRIENDSHIP: SAVANNAH FOR CHRISTMAS

After Sherman's capture of Atlanta in early September of 1864, the remainder of the autumn brought about the supreme test for Grant and Sherman's personal and military relationship. Four months before, Sherman had received Grant's approval for his bold campaign that had moved through Georgia for a hundred miles and resulted in his taking Atlanta. Now, in a letter that Horace Porter carried back to Grant, he sent word that he wanted to march on to the southeast from Atlanta, cut through Georgia for 225 more miles, and capture the great coastal port of Savannah.

Sherman initially presented his plan in a confident, high-hearted way. In his letter of September 20 carried by Porter, he closed a long description of his proposed campaign with these words. "I admire your dogged perseverance and pluck more than ever. If you can whip Lee and I can march to the Atlantic I think Uncle Abe will give us a twenty days' leave to see the young folks."

Despite the recent superb performance of Sherman and his army, Grant was doubtful that this would be the best use of Sherman's forces, and Lincoln and Stanton were even more skeptical about the idea. As Grant and Sherman discussed the strategic situation in the South in a series of letters and telegrams in late September, Grant first proposed that Sherman move south to Mobile and crush the remaining Confederate strength on the Gulf Coast. Sherman soon persuaded him that, as a campaign in itself, the march to Savannah would be feasible, but Grant was worried about what Sherman would be leaving in his rear when he did that. If Sherman headed for the Atlantic Ocean, he would be marching away from John Bell Hood's Army of Tennessee, which would then be opposed only by the Union forces under General George Thomas. Grant feared that Hood, who had been beaten by Sherman, would in his turn be able to defeat Thomas. If that happened, no matter how much progress Sherman was making as he went in the opposite direction southeast of Atlanta, Hood could march his army north into areas that had for some time been under Union control. Hood could move up through Tennessee and Kentucky, and might even reach Cincinnati on the Ohio River. Leaving Hood's army intact was a terrible risk, and one that need not be taken. With this frightening prospect in mind, Grant told Sherman that he could strike out for Savannah, but only after destroying Hood's forces.

Relying on Grant's willingness to hear something more about all this, Sherman argued that Thomas was equal to any threat from Hood, and then he went beyond advocating the purely military aspects of his proposed march to the coast. This, he told Grant, was the chance to break the South's will, its thus far remarkable fighting spirit. If he could march from Atlanta right to the sea, this demonstrated ability to move through the heart of the South on a path of the Union Army's choosing would show everyone, North and South, that night was descending upon the Confederacy. Sherman wanted to convince every adult white Southerner that continuing to fight for the cause of secession would result in personal catastrophe and ruin. "Even without a battle," Sherman now wrote Grant of the dramatic march he wanted to undertake, "the result operating on the minds of sensible men would produce fruits more than compensating for the expense, trouble, and risk." In another letter to Grant, he unveiled his concept of waging war upon everything in his path, the countryside itself, in a harsher fashion than he had been able to do on the way to Atlanta, when he had to face Johnston's troops at every turn. Speaking of Georgia, he said that "the utter destruction of its roads, houses and people will cripple their military resources." He wanted to move ahead and keep going, letting his men and horses live off the land through which they passed, without worrying about what might happen if he had to guard supply lines to his rear: "By attempting to hold the roads, we will lose a thousand men monthly and gain no result." He added this chilling reassurance to Grant: "I can make the march and make Georgia howl."

In his headquarters at City Point, Grant considered all of this, balancing his confidence in Sherman against his responsibility to avoid a disaster that could change the entire course of the war. Fighting Lee in Virginia was supremely hard, taxing the Union's strength and resolve to its utmost. If Hood should get loose, bring his forces north on the inland side of the Appalachian Mountains, and open a new front well to the west of Lee's Army of Northern Virginia, no one could foretell the calamities that might visit upon the Union.

Halfheartedly at first, Grant began to make a great act of faith in his friend Sherman. He started in early October by writing him, "If there is any way of [your] getting at Hood's army, I would prefer that, but I must trust to your own judgement." A few days later, he added, "On reflection, I think better of your proposition." Sherman realized that he still did not have the kind of support from Grant that he needed. He knew that only Grant could convince Lincoln and Stanton to agree with this hazardous strategy—on October 13, Stanton wired Grant that Lincoln was worried that "a misstep now by General Sherman might be fatal to his army"—and he sensed that Grant was still not ready to approve his plan. On November 1, with Hood already moving up toward Chattanooga to confront the Union army under George Thomas, Grant worriedly wrote Sherman, "Do you not think it would be advisable now that Hood has gone so far north, to entirely settle him before starting on your proposed campaign? With Hood's army destroyed, you can go where you please with impunity."

Sherman responded to this on the same day with two telegrams. In the first, he assured Grant that Thomas would be able to stop Hood before he could do any significant damage. In the second, he told Grant that "if I turn back, the whole effect of my campaign will be lost . . . I am clearly of [the] opinion that the best results will follow my contemplated movement through Georgia."

Grant was reluctantly persuaded. Within hours, he gave his approval: "I do not really see that you can withdraw from where you are to follow Hood, without giving up all that we have gained in territory. I say, then, go on as you propose."

The two friends had disagreed and set forth their positions. Grant, Sherman's superior and a man capable of saying no to anything, had decided that Sherman had made his case and agreed to let him go forward, even though Lincoln and Stanton remained doubtful about the movement. Grant knew that the stakes were huge but acted in accordance with his conviction that if a thing was worth doing, it was worth doing wholeheartedly. Five days after giving his approval, when Sherman had his forces ready to head out of Atlanta toward Savannah, Grant wrote him, "Great good fortune attend you. I believe you will be eminently successful, and, at worst, can only make a march less fruitful of results than hoped for."

Sherman's preparations for leaving Atlanta indicated that this march would be unlike anything seen before. He cut his own telegraph lines to the North, as well as the railroad links. For a month, no one, not Grant, not Halleck, not Stanton, was going to be able to find him. Sherman was moving out with sixty-two thousand men, to advance in four huge columns, on a front sixty miles wide; the Confederates in the path of this advance, most of them in understrength cavalry units, were not going to know just where this behemoth was going, let alone be able to stop it. This army was taking a twenty-day supply of food, including three thousand beef cattle they herded along, but as Sherman's columns cut their wide swath through Georgia, they would have no supply lines behind them; in language that profoundly understated the harsh reality to come, Sherman's orders were that "the army will forage liberally on the country during the march."

The night before the Union troops marched out of Atlanta, much of which had earlier been laid waste by the withdrawing Confederates, Sherman ordered the commercial and manufacturing sections of the city to be burned. When he rode from the city at seven o'clock on the morning of November 16, he looked back and saw the results of his orders: "Behind us lay Atlanta, smouldering and in ruins, the black smoke rising high in the air, and hanging like a pall over the city." As for his army, he remembered "the white-topped wagons stretching away to the south" and the troops with their "gun-barrels glistening in the sun . . . marching steadily and rapidly, with a cheery look and swinging pace . . . Some band had, by accident, struck up the anthem of 'John Brown's soul goes marching on'; the men caught up the strain, and never before or since have I heard the chorus of 'Glory, glory, hallelujah!' done with more spirit."

In a diary entry, a sergeant from Iowa captured the esprit de corps of Sherman's men, many of whom had been fighting for more than two years under the man most of them now called "Uncle Billy": "Started this morning early for the Southern coast, somewhere, and we don't care, as long as Sherman is leading us." Other men were less confident. Captain Orlando Poe of Sherman's staff, an engineer officer who was teaching the troops how to tear up Confederate rail lines, looked at this army as it headed into thousands of square miles of the enemy heartland, hoping to reach the coast, and wrote his wife that "this may be the last letter that you ever get from me." As for Sherman's own frame of mind, he felt that he and Grant were working in complementary fashion, toward a common end. He was in command of the largest force acting as light infantry the world had seen, an enormous flying column with which he intended to destroy both the enemy's rear area and its will to fight, while Grant, 450 miles to the northeast, continued to bleed Lee's Army of Northern Virginia to death in a form of trench warfare at Petersburg. (Lincoln put it this way: "Grant has the bear by the hind leg while Sherman takes off its hide.") Adding to the pressure being put on Confederate military resources in Grant's overall Northern theater of operations, Grant's cavalry chief Philip Sheridan had defeated Confederate general Jubal Early's outnumbered forces in Virginia's Shenandoah Valley.

For thirty-one days, no one in Washington knew just where Sherman and his army were or how they were faring. When Sherman's brother Senator John Sherman saw Lincoln one day, he asked if there had been any communications from his brother in Georgia. The recently reelected president answered, "Oh, no, we have heard nothing from him. We know what hole he went in, but we don't know what hole he will come out of."

If Grant was angry about this lack of information, there is no record of it. When Lincoln told Grant that he was concerned about what might be happening to Sherman and his army, Grant answered that he was confident that Sherman would reappear "on Salt Water some place." Grant's biggest worry was the one he had discussed with Sherman weeks before; as he had expected, Hood was marching his Confederate columns north into Tennessee, and it remained to be seen whether George Thomas could stop him from going on up through Kentucky to Cincinnati. At first, it seemed that Grant had been all too right and should have insisted that Sherman "settle" Hood's forces before heading from Atlanta to the coast. For many weeks, Thomas repeatedly delayed executing Grant's orders to attack Hood promptly, citing such reasons as bad weather, which finally brought him a pointed response in a telegram from Grant sent on December 11: "If you delay attack longer the mortifying spectacle will be witnessed of a Rebel army moving for the Ohio River and you will be forced to act, accepting such weather as you find. Let there be no further delay." Four days later, Thomas attacked Hood's twenty-three thousand men at Nashville with his own force of forty-nine thousand in a two-day battle, and, as Sherman had predicted, decisively defeated Hood's army and removed the threat to Tennessee and Kentucky.

On his march, Sherman set the astonishing initial goal of moving his sixty-two thousand men fifteen miles a day and kept to that for a week. No ordinary men could have done this. A soldier from Illinois wrote that his comrades "had been in the service from the beginning and what they did not know about campaigning was not worth inquiring into. Each soldier was practically a picked man. Such was the ratio of casualties that he may be said to be the sole survivor of four men who had set out from Cairo [Illinois] in 1861; all but he having succumbed to disease or death." Sherman's aide Major Henry Hitchcock expanded on this theme of confident pride: "It is a magnificent army of *veterans*, brimful of spirit and deviltry, literally 'spoiling for a fight,' neither knowing nor caring where they are going, blindly devoted to . . . the 'old man[,]' in splendid condition, weeded of all sick, etc., and every man fully understanding that there is no return, no safety but in fighting through."

The "old man" watched over his army like a nervous mother hen, moving around to check his different units at night and "prowling around a camp fire in red flannel drawers and a worn dressing gown." He was also seen, like his men, swimming naked in a river to get himself clean, and on the march he sometimes hiked along beside the enlisted men, talking with them as equals. A major new to Sherman's command described him:

> General Sherman is the most American looking man I ever saw, tall and lank, not very erect, with hair like thatch, which he rubs up with his hands, a rusty beard trimmed close, a wrinkled face, sharp prominent red nose, small bright eyes, coarse red hands; a black felt hat slouched over his eyes . . . field officer[']s coat with high collar and no shoulder stripes, muddy trousers and one spur. He carries his hands in his pockets, is very awkward in his gait and motions, talks continually and with rapidity.

As the army advanced, the men acting as foragers quickly established themselves as an odd elite. Each morning some thirty or forty men from each brigade set out, often on captured horses or mules and frequently using captured carts, some men moving ahead of their massive column and others moving along the flanks. Known as "bummers," their job was to pass through the countryside, taking anything useful that they found on farms or plantations—corn for men and horses, vegetables, livestock—and bring it to the roads on which the main forces were passing, where the regular supply wagons would take charge of what they had stripped from the land. Their skills at finding useful things impressed the blacks on the farms and plantations: one just-freed slave said, "Yankee soldiers have noses like hounds. Massa hid his horses way out dar in de swamp. Some soldiers come along. All at once dey held up dere noses and sniffed and sniffed, and stopped still and sniffed, and turned into de swamp and held up dere noses and sniffed, and . . . went right straight to where de horses was tied in de swamp."

A number of foragers frequently acted for their own profit, sometimes harshly, doing things such as choking an aged plantation owner until he told the soldiers where the family's silver dinner service was concealed. In addition to their foraging, the bummers acted as scouts, directing larger units forward to attack Confederate patrols they had spotted. Occasionally they were able to team up among themselves and rout small parties of enemy horsemen. Both the bummers and the marching rank and file picked up various animals as pets and brought them along: in addition to dogs and cats, there were lambs, raccoons, and hundreds of gamecocks, the last pitted against one another in nightly cockfights.

Other men, including freed slaves who were being paid for their labor, pried lengths of rails loose from railroad ties, heated them in the middle until they were orange in color and soft enough to be twisted, and left them wrapped around trees; these became known as "Sherman Neckties." Occasionally these workers bent the rails into the letters "U" and "S" and placed the "U S" on a hillside for the Southern populace to contemplate. They also became so skilled at rebuilding destroyed bridges and clearing enemy obstructions on the roads that when it became necessary to open closed tunnels, they did it so swiftly that the Confederates began to say that the Yankees had brought their own spare tunnels with them. As for the impression Sherman's advancing columns made on the slaves who became free as they passed, one of them joyously shouted, "Dar's millions of 'em—millions! Is dere anybody left up north?"

The original orders were to restrict the foraging to supplies needed by the army and to avoid entering Southern homes, but a student of the campaign observed that "the distinction between forage and pillage is easily obscured." In addition to the bummers, rank-and-file soldiers began entering the houses of Southern civilians and stealing whatever objects struck their fancy. In one town, a Union officer saw "soldiers emerging from doorways and backyards, bearing quilts, plates, poultry, and pigs." This kind of looting led to confrontations with enraged Southern women, most of whom equaled and often surpassed the most ardent Confederate soldiers in their detestation of the Northerners coming into their neighborhoods. Few among the Union soldiers, to whom one long red road through Georgia looked like another, comprehended the sense of emotional violation, existing quite apart from the issues of secession and slavery, felt by Southerners who saw only an invasion of their land. A man from Iowa was met by a Georgia woman on the porch of her house, and she launched into him: "My husband is a captain in the Confederate army and I'm proud of it. You can rob us, you can take everything we have. I can live on pine straw the rest of my days. You can kill us, but you can't conquer us."

Some of these encounters, including situations in which Union soldiers were not engaged in theft, turned into interesting debates. A major from Illinois found an old woman, the mistress of a plantation, lecturing him that the Northern policy of freeing the slaves would lead to what she called "Amalgamation"—racially mixed children. "The old lady forced it on me," he recalled, "and as there were three or four very light colored mulatto children running around the house, they furnished me an admirable weapon—She didn't explain to my entire satisfaction how her slaves came to be so much whiter than African Slaves are usually supposed to be." When Southern women stared disdainfully at them, one of Sherman's soldiers wrote, "The boys would stir up the female Rebels, just to hear them talk, like the boys at the menagerie stir up the lions just to hear them roar."

Other Union soldiers had more amiable experiences. Brief as some of these meetings with Southern girls were, they made an impression. On the same day, a captain from Ohio met a Miss Glenn, who he noted in his diary was "well dressed polite and agreeable . . . pretty foot and ankle, beautiful complexion," and later encountered two sisters, "one talkative, rebellious but sensible in every other way, both good looking and one finely developed bust, luscious."

It was in Milledgeville, then the capital of Georgia, that things became uglier. Reaching there on the ninth day of their march, the troops saw for the first time some Union soldiers who had escaped from the Confederate prison camp at Andersonville. Starved and sick, with what a colonel from Indiana described as a "wild-animal stare" as they spoke, these living skeletons told tales of their mistreatment that quickly spread through Sherman's ranks. When a Southern woman walked up to a federal soldier on the street in Milledgeville and spat on him, he and his comrades did not touch her but burnt down her house. At the same time, the men became aware of an order from Jefferson Davis to all Confederate officers in Georgia, exhorting them to make "every effort" to obstruct the Union advance, these measures to include "planting sub-terra shells [land mines]." (Sherman's response to this was, when his men suspected land mines had been laid in front of them, to have Confederate prisoners take the risks of digging them up.) On a lighter note, a group of troops spontaneously conducted a mock session in the state's legislative chamber in Milledgeville, voting Georgia out of the Confederacy and back into the Union, and named a committee to punish Jefferson Davis, if he were captured, by kicking him repeatedly from behind.

The army that left Milledgeville was required to move only ten, instead of fifteen, miles a day. One reason for this was the intensification of the manner in which the countryside was being laid waste. Foragers who had begun by rounding up chickens and pigs now decided that wrecking farm equipment and burning barns was in keeping with the idea of destroying the South's ability to raise food. Although many houses were left standing, the next step went from torching a farmer's barn to setting fire to his house, and a lot of bummers took the added time to do that. The headquarters companies of Sherman's major units had brought with them flares that could be shot aloft at night, so that each of the four columns would know where the others were. This was no longer necessary: the location of each advancing corps could be seen by the flames along its route.

There were exceedingly few cases of rape, murder, or beating of civilians, but the original standards of behavior for the march largely vanished. Sherman later wrote: "I know that in the beginning, I too had the old West Point notion that pillage was a capital crime, and punished it by shooting." In that view, confiscating crops and all kinds of food, as well as animals and equipment, was acceptable as long as it was for the good of the army as a whole, but a man was severely punished for stealing for his own profit. As the campaign progressed, this distinction vanished, and Sherman said that he and his officers "ceased to quarrel with our own men about such minor things, leaving minor depredations to be charged up to the rebels who had forced us into the war, and deserved all they got and *more*."

The troops became particularly aggressive when they came to the handsome houses of those who were both slaveholders and the owners of objects they might steal. The mistress of a plantation described the scene as Union cavalrymen entered her house and plundered it. "It is impossible to imagine the horrible uproar and stampede through the house, all of them yelling, cursing, quarreling, and running from one room to another in wild confusion. Such was their blasphemous language, their horrible countenances and appearance . . . their mouths filled with curses and bitterness and lies."

The thousands of freed blacks, most of them determined to stay right with the Union troops they hailed as their liberators, added liveliness and confusion to the daily scenes of the march. The black men walked beside the troops, happy to carry their muskets. At night they cooked spicy dishes and danced around the campfires. Many black girls gave themselves freely to the young troops, and one man noted that "I have seen officers themselves very attentive to the wants of pretty octoroon girls, and provide them with horses to ride."

It was not all levity and licentiousness. An officer from Indiana wrote his wife:

It was very touching to see the vast number of colored women following us with babies in their arms, and little ones . . . clinging to their tattered skirts. One poor creature, while nobody was looking, hid two boys, five years old, in a wagon, intending, I suppose, that they should see the land of freedom if she couldn't. Babies tumbled from the back of mules to which they had been told to cling, and were drowned in the swamps, while mothers stood by the roadside crying for their lost children and doubting whether to continue with the advancing army.

Ironically, Sherman, who was being hailed by the freed blacks as their savior, still saw them as greatly inferior beings and remained opposed to enlisting black men as soldiers. At the moment, he was out of communication with anyone in the North, but he would soon write to Secretary of the Treasury Salmon P. Chase, "The negro should be a free race, but not put on any equality with the Whites," and to an old friend in St. Louis he said in a letter, "A nigger as such is a most excellent fellow, but he is not fit to marry, to associate, or vote with me, or mine." Sherman was far more interested in military victory than in ending slavery, and he worried about how he could continue to feed the increasing masses of freed slaves who insisted on accompanying his troops on their way to the sea. Nonetheless, he had moments of revulsion at things he saw. Coming to a plantation near Milledgeville owned by the Confederate general Howell Cobb, who had been President Buchanan's secretary of the treasury before the war, Sherman inspected the wretched slave quarters and was struck by the pitiful condition of the slaves who greeted him as their hero. He ordered his men to "spare nothing," and the destruction began.

Amid all this, Confederate bullets were still killing a number of Union soldiers as they moved through the countryside. Because this army had no rear bases with hospitals, the wounded had to be carried along day after day in wagons, with no hope of receiving full medical attention until the march ended. Some Union troops were captured in surprise Confederate forays. Two days after leaving Milledgeville, a major in an Illinois regiment wrote in his diary of the determined, punitive frame of mind that "has settled down over the army in its bivouac tonight. We have gone so far now in our triumphant march that we will not balk. It is a question of life and death for us, and the considerations of mercy and humanity must bow before the inexorable demands of self preservation."

For those in Sherman's army who thought about justifying it all, some were shocked when they saw that the backs of some freed slaves were a mass of scars from whippings, but for many the most comforting idea was that relatively bloodless violence, right then, could save much more bloodshed on both sides later. A soldier from Wisconsin said in a letter to his parents, "Anything and Everything, if it will help us and weaken them, is my motto," but another enlisted man probably got closest to the soldiers' deepest feelings when he wrote: "The prevailing feeling among the men was a desire to finish the job; they wanted to get back home."

On December 10, Sherman neared Savannah. He had moved sixty-two thousand men through 225 miles of enemy territory in twenty-four days. His troops could smell but not see the ocean, because Savannah's defender, General William J. Hardee, had flooded the rice fields along the coast, leaving just five causeways running into the city. Sherman decided not to attack Savannah along these exposed perilous approaches but to begin a siege and see if the enemy garrison of some eighteen thousand men would surrender.

Since no one in the North knew just where Sherman and his army were, he could not yet make contact with the federal ships that he was sure were offshore. His men would soon need more to eat, and for the moment they could not get any of the supplies of all kinds that he had been promised would be aboard those vessels.

Fort McAllister, a lightly garrisoned post on the south bank of the Ogeechee River, below the city, protected the city's access to the Atlantic, and its twenty-three cannon denied any invading fleet the opportunity to come close enough to bombard it. Although keeping to his decision not to launch a major attack at Savannah itself, within three days Sherman had one of his divisions ready to storm this fort. Just before sunset on December 13, with his selected division about to make its attack, Sherman was watching from the roof of a rice mill beside the river. A Union steamship came into view down the river and used its signal flags to ask Sherman's staff, "Is Fort McAllister taken?" Sherman signaled in response, "Not yet, but it will be in a minute." Fifteen minutes later, after a tactically perfect assault that cost eleven men killed and eighty wounded, the fort surrendered.

Even after this loss of one of the keys to the city's defenses, Hardee did not give up. Sherman, having brought sixty-two thousand men all the way from Atlanta with a total of only seven hundred killed, wounded, and missing, wanted to spare his men's lives if he could and decided for the moment not to make further attacks but continue the siege in hopes of seeing a white flag run up over Savannah. (Either because his troops could not get there, or because he hoped Hardee would avoid a battle by withdrawing his troops from the city, Sherman left open a route of retreat to the north along one of the causeways that ran through the flooded rice fields.)

Now, with easy access to the many ships that had been waiting for him offshore, Sherman was able to get food and supplies for his army, and from an inland direction he also received the first communications from the North that he had seen in a month. He learned that his son Charles, born on June 11, had died on December 4, making this the second child that he and Ellen had lost in fourteen months. Sherman appears not to have written Ellen immediately, and when he did, his words about this infant he had never seen sounded distant, stoical:

> The last letter I got from you . . . made me fear for our baby, but I had hoped that the little fellow would weather the ailment, but it seems that he too, is lost to us, and gone to join Willy. I cannot say that I grieve for him as I did for Willy, for he was but a mere ideal, whereas Willy was incorporated with us . . . But amid the Scenes of death and desolation through which I daily pass I cannot but become callous to death[.] It is so common, so familiar that it no longer impresses me as of old—You on the contrary surrounded alone by life & youth cannot take things so philosophically but are stayed by the Religious faith of a better and higher life elsewhere[.] I should like to have seen the baby of which all spoke so well, but I seem doomed to pass my life away so even my children will be strangers.

At the same time, Sherman received a disquieting letter from Grant in Virginia, who, once again in the spirit of "keep the ball moving," wanted him to waste no more time in massively besieging or attacking Savannah, which was now effectively cut off from aiding the Confederate cause. Just throw a screen of men around the city and build up a base anywhere near there on the coast, Grant told him, and as soon as we get enough transport ships down to you, embark your army and "come here by water with all dispatch." He explained that he wanted to bring Sherman's army straight to Virginia because "I have concluded that the most important operation toward closing the rebellion will be to close out Lee and his army."

This was not what Sherman wanted, but he began turning captured Fort McAllister into the base that Grant told him to create. Reminiscent of the way that Grant, commanding smaller forces along the Mississippi earlier in the war, had taken advantage of every opportunity that was not specifically prohibited, Sherman decided to try to seize the city before the transports arrived to take his men up the coast to Virginia. With the escape corridor still open north of the city, Sherman began closing in on Savannah.

Everything fell into place for him. On December 21, Hardee used the causeway that Sherman's men had not closed, hurried his defenders out of the city, and fled north across the Savannah River into South Carolina, leaving behind one of the Confederacy's largest concentrations of heavy artillery. Sherman marched into Savannah, in the heart of the South; as had been the case with Atlanta, where Hood evacuated the city, there was no significant capture of enemy troops, but he had successfully completed his epic March to the Sea. The next day he sent a telegram to Abraham Lincoln that said in its entirety: "I beg to present you as a Christmas gift the City of Savannah with 150 heavy guns & plenty of ammunition & also about 25,000 bales of cotton."

Sherman became the Union's man of the hour. The joyous news thrilled the North: strangers on the street stopped each other to cry out, "He's made it! Sherman's at Savannah!" In a headline, the *Chicago Tribune* called him "Our Military Santa Claus." Praise engulfed him. Lincoln wrote:

> My Dear General Sherman:
>
> Many, many thanks for your Christmas gift—the capture of Savannah. When you were about to leave Atlanta for the Atlantic coast, I was *anxious*, if not fearful . . . Now the undertaking being a success, the honor is all yours, for I believe none of us went further than to acquiesce . . .
>
> But what next? I suppose it will be safe if I leave it to you and General Grant to decide.

Grant, adding his praise in a letter to Sherman marked "Confidential," said, "I congratulate you, and the brave officers and men under your command, on the successful termination of your most brilliant campaign . . . the like of which is not read of in past history." He included the somewhat questionable statement that "I never had a doubt of the result," and closed with, "I subscribe myself, more than ever, if possible, Your Friend, U. S. Grant." Writing to his father, Grant underscored his enthusiasm by saying, "Sherman has now demonstrated his great Capacity as a Soldier by his unequalled campaign through Georgia."

The news of Sherman's March to the Sea and its climax resonated in

Europe. The *Edinburgh Review* described it as being among "the highest achievements which the annals of modern warfare record," and the London *Times*, comparing him with the duke of Marlborough, said of his campaign, "military history has recorded no stronger marvel." For many in the South, the inability of Confederate forces to stop a march right through its heart signaled the end of any chance of turning the tide. Even the bravest men quailed at the thought of an enemy army marching upon their homes and families: a Confederate officer wrote that his worries about his family made his "soul to sink in anguish" and his hopes "perish." Southern women remained bitterly opposed to the Northern invasion and hated the often rude and sometimes brutal and thieving incursions into their homes. Many still expressed their hopes for a Confederate victory—when the Northern columns occupied Savannah, Mrs. William Henry Stiles wrote her son William, a Confederate soldier serving in Virginia, "After seeing what we have, *we know* how formidable Sherman's army is . . . Still with General Lee at our head, and with the blessing of the Almighty, we shall not be made slaves to these wretches." But some women who had not done so before began to see that the men they had sent off to war could not save the way of life dear to them all. Allie Travis, of Covington, Georgia, thirty-two miles east of Atlanta, was described by a correspondent traveling with Sherman's army as "very pretty and intelligent." She wrote of the day the Union troops marched through on their way to Savannah, "The street in front of our house was a moving mass of 'blue coats'—infantry, artillery, and cavalry—from 9 o'clock in the morning to a late hour at night." She reflected, "Who can describe our feelings on that morning! All human aid was gone. Prayers for personal safety went up to Heaven from the depths of [a] woman's agonized heart."

For the moment, Sherman's aggressive side seemed to be at rest. From Savannah, on January 2, 1865, he wrote Ellen, "I feel a just pride in the Confidence of my army, and the singular friendship of Genl. Grant, who is almost childlike in his love for me." Sherman had instituted a comparatively courteous and orderly military occupation of Savannah reminiscent of his policies when he served as military governor of Memphis earlier in the war. Writing Ellen again on January 5, he referred to families he had met during his tours of duty in the Deep South as a young officer more than twenty years before: "There are some very elegant people here, whom I knew in Better days and who do not seem ashamed to call on the Vandal Chief. They regard us just as the Romans did the Goths and the parallel is not unjust. Many of my stalwart men with red beards and huge frames look like Giants."

As for how Sherman actually felt about his epic March to the Sea, he also wrote this to Ellen: "I can hardly realize it for really it was easy, but like one who has walked a narrow plank I look back and wonder if I really did it." He added, "People here talk as though the war was coming to a close, but I know better."

At this point Sherman was confronted with an unusual result of his famous march, involving an incident at which he had not been present. On December 9, twelve days before Sherman entered Savannah, the commander of his Fourteenth Corps had some of his troops crossing Ebeneezer Creek near the city on a pontoon bridge. This officer was Brigadier General Jefferson C. Davis (not related to the Confederate president). Davis had a well-known capacity for anger and violence: on September 29, 1862, after his superior officer General William Nelson had criticized him, Davis provoked an argument with Nelson in the lobby of the Galt House in Louisville and had returned with a pistol and mortally wounded him. There were those who thought he should be tried for murder, but he was restored to duty through the intercession of his friend and political patron Governor Oliver P. Morton of Indiana and went on to distinguish himself at Chickamauga and other actions. On the day Davis was crossing Ebeneezer Creek, with Confederate cavalry general Joseph Wheeler's men closing in on the rear of his column, a crowd of black refugees was following just behind the Union troops. As soon as the last of Davis's soldiers crossed the pontoon bridge, he ordered it to be taken down: stranded on the far side, the freed slaves were terrified that the advancing Confederates would kill them for casting their lot with the Northern troops they regarded as being their liberators. They began leaping into the water in an effort to escape by crossing the creek. Most could not swim: despite the efforts many Union soldiers made to save them, an undetermined but significant number of black men, women, and children drowned.

At the time, it had been only one incident in a massive campaign, and Sherman had supported Davis's removal of the pontoon bridge as an act to save his troops from an attack by enemy forces that were right on their heels. In the North, while Sherman remained an immensely popular hero, some in Congress saw the drowning tragedy as demonstrating a cruel indifference to the blacks' fate and as being indicative of Sherman's sometimes expressed views on their racial inferiority. On January 9, Secretary of War Stanton arrived at Savannah aboard the ship *Nevada*; he had been in poor health and this trip was in part supposed to give one of the hardest-working men in the government something of a rest in a warm climate, but he had a number of important matters he wished to discuss with Sherman, and the drowning was uppermost. When Sherman again defended Davis's decision to dismantle the pontoon bridge, Stanton asked Sherman to organize a meeting with representatives of Savannah's black population. Sherman invited twenty men, most of them ministers, to meet with Stanton and was offended when Stanton asked him to leave the room when he finished the questions about the tragedy and turned to soliciting the black leaders' impressions of Sherman.

Sherman need not have worried about the black leaders' view of him. The notes made by Assistant Adjutant General Edward D. Townsend, who had accompanied Stanton from Washington, included, "His conduct and deportment toward us characterized him as a friend and gentleman. We have confidence in General Sherman, and think what concerns us could not be in better hands." After that, developing the document in conference with Stanton, Sherman promulgated his Special Orders No. 15, in which parts of Georgia and South Carolina's Sea Islands were reserved exclusively for black land ownership. Reversing his views at least publicly on having black soldiers in the Union Army, Sherman offered them, as an incentive for enlisting, a guarantee that they would receive their share of the Sea Island lands after the war.

Seemingly satisfied on that issue, although in fact Sherman would do little to implement his order, Stanton discussed the overall conduct of the war with Sherman, pointing out among other things the desirability of bringing the war to an end quickly because the federal government was nearly bankrupt. He also made the argument, with which Grant had already agreed, that bringing more black troops into the army and using them for garrison duty would free experienced white regiments to participate in the offensives to end the war.

These meetings between the tall, rangy, gesturing Sherman with his short red beard, and the stocky five-foot-eight intense fifty-year-old Stanton with his long graying beard and small steel-rimmed spectacles brought together two men with complicated personalities. Stanton, who had suffered from asthma his entire life, had endured personal suffering that exceeded even Sherman's loss of his beloved son Willy. At the age of twenty-seven, when Stanton was a rising lawyer, his year-old daughter Lucy died; three years later, when his wife, Mary, suddenly died of a "bilious fever," in his grief he came close to insanity, leaving his room night after night carrying a lamp as he searched the house, crying out, "Where is Mary?" Stanton had always been fond of and proud of his younger brother Darwin, whom out of the profits from his hardworking law practice he had helped send to Harvard to study medicine, and whom he was also able to assist in being elected to the Virginia legislature; in 1846, two years after his wife died, Dr. Darwin Stanton committed suicide by cutting his throat. It was not clear whether Stanton came upon the scene himself, but an account of the death written by a doctor said, "The blood spouted up to the ceiling," and Stanton ran into the woods in the night, with his law partner and other friends searching for him until they found him and were able to lead him home. From that time on, Stanton had become outwardly colder and more hardworking and efficient. He went on to remarry and became an important lawyer and politician, serving as President James Buchanan's attorney general and then returning to his private practice of law in Washington until Lincoln asked him to serve as his secretary of war.

Of these two leading figures of the Union war effort now meeting in Savannah, Stanton had the more difficult wartime role to play. As a soldier, Sherman's objective, like Grant's, was to defeat the enemy. Serving as Lincoln's highly effective secretary of war, Stanton also clearly had victory as his objective, but he found himself in the midst of the frequent tension between Lincoln and the Radical Republicans. Stanton had great quiet admiration and sympathy for Lincoln, who had to put down the Confederate rebellion and yet wished to impose a gentler peace than the one the Radicals wanted. In addition to that, as president, Lincoln was leader of his Republican Party, which had very nearly foundered during the election year just past. Trying to be president of all the people, not just Republicans and Democrats but *all* Americans— whites, blacks, the people not only of the North but also of the South when its seceded status ended—Lincoln was engaged in the greatest balancing act in American history. Stanton saw and understood all that, and had indeed thrown much of his great energy and administrative skill into being an important initiator and coordinator of many aspects of Lincoln's 1864 presidential campaign. As a member of Lincoln's cabinet, he was fulfilling his duty to carry out the commander in chief's policies, but as the politically astute creature he was, Stanton had also kept on good terms with the leading Radicals. (A measure of Stanton's adroitness was that, when Lincoln had named him secretary of war, every faction in Congress felt that he was the man to further their agendas.)

So it was that the two men conferring in Savannah had at least some identical interests. They wanted to end the war quickly; Stanton, who was a friend of Sherman's father-in-law, Thomas Ewing, was talking with the general who seemed destined to play a large role in bringing that about, and Sherman knew that Stanton was ready to throw all the available military resources into the effort. After four days in Savannah, Stanton returned to Washington.

In fact, both Sherman and Stanton remained wary of each other. Before Stanton arrived, Halleck had warned Sherman that Lincoln himself was being urged to punish Sherman not only for the Ebeneezer Creek incident but also for his views on slavery, and even on his way back to Washington Stanton had a wire sent to Grant asking to meet "so as to communicate other matters that cannot safely be written"—presumably Sherman's political volatility, both as to racial remarks and his ambivalence toward white Southerners, in which he mixed his fire-and-sword policy with fond memories of his prewar experiences in the South, and his evident willingness to extend softer peace terms than those the Radicals relentlessly sought. As for Sherman, he mistakenly felt that he had brought Stanton close to his view that yes, the slaves should be freed, but that they were of an inferior race that, even though they were now in the Union ranks, would never make as good soldiers as white men could. On January 15, two days after Stanton left, Sherman wrote Ellen, "Mr. Stanton has been here and is cured of that Negro nonsense."

What Sherman may not have fully grasped, despite the political knowledge available to him both through Halleck and his brother the senator, was that Stanton had been sympathetic to the abolitionist cause from a time long before the war and had a deep distrust of West Pointers—a feeling shared by many of the Radicals, who felt that the officers of the Regular Army formed a clique with little interest in the values of a democracy and comprised a group that could seize and hold despotic power. There was also the health-draining pressure that Stanton was feeling from the demands of his position: in addition to his constant responsibilities as one of the key figures in the prosecution of the war, by the end of Lincoln's presidential campaign, Stanton had come down with a combination of chills and fever that had kept him in bed for three weeks, during which he ran his part of the war from his house. High-strung and driven as always, Stanton was more of an enigma than many of his governmental associates knew. While staying on good terms with the Radicals, he had a great unspoken affection for Lincoln, who would soon memorably proclaim in his Second Inaugural Address his policy of acting "With malice toward none; with charity for all," and later say of the prostrate South, "Let 'em up easy."

At this point, with the war yet to be won, Lincoln was still formulating his thoughts as to the terms on which the seceded states were to be re-admitted to the Union and the way the freed slaves were to be given their civil rights. Stanton saw in Lincoln the indispensable leader who was guiding the nation through the maelstrom. Like Sherman, Stanton had a great desire for order, an impulse almost constantly thwarted by the real-ities of the war and wartime politics, and Stanton was ready to see a po-litical dissident as an outright traitor. Just where all this was taking Stanton's tendency to mistrust Sherman would become dramatically ap-parent within a few months. When that happened, under circumstances that at the moment seemed unimaginable, Grant would extricate Sher-man from the crisis he created for himself.

13

THE MARCH THROUGH THE CAROLINAS, AND AN ADDITIONAL TEST OF FRIENDSHIP

Sherman's widely praised capture of Savannah still left unanswered Lincoln's question, "But what next?" Grant and Sherman differed on this and, as was the case before Sherman set out from Atlanta for Savannah, each man held reasonable views and put them before the other. Grant, still locked in daily heavy combat with Lee at Petersburg that was costing many Northern lives, continued to feel that if he could have ships land Sherman's splendid army near him on the Virginia coast, between them they could swiftly "close out Lee and his army." That would mean the fall of Richmond, the Confederacy's capital. The Confederate forces under Beauregard in North Carolina would still be in existence, but Beauregard, hugely outnumbered at that point, might well surrender, and the war could be over then and there.

Sherman saw it differently. Trusting in himself and his men—he wrote Grant, "I don't like to boast, but I believe this army has a confidence in itself that makes it almost invincible"—Sherman wanted to turn his army north and make another march, up through the Carolinas, continuing to disembowel the South and destroy its will and capacity to make war. This northward march—it was 270 miles in a straight line between Savannah and North Carolina's capital of Raleigh, but the real distance to be covered through swamps and on terrible roads was more than 400—would keep Beauregard, soon to be replaced by Joseph E. Johnston, from moving up to Virginia to reinforce Lee. Sherman's plan was to defeat the enemy in the Carolinas wherever they were, then move on and attack Lee's rear while Grant smashed at his front, and bring the war to an end that way.

Although Grant and Sherman laid out their differing points of view on Sherman's next move in a quick series of telegrams and letters, this time a logistical reality decided the matter. Grant found that there were not enough ships available to bring Sherman's army up to the Virginia coast quickly enough to justify his strategy. On December 27, he approved Sherman's plan for his inland march.

Ulysses S. Grant was still Sherman's superior, but both of them knew that Sherman's brilliant slashes through the South, taking important cities and costing few casualties, were making him greatly popular throughout the North, while Grant remained the general under whose direction the Army of the Potomac was losing many thousands of men every month. On the last day of 1864, with his army unable to start its northward march until widespread flooding in the Carolina coastal lowlands subsided, Sherman wrote a most tactful letter to Grant. Without referring to the change in the public's perception of the two of them, he said, "I am fully aware of your friendly feelings towards me, and you may always depend on me as your steadfast supporter. Your wish is Law & Gospel to me and such is the feeling that pervades my army." At the same time, possibly having this letter from Sherman in mind, Grant wrote Julia, "How few there are who when rising to popular favor would stop to say a word in defence of the only one between himself and the highest in command. I am happy to say that I appreciated him from the first feeling him to be what he is proven to the world he is." During December, Grant had helped to start the Sherman Testimonial Fund of Ohio, which was collecting contributions from businessmen to give to Sherman, who had saved little money and had no house, to help him buy a house for his family when his life became more settled. In sending his own contribution of five hundred dollars, Grant wrote the committee, "I can not say a word too highly in praise of General Sherman's services from the beginning of the rebellion to the present day . . . Suffice it to say that the World[']s history gives us record . . . of but few equals. I am truly glad for the movement you have set afoot and of the opportunity of adding my mite in testemonial [sic] of so good and great a man."

The bond between Grant and Sherman was soon to be brought under pressure again. Around the time Sherman commenced his march into South Carolina at the beginning of February 1865, rumors started to circulate that Sherman was going to be promoted to lieutenant general, making him equal in rank to Grant. It would then be possible for Sherman to be named general in chief, replacing Grant as the Union's top military leader. A bill for Sherman's promotion was introduced in Congress; as soon as he heard of it, Sherman wrote to his brother the senator, stating that he wanted the effort stopped: "I will accept no commission that would tend to create a rivalry with Grant. I want him to hold what he has earned and got. I have all the rank I want." He also wrote Grant about his feelings on the matter, telling him, "I would rather have you in command than anyone else [and] I should emphatically decline any commission calculated to bring us into rivalry." Grant replied to Sherman, "I have received your very kind letter in which you say that you would decline, or are opposed to, promotion. No one would be more pleased at your advancement than I, and if you should be placed in my position and I put subordinate it would not change our personal relations in the least. I would make the same efforts to support you that you have ever done to support me, and would do all in my power to make our cause win."

That cause was nearer to being won than either Grant or Sherman realized. At the beginning of March, by which time Sherman was well on his way up through the Carolinas, Grant received a letter from Robert E. Lee. It developed that, during a meeting in Virginia under a flag of truce to exchange political prisoners, Union general Edward Ord had found himself talking with Grant's old friend and West Point classmate, Julia's cousin Confederate general James Longstreet. Ord and Longstreet were also good friends from the prewar army and, with the business of exchanging the prisoners completed, they began to discuss the possibilities for holding peace talks. After Ord told Longstreet that a first step might be for Lee and Grant to meet, Longstreet took the suggestion to Lee, who wrote Grant about what he termed "the possibility of arriving at a satisfactory adjustment of the present unhappy difficulties." Lee added, "Sincerely desiring to leave nothing untried which may put an end to the calamities of war, I propose to meet you at such convenient time and place as you may designate." Grant immediately forwarded this to Stanton, who sent back this equally prompt reply:

The President directs me to say that he wishes you to have no conference with General Lee unless it be for the capitulation of Gen. Lee's army . . . He instructs me to say that you are not to decide, discuss or confer upon any political question. Such questions the President holds in his own hands.

Nothing came of the matter, but the fact that Lee made this overture demonstrated the deterioration of the Confederacy and its morale. Many men were deserting from Lee's Army of Northern Virginia, but Grant, still locked in combat with Lee at Petersburg, remained wary of his remarkable opponent. He knew that Lee could do to him what he had done to Lee at Cold Harbor: move out overnight, and in this case head south to link up with Joseph E. Johnston. Of these days in March of 1865, Grant wrote that "I was afraid, every morning, that I would wake from my sleep to hear that Lee had gone, and that nothing was left but his picket line. I knew he could move much more lightly and more rapidly than I, and that, if he got the start, he would leave me behind so that we would have to fight the same army again further south—and the war might be prolonged another year."

As for what was going on "further south," after treating the residents of Savannah gently, Sherman and his army had entered South Carolina in an increasingly vindictive frame of mind toward the state they felt had begun the war and started causing the deaths of their comrades. Just as they were leaving Georgia, Sherman told one of his division commanders, Henry W. Slocum, "Don't forget that when you cross the Savannah River you are in South Carolina . . . The more of it you destroy the better it will be." Speaking of South Carolina, he wrote Halleck in Washington that "I almost tremble at her fate, but feel that she deserves all that is in store for her." (Halleck had written Sherman, "Should you capture Charleston, I hope that by *some accident* the place may be destroyed," and when as the campaign began, Sherman's cavalry leader Brigadier General Judson Kilpatrick asked him, "How shall I let you know where I am?" Sherman replied, "Oh, just burn a barn or something. Make smoke like the Indians do.")

Once again, there was skepticism about the outcome of a march that flouted the conventional military belief that an advancing army must have lines of supply and communication extending behind it to bases in the rear. The British *Army and Navy Gazette* said, "If Sherman has really left his army in the air and started off without a base from Georgia to South Carolina, he has done either one of the most brilliant or one of the most foolish things ever done by a military leader." In any event, Sherman and his army were on their way. Putting down logs to make roads through huge swamp areas that the Confederates had considered impassable, they occasionally engaged in short battles that caused their enemies to fall back before them.

Looting and burning more than they had in Georgia, Sherman's columns moved through South Carolina along different routes. Near Barnwell, Mrs. Alfred Proctor Aldrich, mistress of a plantation named The Oaks and a woman who had a husband and two sons in the Confederate Army, braced herself for the arrival of Sherman's men.

> The first of the soldiers who rushed into the house seemed only intent upon procuring food, and . . . ate like hungry wolves.
>
> So soon, however, as they were satisfied, their tramp through the house began. By this time they were pouring in at every door, and without asking to have bureaus and wardrobes opened, broke with their bayonets every lock, tearing out the contents, in hunting for gold, silver, and jewels, all of which had been sent off weeks before. Finding nothing to satisfy their cupidity so far, they began turning over mattresses, tearing open feather-beds, and scattering the contents in the wildest confusion.

After the troops found and drank some bottles of whiskey, "work of destruction began in earnest. Tables were knocked over, lamps with their contents thrown over carpets and mattings, furniture of all sorts broken, a guitar and violin smashed." For ten days, as different units of Sherman's army passed through, camping on her plantation at night, Mrs. Aldrich tried to save her house. Occasionally, Union officers and enlisted men came to her aid, the officers ordering off groups of marauding troops. One night an enlisted man from Ohio named McCloskey appointed himself as a sentry; leaning his rifle against the door, he said to Mrs. Aldrich and one of her young female relatives who had her terrified children with her, "Ladies, it makes my heart sick to see this. I never approved of fighting your people, and would not volunteer for the war, but lately I have been drafted into a new regiment. I have no family of my own, but my mother and sisters are as little in favor of this trip as I am. I can't bear to see women and children ill used."

Different efforts were made by individuals or small groups to start putting the house to the torch, but timely interventions by other Union soldiers combined with Mrs. Aldrich's own steadfast courage to save her house. At times she simply faced down some of the intruders, and at one point she shamed Union General David Hunter, whose tent was pitched on her lawn, into ordering some men to extinguish a fire that had just been set in her corn house. Eventually the corn house was burnt to the ground, as were the plantation's stables, and the books from the library were carried off. The house survived, but this was the scene that now surrounded it: "My beautiful avenue of oaks had been ruthlessly cut down or killed by camp fires near the gates. The park fence was burned up, the large entrance gate cut down, and the undergrowth scorched as black as midnight."

After Sherman's troops moved on, it was a few days before Mrs. Aldrich went into little nearby Barnwell. "I do not remember the day our town was burned, or the division that accomplished it, but I do remember the spectacle presented the first time I beheld its ruins. All the public buildings were destroyed. The fine brick Courthouse, with most of the stores, laid level with the ground, and many private residences, with only the chimneys standing like grim sentinels; the Masonic Hall in ashes."

Barnwell had been a small place. Soon Sherman's army would arrive at a bigger one. On February 17, the largest of Sherman's columns, led by him, came to Columbia, South Carolina's capital. As he entered the city, accompanied by its mayor, who had ridden out to meet Sherman and assure him that he would encounter no opposition, he came upon an already chaotic scene. The retreating Confederates had burnt down the railroad station and a warehouse, which still had some flames among the ruins. The Southern soldiers had done some looting as they left, and broken furniture and scattered household objects littered the sidewalks and streets. Many bales of cotton had been ripped open, so that their contents would scatter and become useless. There was a high wind, and wisps of cotton were flying about in a way that reminded Sherman of a "Northern snow-storm." A number of these fragments were catching fire from the smoldering buildings. Soon the streets filled with black people tumultuously greeting the troops, and in a short while many of Sherman's soldiers had been given liquor or had stolen it, and became increasingly drunk.

That night, as the winds continued, Sherman saw that "the whole air was full of sparks and flying masses of cotton, shingles, etc. some of which were carried by the wind for four or five blocks, and started new fires." He ordered an entire division of his troops to start fighting the spreading blaze, but while they did this, many blacks and drunken Union soldiers wantonly started other fires. (At least one Union officer said that even many of the sober and disciplined troops, ordered to fight the fire, ceased to do so whenever the officers' backs were turned. When a particularly disciplined brigade was ordered to round up the drunk and disorderly soldiers on the streets, one of the unit's officers said that they "very frequently had to use force, and many men would not be arrested, and were shot. *Forty* of our men were killed this way, many were wounded, and several dead drunk men were burned to death.") Only at four in the morning, when the wind stopped, could the fire be brought under control. By then, a third of the city was in ashes.

The burning of Columbia became the subject of endless argument and investigation. It entered the Southern psyche as a deliberate, organized effort to burn an entire city to the ground, after its military defenders had left and it had surrendered and was clearly offering no resistance. Many of its residents had certainly seen Sherman's soldiers setting fires. General William B. Hazen, whose division furnished the brigade that began shooting their drunken fellow Union soldiers who resisted arrest, took the position that "no one ordered it, and no one could have saved it." Sherman's attitude seems to have fallen somewhere between callous indifference and vengeance: he later said defiantly, "If I had made up my mind to burn Columbia, I would have burnt it with no more feeling than I would a common prairie dog village, but I did not do it." Two weeks after the conflagration, a colonel who had not been there heard Sherman say in an informal conversation, "Columbia!—pretty much all burned, and burned *good!*"

By the time Sherman marched one of his principal columns toward the outskirts of Goldsboro, North Carolina, in the state's interior, forty-five miles southeast of Raleigh, his men thought that their "Uncle Billy" was nearly superhuman. He felt the same way about them: as he had watched the 104th Illinois stride into Fayetteville after marching through fifteen miles of thick mud in five hours, he said, "It's the damndest marching I ever saw," and he noted that fewer men in his army were sick on the march than when they were in relatively permanent camps. As he moved up through North Carolina, Sherman's confidence grew: in a report that he sent to Grant on March 22 concerning everything his army had done the previous day, he referred to three of his generals: "Our combinations were such that Schofield entered Goldsboro, from New Bern, Terry got Cox's brigade with pontoons laid and a bridge across [Mill Creek] and Entrenched, and we whipped Joe Johnston on the same day."

"Whipped" suggests a greater victory than what took place. At the Battle of Bentonville to which Sherman referred, his troops inflicted more casualties than they sustained, but he failed to press home an initially successful attack on Johnston's left flank, and when Johnston, who had replaced Beauregard, counterattacked and then extricated his forces from the battlefield with his usual defensive skill, Sherman did not pursue him.

Here was a major difference from Grant's behavior. After every battle, Grant did everything to "keep the ball moving." Indeed, for Grant, that further effort to pursue, to exploit whatever had been gained, was seemingly a reflex action, a part of the battle itself. Perhaps, despite Sherman's admiration for Grant, Sherman had been influenced more than he knew by the French military thinker Jomini, whose preference for winning by maneuver rather than frontal attack Halleck emulated. Sherman often said that he wanted to minimize his casualties, and he did, but at Bentonville he missed the opportunity to deal Johnston a blow that might have shortened the war and in the process spared both sides suffering yet to be endured.

Nonetheless, Sherman was proving himself a master of maneuver. Moving on after Bentonville, the men with Sherman felt themselves to be part of an irresistible northward march. As they came in sight of the houses of Goldsboro on the afternoon of March 22, they saw, in what seemed a remarkable piece of military choreography, a heartening sight: "A locomotive train came thundering along from the Sea 96 miles distant loaded with shoes, & pants, & clothing as well as food."

The following day, when they all entered Goldsboro, they found Brigadier General John Schofield waiting for them with his Twenty-third Corps. (At Atlanta, when Sherman wanted to know what to expect from his new opponent John Bell Hood, it was Hood's West Point classmate Schofield who told him that Hood would attack within forty-eight hours—an estimate that Hood undercut when he attacked the next day.) Here at Goldsboro, within twenty-four hours, eighty thousand men of Sherman's army, an army that had been moving through North Carolina along several routes, some that included swamps, were reassembled in one gigantic encampment. All of them had marched 330 miles or more since leaving Savannah on their different missions, some of them covering the distance in as little as twenty-one days.

Sherman's entire army had become men whose marches rivaled those of the Roman legions. From the time they had left Meridian, Mississippi, after the Vicksburg campaign, his forces had traveled more than two thousand miles. Coming into Goldsboro, more than half the men had worn out their shoes and were walking on calloused bare feet, and the uniforms of most of the soldiers who proudly swung past Sherman had rotted into rags. When a Union general who was beside him said of the passing troops, "Look at those poor fellows with bare legs," Sherman, whose own uniform was in little better condition, shot back with, "Splendid legs! Splendid legs! I'd give both of mine for any one of 'em."

As the army paused briefly at Goldsboro to rest, even some Southerners were ready to give Sherman credit for the unconventional strategy that had brought his army so far. At this point, after nearly four years of war, *The Richmond Whig* said, "Sherman is simply a great raider. His course is that of a bird in the air. He is conducting a novel military experiment and is testing the problem whether or not a great country can be conquered by raids."

Despite Sherman's failure to pursue after Bentonville, in a recent letter to Grant he had appropriated Grant's own phrase, "keep the ball moving," and was thinking hard about his role in what he had no doubt was the impending end of the war. He wanted to be in at the kill, not only defeating Johnston but also sharing in the defeat of Lee. (His soldiers shared the same feeling: a sergeant from Iowa wrote home that "it is the talk of the Boys now that our next moove [*sic*] will be in the direction of Richmond, but the boys say it is hard to tell which way Crazy Bill will go for he goes wherever he wants and the rebs can[']t help themselves.")

Sherman had no way of knowing that Grant, also certain that the end was near, had come to think that it would be better for the postwar political situation if his Eastern forces defeated their old adversary Lee by themselves. Even though he was a man of the West himself, Grant felt that if Sherman's men completed their remarkable series of campaigns by coming up from North Carolina into Virginia to take a large part in defeating Lee's Army of Northern Virginia, "It might lead to disagreeable bickering between members of Congress of the East and those of the West" as to which area of the nation deserved credit for winning the war. (Grant was to say that when he spoke of his concern about this to Lincoln, the president considered it valid but told Grant that he "had never thought of it before, because his anxiety was so great that he did not care where the aid came from so [long as] the work was done.")

With the subject of the war's final strategic moves on his mind, Sherman wrote Grant on March 23 that "if I get the troops all well placed, and the supplies working well, I might run up to see you for a day or two, before diving into the bowels of the Country again." The following day he added in another letter that "I think I see pretty clearly how in one more move we can checkmate Lee, forcing him to unite Johnston with him in the defense of Richmond, or by leaving Richmond to abandon the cause. I feel certain that if he leaves Richmond[,] Virginia leaves the Confederacy."

There is no record of Grant's response to Sherman's "I might run up to see you for a day or two," but on March 25, when his engineering troops finished repairing the torn-up rail line from Goldsboro to New Bern, Sherman left his army under the command of General Schofield and started for Morehead City, a port on the North Carolina coast nearly a hundred miles away. According to a reporter from *The New York Herald*, when Sherman stopped overnight at New Bern, sixty miles southeast of Goldsboro, some of his off-duty soldiers saw him walking down the street and enthusiastically "rushed around him as if they were going to tear him to pieces and all the while calling for a speech." Sherman said only this to them: "I'm going up to see Grant and have it all chalked out for me and then come back and pitch in. I only want to see him for five minutes and won't be gone but two or three days." At Morehead City the following day, Sherman embarked on the swift steamer *Russia*, a captured Confederate blockade-runner. Writing Ellen from the ship as it moved north, he told her, "There is no doubt we have got the Rebels in a tight place and must not let them have time enough to make new plans . . . I will now concoct with Grant another plan." In closing he said, "The ship is pitching a good bit, we are just off Hatteras, and I cannot write more."

Heading north at sea, moving toward his friend and military superior, Sherman was coming to an almost symmetrically placed point in the plans that he and Grant had made in that hotel room in Cincinnati a year before. They had not seen each other since, but in the meantime they had indeed lived Grant's dictum, which Sherman expressed as, "He was to go for Lee, and I was to go for Joe Johnston." They had done that. The remaining question was still the one Lincoln had asked after the capture of Savannah: "But what next?"

14

GRANT, SHERMAN, AND ABRAHAM LINCOLN
HOLD A COUNCIL OF WAR—AND PEACE

Sherman's destination, Grant's busy headquarters at City Point, was on the same side of the Appomattox River as besieged Petersburg, nine miles away. It was at just that time a particularly interesting and dramatic place to be. President Lincoln enjoyed getting out of Washington and being with the troops, and his son Robert, who had graduated from Harvard the year before, was now serving as a captain on Grant's staff. Lincoln and his wife had recently arrived for an extended stay. On the same day that Sherman left his men at their inland encampment at Goldsboro in North Carolina, Lincoln at City Point had been taken to a hill near the Petersburg front to watch the battle for Fort Stedman, an effort by Lee to break and weaken Grant's line that cost the Confederacy more than four thousand casualties in one day.

During this presidential visit, it was Julia Grant's misfortune to have to deal with Lincoln's mentally unstable wife, who frequently had the idea that every woman was trying to steal her husband. In addition, Mary Todd Lincoln insisted on having every kind of privilege accorded her and often saw slights where none were intended. The Lincolns stayed aboard a handsome ship, the *River Queen*, which brought them down from Washington and was anchored in the river. Grant had at his disposal a smaller vessel, the *Mary Martin*, a fast little steamship that was frequently tied up at a dock near headquarters. The first time the Lincolns came ashore, the *River Queen* was brought alongside Grant's *Mary Martin*, and the Lincolns walked across the *Mary Martin*'s decks to the dock, where they were greeted by Grant and Julia. That happened only once: Mrs. Lincoln let it be known that she did not want to have to cross another ship's deck to come ashore, and while soldiers were dying some miles away, Grant's vessel was moved out into the river every time she wanted to come ashore, so that she could step straight ashore from the presidential ship. A most unfortunate outburst occurred when, sitting beside Julia in an ambulance being used as a carriage at a large military review, she saw the beautiful wife of General Edward Ord, a stylish woman who was a superb equestrienne, riding her horse in a party of generals and other notables that included Lincoln. When Mrs. Ord was brought alongside the makeshift carriage to be presented to Mrs. Lincoln, her horse wheeled and carried her off in pursuit of the group that included Lincoln. An officer tried to explain to the jealous Mrs. Lincoln that Mrs. Ord's horse was trained to stay near General Ord. The man added, trying to be helpful and pleasant, that the horse "will not let the lady leave her husband's side. I would recommend that you get one just like it. If you would like . . . I will try to get him for you; he is just what you want." Mary Todd Lincoln took this as a warning that Mrs. Ord might steal Abraham Lincoln from her if she did not get a horse and ride right beside him and angrily cried out, "What do you mean, Sir?" It took all of Julia's tact and good sense to soothe her, and even then Mrs. Lincoln struck out at Julia with, "I suppose you think you'll get to the White House yourself, don't you?"

Lincoln, on the other hand, seemed to be calmer, even within the sound of the cannon firing back and forth at besieged Petersburg, than in Washington. He enjoyed riding Grant's big horse Cincinnati—Grant let no one else ride his favorite mount—but Lieutenant Colonel Horace Porter of Grant's staff remarked that he seemed sad and tired, and described something that happened in one of the headquarters tents.

Three tiny kittens were crawling about the tent at the time. The mother had died, and the little wanderers were expressing their grief by mewing piteously. Lincoln picked them up, took them on his lap, stroked their soft fur and murmured: "Poor little creatures, don't cry; you'll be taken good care of," and turning to Bowers [a colonel of Grant's staff], said: "Colonel, I hope that you will see that these motherless little waifs are given plenty of milk and treated kindly." Bowers replied: "I will see, Mr. President, that they are taken in charge by the cook of our mess, and are well cared for."

Several times during his stay Mr. Lincoln was found fondling these kittens. He would wipe their eyes tenderly with his handkerchief, stroke their

smooth coats, and listen to them purring their gratitude to him. It was a cu-
rious sight . . . upon the eve of a great military crisis in the nation's history, to
see the hand which had affixed the signature to the Emancipation Proclamation . . . tenderly caressing three stray kittens.

On the morning of March 27, Grant's headquarters received the
news that Sherman had arrived at Fortress Monroe, on the Virginia
coast, and that the *Russia* was proceeding up the James River to City
Point. The reunion about to take place came not only at a critical point
in the war but also at a critical moment in each of their careers. Sherman's far-ranging campaigns had made him so famous that there were
now those who thought that when victory came he might be made some
sort of American dictator, or at least become president. (Of the latter
idea, Sherman wrote to a friend, "You may tell *all* I would rather serve
4 years in the Singsing Penitentiary.")

In contrast with the image Sherman had gained of being the all-
conquering general, Grant was embedded in the public mind as the
bulldog who knew only how to throw regiments straight ahead in his
constant bloody battles with the equally tenacious Lee. In fact, like Sherman and Lee, Grant was a great practitioner of the war of movement: he
had demonstrated that skill in his Western campaigns and in his slipping
115,000 men away from Lee's front after Cold Harbor to reappear swiftly
before Petersburg. In the last year, however, both Grant and Lee had
become prisoners of the Confederate determination to hold the two
neighboring cities of Richmond and Petersburg at all costs, a political
decision that required Lee to defend that area. Grant's goal was to destroy Lee's army rather than to capture Richmond, but to do the one
thing, it appeared that he would have to do the other as well.

Whatever the public perception of him, Grant was doing much more
than presiding over the war of attrition against Lee's Army of Northern
Virginia: as general in chief, in the days just before Sherman arrived
he sent specific operational orders to the commander at Knoxville,
made recommendations for a reorganization of the command structure
for Arkansas, Missouri, and Kansas, suggested a campaign be made
in northeast Texas to subdue the remaining Confederates there, and
dealt with matters concerning the United States Army as a whole that involved promotions, prompt pay for black soldiers, and the exchange of
prisoners.

Grant's aide-de-camp Porter stood nearby as Sherman arrived aboard the *Russia* at three in the afternoon.

> General Grant and two or three of us who were with him started down the wharf to greet the Western commander. Before we reached the foot of the steps, Sherman had jumped ashore and was hurrying forward with long strides to meet his chief. As they approached Grant cried out, "How d'you do, Sherman?" "How are you, Grant!" exclaimed Sherman; and in a moment they stood upon the steps, with their hands locked in a cordial grasp, uttering earnest words of familiar greeting. Their encounter was more like that of two school-boys coming together after a vacation than the meeting of two actors in a great war tragedy.

Grant walked Sherman up to his headquarters, where Julia greeted Sherman warmly, and members of Grant's staff crowded around. Porter said, "Sherman then seated himself with the others . . . and gave a most graphic description of the events of his march through Georgia. The story was the more charming from the fact that it was related without the manifestation of the slightest egotism. Never were listeners more enthusiastic; never was a speaker more eloquent. The story, told as he alone could tell it, was a grand epic related with Homeric power." Another officer noted Sherman's appearance: "his sandy whiskers closely cropped . . . sharp twinkling eyes, long arms and legs, shabby coat, slouch hat, his pants tucked into his boots." A man from Massachusetts who first saw him during this visit observed that "his features express determination, particularly the mouth, which is wide and straight with lips shut tightly together . . . a very remarkable-looking man such as could not be grown out[side] of America—the concentrated essence of Yankeedom."

After an hour of Sherman's accounts, Grant said, "I'm sorry to break up this entertaining conversation, but the President is aboard the *River Queen*, and I know he will be anxious to see you. Suppose we go and pay him a visit before dinner." About an hour after that, the two generals returned to Grant's log cabin. This first visit had been largely a courtesy call; Lincoln had initially expressed concern that Sherman was not with his army but then relaxed and listened intently to Sherman's stories of his recent campaigns. It was understood that Grant and Sherman were to meet with Lincoln the following morning for a conference on the military and political strategy to be followed in ending the war.

Porter had been talking with Julia as he waited for his chief to reappear and witnessed what came next. Julia had prepared the two generals some tea, and as they entered, she asked her husband, "Did you see Mrs. Lincoln?" Grant, taken aback, said "Oh," and sheepishly added, "We went rather on a business errand, and I did not ask for Mrs. Lincoln." Sherman chimed in with, "And I didn't even know she was on board."

"Well, you are a pretty pair!" Julia said, chiding them. "I do not see how you could have been so neglectful. Now, you have got your foot in it."

Grant replied contritely, "Well, Julia, we are going to pay another visit in the morning, and we'll take good care then to make amends for our conduct today."

"And now," Sherman suggested to Grant as they settled down with their tea, "let us talk further about the immediate movements of my army."

At this point Julia said, "Perhaps you don't want me here listening to all your secrets."

Sherman smiled at Julia and asked, "Do you think we can trust her, Grant?"

"I'm not so sure about that, Sherman," Grant answered lightheartedly. This led to some banter about Julia's trustworthiness, and Sherman said, "Now, Mrs. Grant, let me examine you, and I can soon tell whether you understand our plans well enough to betray them to the enemy."

"Very well," Julia replied. "I'm ready for all your questions."

Porter described what happened next. "Then Sherman turned his chair squarely toward her, folded his arms, assumed the tone and look of a first-class pedagogue, and, in a manner which became more and more amusing as the conversation went on, proceeded to ask all sorts of geographical questions about the Carolinas and Virginia."

In fact, Julia had studied the big headquarters maps with great interest during her sometimes weeks-long visits to City Point and knew a great deal about the territory in which the Union forces were operating. She decided to answer "wide of the mark," as she later put it. Porter said, "When asked where a particular river in the South was, she would locate it a thousand miles away, and describe it as running up stream instead of down; and when questioned about a Southern mountain she would place it somewhere in the region of the north pole. Railroads and canals were also mixed up in interminable confusion."

After some minutes of what Julia described as "throwing dust in Sherman's eyes," with Grant enjoying it all greatly, Sherman turned and said, "Well, Grant, I think we can trust her." Then he said to Julia, "Never mind, Mrs. Grant; perhaps some day the women will vote and control affairs, and then they will take us men in hand and subject us to worse cross-examinations than that."

Grant suddenly spoke up. "Not if my plan of female suffrage is adopted."

"Why, Ulys," Julia said, "you never told me you had any plans regarding that subject."

"Oh, yes," Grant continued. "I would give each married woman two votes; then the wives would all be represented at the polls, without there being any divided families on the subject of politics."

Then Grant and Sherman got down to business, in what Julia called "a long talk of troops and movements." They both knew the various possibilities—the war could end with Sherman defeating Johnston in North Carolina and Grant defeating Lee in Virginia, or Lee might make a sudden march south with the remains of his Army of Northern Virginia to link up with Johnston for a final combined stand, or Johnston might be able to come north to aid Lee. Sherman now said that whatever movements Johnston might make, he felt he could march north immediately to join in a final defeat of Lee. "Grant, if you want me to help you, I can come up. Yes, I can manage it, I'm sure." Grant answered, "No, I can manage everything myself. You hold Joe Johnston just where he is. I do not want him around here."

When dinner was announced, Sherman took Julia into the headquarters mess on his arm. After all the other dinner guests had gone and Grant and Sherman continued to talk, with Grant's staff officer Horace Porter still present, thirty-four-year-old Major General Philip Sheridan arrived near midnight, explaining that he was late because the train bringing him had been derailed. In the sixteen months since Grant commended him for his "prompt pursuit"—indeed, the only effective pursuit—of the enemy after the taking of Missionary Ridge at Chattanooga at the end of November 1863, Sheridan had risen to hold a unique place within the Union Army. From being an able, aggressive commander of foot soldiers who had experience with mounted troops, he had become the leading cavalry general whose seven-month-long Shenandoah Valley Campaign against Jubal Early laid waste the fertile valley to the point that Sheridan said, "A crow could not fly across it without carrying his rations with him." His campaign produced the important victories at Winchester, Fisher's Hill, and Waynesboro, but his most dramatic day came at Cedar Creek on October 19, 1864. When Sheridan, returning from a conference in Washington, learned at Winchester in the morning that his army had been routed in a surprise attack eight miles to the south at Cedar Creek, he raced there on his horse Rienzi. Arriving amid his demoralized retreating troops at about ten-thirty a.m., he leapt Rienzi across an improvised barricade made of fence rails. Facing his startled soldiers, he bellowed, "Men, by God, we'll whip them yet! We'll sleep in our old tents tonight!"

From that moment, aided by the Confederate delay in trying to exploit their original success, the Union front began to stabilize; when one of his generals said he could organize his units to cover an orderly retreat, Sheridan exclaimed, "Retreat, hell! We'll be back in our camps tonight!" Reorganizing his scattered front, at noon he cantered all along the line, swinging his hat in his right hand, his head bared so that his men could see that it was indeed their commander, and was greeted with a roar as he came by. At four o'clock that afternoon, two hundred of Sheridan's buglers sounded the signal to attack, and within ninety minutes the Confederates were swept from the field, pursued by some of Sheridan's cavalrymen until long after dark. Grant praised Sheridan for "turning what bid fair to be disaster into glorious victory."

A week before the night he walked into Grant's headquarters at City Point, Sheridan had finished a daring month-long raid deep into Virginia, destroying railroads east of the Shenandoah Valley after finishing his valley campaign. The five-foot-five West Pointer, now the best-known Union general after Grant and Sherman, with George Gordon Meade being next due to his victory at Gettysburg, had proved to be the North's bold and successful answer to the brave and dashing horsemen of the Confederacy. As he did with Sherman, Grant treated this commander of large and powerful mounted columns almost as an equal: he listened to Sheridan's ideas and sometimes was persuaded by them, but his word was final.

Like Sherman, Sheridan wanted his share of the final victory. Having torn apart much of the western side of Virginia, he had decided to "join General Grant in front of Petersburg . . . Feeling that the war was nearing its end, I desired my cavalry to be in at the death." Grant had urged him to come ahead, and the muscular little cavalryman now had nine-teen thousand of his troopers, and their horses, encamped south of Petersburg, poised to go into action.

There would be different versions and interpretations of what Grant had previously told Sheridan concerning the role that his two cavalry divisions were to play in what they both knew was to be the last phase of the war. Grant had written Sheridan that he was to cut the Southside Railroad and the Richmond and Danville Railroad, which were Lee's last means of moving south from Petersburg and Richmond to link up with Joseph E. Johnston in North Carolina. After that, Grant told Sheridan, he could either return to Grant's front facing Lee in the north, "or go on to Sherman, as you deem most practicable." (In his memoirs, Grant offered the questionable explanation that any talk of Sheridan going to serve with Sherman at this point was simply a "blind," apparently to prepare an excuse; if Sheridan's raid against the railroads failed, they both could simply claim that the real objective of the movement was to link up with Sherman's army.)

That was the background against which a midnight talk among Grant, Sherman, and Sheridan now took place. Sherman started exploring the idea that the best use of Sheridan's two cavalry divisions would indeed be to leave the Petersburg-Richmond area, heading immediately toward his army, so that by being farther south Sheridan would be in a better position to cut off Lee if he made a quick retreat to join Johnston's army in North Carolina. Grant's aide-de-camp Porter was still present and noted that "Sheridan became a good deal nettled by this, and argued earnestly against it; but General Grant soon cut short the discussion by saying that it had been decided that Sheridan was to remain with our army, then in front of Petersburg."

Everyone then went to bed, but Sherman persisted: waking Sheridan well before dawn, he tried again to persuade him to move his cavalry divisions so that they were at least somewhat better positioned to come to North Carolina. This time, without Grant being present, Sherman got a taste of the powerful personality that had been directing destructive assaults in the Shenandoah Valley that compared with what Sherman had wreaked in his own famous campaigns. Sheridan, who knew he would be leading his columns to a jumping-off point within a few hours and wanted to sleep a little longer, bluntly told Sherman that they both had heard what Grant said. Orders were orders: Sheridan was staying in Virginia with Grant to deal with Lee, and Sherman was supposed to go back to North Carolina and catch up to Joseph E. Johnston.

Later in the morning, after General Meade and other leaders of Grant's Eastern army came to visit with Sherman, Grant and Sherman were joined by their colleague from the Vicksburg campaign, Admiral David Dixon Porter, whose ships would be supporting the final land offensives. The three men presented themselves aboard the *River Queen* for what was to be a momentous conference with Abraham Lincoln, and the president greeted them and conducted them to the after cabin. Mindful of their breach of etiquette the day before, Sherman said that, "After the general compliments, General Grant inquired after *Mrs.* Lincoln, when the President went to her state-room, returned, and begged us to excuse her, as she was not well."

Getting down to business, Grant reported, in Sherman's words, that "at that very instant of time, General Sheridan was crossing James River from the north, by a pontoon bridge below City Point; that he had a large, well-appointed force of cavalry, with which he proposed to strike the Southside and Danville Railroads, by which alone General Lee, in Richmond, supplied his army; and that, in his judgment, matters were drawing to a crisis."

Sherman, still wanting some part in defeating Lee's Army of Northern Virginia, remarked that his own army in North Carolina "was strong enough to fight Lee's army and Johnston's combined," if Grant could quickly support him. He added that in any case, if Lee would stay in his defensive positions in front of Richmond and Petersburg for another two weeks, he could vanquish Johnston in North Carolina and march into Virginia to help Grant finish Lee. Grant and Sherman both indicated that they expected there would be one more major battle. His voice filled with emotion, Lincoln said that there had been "blood enough shed" and asked if "another battle could not be avoided." Sherman replied that the answer to whether a big final battle would occur lay with the enemy and that he thought Jefferson Davis and Lee would bring on "one more desperate and bloody battle."

With Grant and Sherman assuring Lincoln that they were ready for any military eventuality and could soon defeat the armies of both Lee and Joseph E. Johnston, the conversation turned to what should happen when the final surrenders occurred. From this point on, Admiral Porter wrote in his notes of the meeting that Grant "sat smoking a short distance from the President" and had little more to say. Sherman said of his own part that he asked such things as, "What was to be done with the rebel armies when defeated? And what should be done with the political leaders, such as Jeff. Davis, etc.? Should we allow them to escape, etc.?"

Now Lincoln started to speak at some length. "When at rest or listening," Sherman recalled, "his arms and legs seemed to hang almost lifeless, and his face was care-worn and haggard; but the moment he began to talk, his face lightened up, his tall form unfolded." In answer to what to do with Jefferson Davis, Sherman noted that Lincoln said Davis "ought to clear out, 'escape the country,' only it would not do for him to say so openly." Lincoln often used homely little stories as political parables. Now he told of a man who offered a teetotaler some lemonade but pointed out to his guest that it would be "more palatable if he were to pour in a little brandy." Lincoln finished the anecdote with the teetotaler's reply: "His guest said, if he could do so 'unbeknown' to him, he would not object." Sherman later wrote, "From which illustration I inferred that Mr. Lincoln wanted Davis to escape, 'unbeknown' to him."

As Grant continued to smoke a cigar and listen, Lincoln and Sherman proceeded to talk about what terms of peace should be offered to the states of the Confederacy. With this, the meeting came to a moment of supreme significance: four years before, the legislative bodies representing eight million white Southerners had withdrawn their respective states from the United States of America, to create a separate nation, the Confederate States of America. That effort, considered to be treason by many in the North, had failed, at a cost of more than half a million American lives. On what conditions were those rebellious states to reenter the national political framework from which they had removed themselves? As Sherman recalled what Lincoln now said, "In his mind he was all ready for the civil reorganization of affairs at the South as soon as the war was over . . . As soon as the rebel armies laid down their arms, and resumed their civil pursuits, they would at once be guaranteed all their rights as citizens of a common country; and that to avoid anarchy the State governments then in existence, with their civil functionaries, would be recognized by him as the government *de facto* till Congress could provide others."

This account of Lincoln's postwar intentions suggests more crystallization of his specific ideas than was seen at this time by some of his governmental colleagues, but the president certainly had a deep desire for the fighting to end. Porter later stated that his impression during the meeting was that Lincoln "wanted peace on almost any terms," and Sherman was soon to prove that he thought Lincoln was giving him a virtual carte blanche in bringing the war to a close in his area of operations. Grant, who sat silent throughout this part of the discussion, may have been less sure about the degree of latitude that Sherman thought he had: less than a month before, when Grant had forwarded to Secretary of War Stanton the overture sent him by Robert E. Lee, he had swiftly been told that Lincoln would authorize him to accept a military surrender but directed Grant that "you are not to decide, discuss or confer upon any political question. Such questions the President holds in his own hands."

Near noon, after a conversation of an hour and a half, Lincoln escorted Grant and Sherman to the gangway of the *River Queen*. Admiral Porter had arranged that Sherman would return to North Carolina aboard the *Bat*, an even faster ship than the one that had brought him to City Point. At that moment of harmonious parting among the participants in the meeting, no indication of any misunderstanding about policy seemed to exist. After being in Lincoln's presence, Sherman had this impression of him: "Of all the men I ever met, he seemed to possess more of the elements of greatness, combined with goodness, than any other." Grant had the same reaction, describing Lincoln as "incontestably the greatest man I have ever known."

Grant and Sherman had been reunited at City Point. When they saw each other again, it would be in the midst of an altogether unforeseen crisis.

15

"I NOW FEEL LIKE ENDING THE MATTER": GRANT'S FINAL OFFENSIVE

The next morning, with Sherman on his way back to his army in North Carolina, Grant prepared to leave City Point for the Petersburg front, to launch the big push that he hoped would end the war. His aide Porter saw Ulysses and Julia kissing repeatedly as they parted at the door of their cabin; Grant walked away a few yards, turned, came back, and they kissed several more times. "She bore the parting bravely," said Porter, who was accompanying Grant to the front, "although her pale face and sorrowful look told of the sadness that was in her heart."

Lincoln was still at City Point, and he walked Grant and a number of his aides down to the railroad line that would take them the short distance west to the network of Union positions near Petersburg. Porter noted, "Mr. Lincoln looked more serious than at any other time since he had visited headquarters. The lines in his face seemed deeper, and the rings under his eyes were of a darker hue. It was plain that the weight of responsibility was oppressing him." At the train, Lincoln shook hands with Grant and his staff officers who were going to the front with him, and stood watching as they climbed aboard the last car and looked down at him from its rear platform. "As the train was about to start we all raised our hats respectfully," Porter recalled. "The salute was returned by the President, and he said in a voice broken by an emotion he could ill conceal: 'Good-bye, gentlemen. God bless you all! Remember, your success is my success.'" The train moved off. After a few minutes Grant said to several of his officers who were gathered around him, "I think we can send him some good news in a day or two."

Setting up his headquarters right opposite the Confederate lines at besieged Petersburg on March 29, Grant had a solid plan and a large superiority in numbers, but in addition to that he was once again acting on instinct. Sherman, back with his reequipped army, would soon write Ellen that on April 10 "I will haul out for Raleigh," forty-five miles north of Goldsboro, to come to grips with Johnston. He thought he still had time to finish off Johnston and then march north into Virginia to realize his cherished idea of helping Grant finish Lee; he shared with John Sherman his estimate that "the next two months will demonstrate whether we can manoeuvre Lee out of Richmond and whip him in open battle."

Grant sensed that things were about to move much faster than that. He never forgot that Lee, while seemingly so determined to continue to hold Petersburg and Richmond, might decide that, slim as the South's military chances now were, they might be bettered if he made a sudden break for it, evacuating both cities and moving quickly to link up with Johnston in North Carolina. In fact, Lee's recent failed attack on Fort Stedman had been an effort to make Grant send reinforcements to the attacked fort from the left end of his line, to the south of the city, weakening the Union positions at just the place Lee might be able to cut through in his army's escape if he chose to make that move.

The offensive Grant had been starting was somewhat tentative. He had been maneuvering various divisions to the south and slightly to the west below Petersburg, to block a retreat toward North Carolina by Lee, but his aggressive instincts reappeared. Within a few hours of getting the feel of the situation at his advanced headquarters at Gravelly Creek, he decided to try to knock Lee out of the war in an all-out drive. "I now feel like ending the matter," he telegraphed ahead to his cavalry chief Sheridan. "We will all act together as one army until it is seen what can be done." The following day, March 30, he wired to Lincoln, who was still at City Point, that "our troops have all been pushed forward."

Robert E. Lee was not through. He remained as aggressive and tenacious as Grant, and as imaginative in maneuvering as was Sherman. Although he had long before this said privately that Confederate defeat was "a mere question of time," he had recently written to his invalid wife in Richmond that "I shall . . . endeavor to do my duty and fight to the last." In Lee's view of his responsibilities and constraints, he was the Confederate mirror image of what had been said to Grant earlier in the month, when Grant forwarded on to Washington Lee's overture to end the fighting. Lee felt empowered to surrender his army rather than see it annihilated, but he believed that when it came to a political settlement of the war, it could be said of his president, Jefferson Davis, just what Grant had been told concerning Lincoln: "Such questions the President holds in his own hands."

At this point, Davis was still in Richmond, still meeting with the cabinet of the Confederate States of America, still conducting that government's business and showing no disposition to sue for peace. Lee had told Davis that he would give him warning if Petersburg was about to fall and Richmond must be evacuated; at the end of March, Lee estimated that the Confederate administration still had ten or twelve days in which to wind up its affairs and leave. Despite the rapidly deteriorating state of his battered army and the many Union divisions arrayed against him, Lee stayed alert for any opportunity to strike the enemy and continued to watch for Grant to make a misstep, leave a gap between divisions, expose a strung-out column on the roads. He ordered his shrunken forces to be ready to take the offensive; like Grant, he was waiting "until it is seen what can be done."

As events would show, Grant had understood the situation intuitively and needed to have Sheridan and his cavalry as his spearhead, constantly ready and flexible as he attacked Petersburg from the east and at the same time tried to cut the Confederate railroad line that ran southwest of the city. In fact, despite his plan to throw everything he had at Lee, Grant needed Sheridan to stiffen his spine; when early in the movements around Petersburg the spirited cavalry leader read an uncharacteristically pessimistic message from Grant, telling him that "the heavy rain of today will make it impossible for you to do much until it dries up a little," he leapt on his horse Pacer and headed straight through the mud for Grant's headquarters at Gravelly Creek. There, in a cornfield that had become a swamp, he found Grant in his tent, arguing with his chief of staff Rawlins, who was trying to persuade Grant not to pause in making his attacks. Sheridan quickly realized that the rest of Grant's staff, uneasily gathered around a campfire in the rain, also wanted their chief to push on; after "pacing up and down . . . like a hound in the leash," as Horace Porter described it, Sheridan got Grant aside and apparently began laying into him with the same force he had used in telling Sherman that he was going to stay with Grant and not go to North Carolina.

Within twenty minutes, Grant said, "We will go on," and Sheridan rode back to his command.

By the next night, it was Sheridan who needed encouragement and support. When Grant read a report from Sheridan sent from southwest of Petersburg, saying that he was facing a Confederate infantry division commanded by General George Pickett and intended to "hold on to Dinwiddie Court House until I am compelled to leave," Grant immediately started a reinforcement of ten thousand horsemen and twenty thousand foot soldiers moving through the darkness to strengthen Sheridan's position. The Union effort to take Petersburg started up yet again; by the morning of Sunday, April 2, when a total of sixty-three thousand federal infantrymen began attacking his lines, Lee realized that his eighteen thousand defenders—some estimates placed the number as low as twelve thousand—would be overrun if they remained in their defensive positions. (In the account of the day given by Confederate major Giles B. Cooke, Lee is said to have turned at one point to one of his officers and said, "Well, Colonel, it has happened as I told them it would at Richmond. The line has been stretched until it is broken.") Composed as always, Lee matter-of-factly began issuing orders that Petersburg must be evacuated by nightfall.

With this last day in the city still ahead of him, on this Sunday morning Lee dictated a telegram to be sent to Confederate Secretary of War John C. Breckinridge in Richmond, reporting the situation and ending with, "I advise that all preparations be made for leaving Richmond tonight." This message was brought to Jefferson Davis, who was attending a morning service at St. Paul's Church. After reading it as he sat in his pew, the Confederate president rose and quietly walked out of the church. At eleven that night, Davis and most of his cabinet left Richmond on a special train, heading 120 miles southwest to Danville, Virginia. Their hope was that Lee might also be able to retreat to Danville, which was right on the North Carolina border seventy miles north of Joseph E. Johnston's army, and that with those combined forces they could carry on the war.

While the message sent by Lee was producing the Confederate government's departure from its capital, his men put up a remarkable last day's defense before their survivors left Petersburg. At Fort Gregg, two miles southwest of the city, some four hundred to six hundred men held off several thousand attacking federal troops for more than an hour; during the final fifteen or twenty minutes of hand-to-hand fighting on the parapet of the earthworks, the battle flags of six Union regiments were seen in the midst of the struggle. Among those Confederates killed during the day was Lee's veteran commander Lieutenant General A. P. Hill, who was shot down when he and a sergeant, out on their horses, encountered two Union soldiers who answered their demand that they surrender by firing their rifles at them. The long, brutal last day at Petersburg ended after dark: riding along with his troops on Traveller, Lee finally left the city he had defended for so long, crossing the bridge to the north bank of the Appomattox River.

Lee's final retreat was under way, with Grant's divisions hemming in his heavily outnumbered columns and cutting them up any time the hungry, weary Confederates tried to turn and make a stand. On April 5, Grant, who was moving along right behind the forward elements of the Union advance, sent Sherman a telegram in cipher, saying, "All indications are that Lee will attempt to reach Danville with the remnant of his force . . . If you possibly can do so push on from where you are and let us see if we cannot finish the job with Lee's and Johnston's Armies." Although the taking of Richmond had always held a symbolic significance because it was the Confederate capital, Grant reinforced his long-held view that the taking of cities and towns was secondary: "Rebel Armies now are the only strategic points to strike at." The next day, in an action at Sayler's Creek during which Union cavalry under General George Armstrong Custer blocked one of Lee's lines of retreat, six Confederate generals were captured, including Lee's son Custis.

Although the message from Grant urging Sherman to "push on from where you are and let us see if we cannot finish the job" was sent by wire, there were no telegraph lines running through Virginia that Union forces could use, so, going by wire to City Point, then on a ship to Morehead City, North Carolina, and then being copied and sent on by wire to Sherman's headquarters, it did not reach Sherman until three days later. As soon as he received it on April 8, Sherman wrote Grant that "I am delighted and amazed at the result of your move to the South of Petersburg, and Lee has lost in one day the Reputation of three years, and you have established a Reputation for perseverance and pluck that would make Wellington jump out of his Coffin." Thinking that his wish to be in at Lee's surrender was about to come true, he added, "It is to our interest to let Lee & Johnston come together just as a billiard player would nurse the balls when he has them in a nice place."

By the time this letter from Sherman reached Grant's constantly moving forward headquarters, Grant's forces had surrounded the shattered remnants of Lee's Army of Northern Virginia at Appomattox Court House, ninety miles southwest of Richmond, making it impossible for them to move farther in any direction. Lee's men were ready for one last battle, but Lee saw that it would have no military effect. Without further fighting, Lee surrendered his army in that suddenly quiet countryside on Palm Sunday, April 9, 1865.

The events of that day would come down through American history in detail, including the fact that at Appomattox there were many reunions between West Pointers who had fought on different sides— Grant had the opportunity for a friendly talk with Julia's cousin James Longstreet—but Grant's hour of victory was filled with mixed feelings. He recalled that when he received Lee's letter saying that "I therefore request an interview" for the purpose of surrendering, he had been suffering from a "sick headache, but the instant I saw the contents of the note I was cured." Grant added that he suddenly felt "quite jubilant," but that when he entered the parlor of the house where Lee stood waiting for him, wearing a gray dress uniform, he became "sad and depressed." Shaking hands and sitting down with the gracious and dignified Lee, whose face betrayed no emotion—Grant called it "an impassible face"— he was ambivalent. "I felt like anything rather than rejoicing at the downfall of a foe who had fought so long and valiantly, and had suffered so much for a cause, though that cause was, I believe, one of the worst for which a people ever fought, and one for which there was the least excuse."

After some minutes of Grant trying to ease the situation by talking about the Mexican War, to which Lee responded in a polite, abstracted fashion, Lee reminded Grant of the reason for their meeting. Grant called for pen and paper, and wrote down his terms for surrender. As was his practice, he knew what he wanted to say and wrote the articles of surrender without stopping, expressing himself clearly and succinctly.

When Lee read the surrender terms, he mentioned something that Grant did not know. Unlike the situation in the Union Army, the many thousands of horses in the Confederate Army belonged to the men who had brought them into the service with them. As the surrender terms now stood, only the Confederate officers could take their horses home, and all the other horses would be held as captured enemy property.

Lee simply mentioned it; he did not beg, and he did not have to. Ulysses S. Grant had learned a lot, peddling firewood on those cold streets in St. Louis during his prewar years. He saw it in an instant. "I take it," he said to Lee, "that most of the men in the ranks are small farmers, and as the country has been so raided by the two armies, it is doubtful whether they could put in a crop to carry themselves through the next winter without the aid of the horses they are now riding." As Lee remained silent, Grant said that he would arrange "to let all the men who claim to own a horse or mule take the animals home with them to work their little farms."

His face remaining impassive, Lee saw all the rest of it; if these horses and men could go home now as the April planting season began in the South, it could make the difference between full stomachs and near starvation for the children of the soldiers for whom he was negotiating. He said quietly, "This will have the best possible effect upon the men." Lee, who had ridden to the meeting on Traveller, four years earlier had been offered the field command of the army to whose present general in chief, Ulysses S. Grant, a former captain, he was now surrendering; arriving for this meeting, he had not known what to expect. Now he knew. Grant was feeding his men; Grant had arranged for everyone to be paroled and not be marched off to prison camps; Grant was letting them take their horses home. Referring to the matter of the horses but speaking in words that covered more than that, Lee said to Grant, "It will be very gratifying and will do much toward conciliating our people." Lee signed the surrender document. It was done.

Grant knew that many thousands of hungry Confederate soldiers were within a mile of where they sat, waiting for the results of this meeting. As soon as Lee had signed, Grant said, "I will take steps at once to have your army supplied with rations. Suppose I send over twenty-five thousand rations, do you think that will be a sufficient supply?"

"Plenty," Lee answered. "Plenty." He spoke as if overcome by this evidence of the resources of the army that had finally hammered him down. "An abundance." Lee paused. "And it will be a great relief, I can assure you."

Before coming, Lee had rejected the idea put forward by some of his diehard officers that his army should slip away in groups and become guerrilla bands operating throughout the South, a development that could have prolonged the fighting for months and even years. Although Lee had said earlier in the day, "Then there is nothing left me but to go and see General Grant, and I would rather die a thousand deaths," he had already decided to live the rest of his life in dignified acceptance of defeat, recognizing, as he soon said to an embittered Confederate widow, that "we are all one country now," and urging all his veterans to take up peaceful pursuits and get on with their lives. For himself, living an example that was an impressive and stabilizing influence for both North and South during the difficult years ahead, Lee would take on the presidency of Washington College, a small war-torn school in western Virginia, and rebuild it into the fine institution that upon his death in 1870 would be renamed Washington and Lee University. While he was still at Appomattox, there was only one thing Lee felt he could not do: in a final brief meeting between Grant and Lee the next day, before they both left the area, Grant told Lee that "there was not a man in the Confederacy whose influence with the soldiers and the whole people was as great as his" and suggested that, in addition to having surrendered his own Army of Northern Virginia, Lee could also "advise the surrender" of the other Confederate forces scattered throughout the South. Lee gave him the same answer that Grant had received concerning Lincoln's powers: only Jefferson Davis could negotiate on behalf of the Confederacy as a whole. Grant realized that this meant the war might go on for a time, and that Sherman would have to deal with Johnston down in North Carolina, but he did not press the matter. "I knew," he said of that moment with Lee, "that there was no use to urge him to do anything against his ideas of what was right."

After the close of the meeting at which the surrender terms were signed, Grant, lost in thought, came down the steps of the house where he and Lee had conferred. Suddenly he realized that this was Lee in front of him, mounted on Traveller, sitting erect and turning his horse's head as he started out of the yard, on the way to tell the men who had followed him through so much, and were still willing to fight on, that it was all over. Many Union officers were moving around in the yard, eager to mount their horses and get back to their commands with the news that the surrender was official.

Grant stopped and took off his hat. The yard became silent; every Union soldier there removed his hat and came to attention. Robert E. Lee lifted his hat once and passed through the gate, a man in a gray uniform riding a gray horse. For the remaining five years of his life, he never allowed a word against Ulysses S. Grant to be spoken in his presence.

16

THE DAYS AFTER APPOMATTOX: JOY AND GRIEF

Late on April 9, Grant's headquarters sent Sherman a telegram that included a copy of the surrender terms, but, partly because of the continuing need for telegrams to travel part of the way to North Carolina by ship and partly because Sherman stuck to his plan to "haul out towards Raleigh" on April 10, it took three days for the message to catch up to his headquarters in the field, between Goldsboro and Raleigh. At five in the morning on April 12, Sherman sent off a telegram to Grant that began, "I have this moment received your telegram announcing the Surrender of Lee's Army. I hardly Know how to express my feelings, but you can imagine them. The terms you have given Lee are magnanimous and liberal. Should Johnston follow Lee's example, I shall of course grant the Same. He is retreating before me on Raleigh, and I shall be there tomorrow. Roads are heavy [muddy], but under the inspiration of the news from you, we can march 25 miles a day."

The word of Appomattox spread through Sherman's sleeping regiments at dawn, with horsemen cantering through the bivouacs as they shouted, "Lee's surrendered!" Regiment after regiment exploded with joy. A man from Minnesota wrote, "I never heard such cheering in my life. It was one continuous roar for three hours." Muskets were fired in the air; fifes and drums played "Yankee Doodle," and Sherman sent this statement to his units: "Glory to God and our country, and all honor to our comrades in arms towards whom we are marching. A little more labor, a little more toil on our part, the great race is won, and our Government stands regenerated, after four long years of war."

The festivities continued into the night. Theodore Upson, a long-serving sergeant of the 100th Indiana, left this account of how things went at an encampment that evidently received the news late in the day.

> We had a great blowout at Hd Quarters last night . . . Gen Woods came out saying "Dismiss the guard, Sergeant, and come into my tent." I thought he was crazy or some thing, so asked for what reason. He said, "Don't you know Lee has surrendered? No man shall stand guard at my Quarters tonight. Bring all the guards here" . . . Officers were coming from every direction . . . He had a great big bowl sitting on a camp table. The General handed me a tin cup. "Help yourself," he said . . . I never had drank liquor, and I did not know what it would do to me.
>
> After a while a band came. They played once or twice, drank some, played some more . . . then they played again but did not keep very good time. Some of them could not wait till they got through a tune till they had to pledge [toast] Grant and his gallant Army, also Lee and his grand fighters . . . The Band finally got so they were playing two or three tunes at once . . .
>
> General Woods shook my hand and said he would promote me, that I could consider myself a Lieut. After a little more talk . . . he made me a Captain, and I might have got higher than that if the General had not noticed the Band was not playing . . . He found the members seated on the ground or anything else they could find, several on the big bass drum . . . He got the big drum, other officers took the various horns and started through the camps—every fellow blowing his horn to suit himself and the jolly old General pounding the bass drum for all it was worth.
>
> Of course we all followed and some sang, or tried to sing, but when "Johnny Comes Marching Home Again" and "John Browns Body" or "Hail Columbia" and the "Star Spangled Banner" are all sung together they get mixed so I don't think the singing was a grand success from an artistic point of view at least.

As the celebrations wound down, Sherman was trying to get in touch with Joseph E. Johnston, to see if he could bring the war to an end on his front. As his men approached North Carolina's capital of Raleigh, with Johnston's rear guard clearly trying to avoid contact, Sherman issued orders to cease laying waste the enemy's country: "No further destruction of railroads, mills, cotton and produce, will be made without the specific orders of an army commander, and the inhabitants will be dealt with kindly, looking to an early reconciliation."

After an agreement had been reached with Raleigh's mayor that there would be no resistance—Johnston's forces had been withdrawn to the north and northwest—Sherman's advanced troops entered the city. Only one remaining diehard, wearing a blue uniform that made him suspect as a spy, fired revolver shots at them on his own initiative, and he was captured and quickly executed on the grounds that he had violated the surrender agreement.

Then Sherman ordered his army to parade through the streets, bands playing. This display of force was intended to impress the city's residents with the futility of further resistance, and it did. Major General Henry Slocum's chief of staff, thirty-five-year-old Union general Carl Schurz, the German-born antislavery orator who had ahead of him a career as a nationally prominent newspaper editor and political figure, said of the endless columns, "As far as the eye can reach is a sea of bayonets." A young Southern woman who was watching wiped her eyes with a hand-kerchief and said through her tears, "It is all over with us; I see now, it is all over. A few days ago I saw General Johnston's army, ragged and starved; now when I look at these strong healthy men and see them com-ing and coming—it is all over with us!" (A small boy who had been told that all Yankees were devils said that he had been watching them pass all morning and hadn't seen one with horns on his head.) Sherman's troops were on their best behavior, far different from their actions and attitude in Georgia and South Carolina. The *Raleigh Daily Standard* soon com-mented on "the gentlemanly bearing of the officers and men. From General Sherman to the humblest private, we have witnessed nothing but what has been proper and courteous."

On April 14, Sherman received a letter from Johnston, asking for "a temporary suspension of active operations . . . to permit the civil authori-ties to enter into the needful arrangements to terminate the existing war." This request for a truce was just what Sherman wanted, and he replied that the two armies should remain apart and that "I undertake to abide by the same terms and conditions as were made by Generals Grant and Lee . . . I will add that I really desire to save the people of North Carolina the damage they would sustain by the march of this army through the central or western parts of the State."

Sherman felt that, after his meeting with Grant and Lincoln at City Point, he understood exactly what he was empowered to do, and that it was now up to Johnston to consult with "the civil authorities" to which he had referred in his letter. In a telegram to Grant, with a copy for Stanton, Sherman set forth the text of his exchange of letters with Johnston and said that he thought this "will be followed by terms of capitulation." He repeated his intention of following Grant's items of surrender agreed upon at Appomattox and added that he would "be careful not to complicate any point of civil policy." He closed by saying he expected that, once Johnston received the Confederate governmental decision to surrender, "all the details are easily arranged."

Joseph E. Johnston moved as quickly as he could, visiting scattered and confused Confederate officials in various locations in North Carolina. He finally tracked down Jefferson Davis in Charlotte, only to find the Confederate president talking about raising new Confederate armies and fighting on. Johnston bluntly told Davis that "it would be the greatest of human crimes for us to continue the war" and added that "the only function of government left in his possession" was that of agreeing to negotiations for peace. Davis did not give Johnston authority to do more than surrender his own army, but a meeting between Sherman and Johnston was finally set for April 17.

While Grant had been closing out the military side of the war, Lincoln had visited the enemy capital soon after its defenders evacuated the city. Alerted by Lee, Jefferson Davis and his cabinet had left Richmond the night of Sunday, April 2, and within hours fires were set to destroy the Confederacy's military records; warehouses filled with cotton, tobacco, and military supplies were also put to the torch. These soon burnt out of control, and other fires started by looters merged into an inferno that engulfed most of Richmond. Everything went up in flames: factories, hotels, business offices, residences. The following morning, federal cavalrymen rode into the still-flaming city, followed by white infantrymen led by a band playing "The Girl I Left Behind Me" and black troops marching to the tune of "Dixie."

On the same day, Lincoln went from City Point to just-evacuated Petersburg, visited with Grant for an hour and a half, and returned to City Point. (Neither man left a record of what they discussed.) The following morning, April 4, the president left City Point for Richmond aboard the USS *Malvern*. Coming ashore accompanied by Admiral Porter and a squad of sailors armed with carbines, he was welcomed by Major General Jacob Weitzel, in charge of the Union forces occupying the smoldering enemy capital. As Lincoln walked through cheering crowds of blacks, many of these freed slaves fell to their knees before him, crying out that he was a "Messiah." Asking them to get back on their feet, Lincoln told them, "Don't kneel to me. You must kneel to God only and thank him for your freedom."

After visiting the Confederate White House that Jefferson Davis had occupied until forty-eight hours before, Lincoln was driven through the city in a carriage and met with John A. Campbell, a former justice of the United States Supreme Court who had served as the Confederacy's assistant secretary of war. The president had last seen Campbell in early February, at the unsuccessful talks that became known as the Hampton Roads Peace Conference, held aboard Lincoln's steamer *River Queen* in the waters off Fort Monroe on the Virginia coast. On that occasion, Lincoln and his secretary of state William Seward had met with Campbell, Confederate vice president Alexander H. Stephens, and the Confederacy's secretary of state R.M.T. Hunter, only to find that the South's representatives still would not consider any peace that did not leave the Confederacy as an independent nation entitled to practice slavery.

As he left Richmond, Lincoln said to General Weitzel, speaking of the defeated Southerners, "If I were in your place, I'd let 'em up easy, let 'em up easy." Back at City Point two days later, knowing that Jefferson Davis was still fleeing, Lincoln wrote Weitzel: "It has been intimated to me that the gentlemen who have acted as the Legislature of Virginia, in support of the rebellion, may now . . . desire to assemble at Richmond, and take measures to withdraw the Virginia troops, and other support from resistance to the General [federal] government. If they attempt it, give them permission and protection." With this, Lincoln was recognizing for the moment at least the legitimacy of Virginia's Confederate legislature to act on behalf of Virginia in cooperating with the federal government—something that would soon bring him once again into a confrontation with the Radicals, who felt that a legislature that had voted to secede from the Union had no authority whatever, and that the only government suitable for a defeated Confederate state was a military occupation.

Everyone who saw Lincoln at this time, in Washington and anywhere else, was struck by the sorrow marked on his face. The poet Walt Whitman, who profoundly admired Lincoln and had spent the war working with the wounded in military hospitals in the Washington area, had recently described Lincoln in a letter to a friend he had met in the course of his work, a nineteen-year-old Confederate captain from Mississippi who lost a leg at Fredericksburg and was taken to Washington as a prisoner: "He has a face like a hoosier Michael Angelo, so awful ugly it becomes beautiful, with its strange mouth, its deep cut criss-cross lines, and its doughnut complexion." In addition to bearing the enormous responsibilities of the convulsive national crisis for the past four years, Lincoln had truly meant it when in his Second Inaugural Address on March 4 he had said, "With malice toward none; with charity for all." Above even his constant concern for his soldiers and their families—in his speech he pledged "to care for him who has borne the battle, and for his widow, and his orphan"—he loved the entire nation, the South as well as the North, and was painfully aware of the passions bred by four years of war. On the one hand, he had such Radicals as Congressman George W. Julian of Indiana, who three years before had said, looking ahead to a Union victory, "Let us convert the rebel States into conquered provinces, remanding them to the *status* of mere Territories, and governing them as such in our discretion." Since then Julian had developed that view into this, from a speech he made on the House floor on February 7, less than a month before Lincoln's "With malice toward none" address: "Both the people and our armies . . . have been learning how to hate rebels as Christian patriots ought to have done from the beginning." On the other hand, Lincoln was equally aware of Southern emotions such as those expressed after the surrender at Appomattox, when a Union general told Confederate general Henry A. Wise that he hoped there would be good relations between North and South. Wise, a former governor of Virginia, replied, "There is a rancor in our hearts that you little dream of. We hate you, sir."

Apart from the great issues that weighed upon him, Lincoln, like Sherman, had experienced personal tragedy during the war. A year and a half before the Shermans' son Willy died at the age of nine, the Lincolns' eleven-year-old son William Wallace Lincoln had died of malaria in Washington. (Jefferson Davis and his wife Marina also lost a child when their five-year-old son Joseph Evan Davis fell to his death from a porch fifteen feet high at the Confederate White House in Richmond in 1864.)

With all his cares, Lincoln spoke of himself as "very greatly rejoiced" by the news of Lee's surrender to Grant at Appomattox on April 9. The day after that, talking to a crowd in Washington that had brought along two or three bands to serenade him, Lincoln said, "I have always thought 'Dixie' one of the best tunes I have ever heard. Our adversaries over the way attempted to appropriate it, but I insisted . . . that we fairly captured it." As the crowd applauded, Lincoln added, to more applause and laughter, "I presented the question to the Attorney General, and he gave it as his opinion that it is our lawful prize."

The next evening, as he made a much longer speech, the public saw a different Lincoln, thoughtful and intense. Here was the eve of peace, but throughout the war he had never been able to get a consensus in Congress for his specific plans for the postwar treatment of the South, and had opposed the Radicals in such a way that they could not enact their harsher views into law. His mood at the moment seemed to match the weather. It was a misty night; reading by the light of a candle held by an aide, Lincoln began speaking to an enthusiastic crowd that had gathered beneath the shadowy second-floor balcony window of the White House, expecting to hear a victory speech. He began as his listeners expected him to, saying, "We meet this evening, not in sorrow, but in gladness of heart," and added, paying tribute to the recent successes of "General Grant, his skillful officers, and brave men," that "no part of the honor, for plan or execution, is mine."

Despite Lincoln's saying that he hoped for "a righteous and speedy peace," those who wanted a rousing speech soon found themselves disappointed. They were listening to a scholarly soliloquy, in which Lincoln revealed himself as still wrestling with the moral problems and political realities that lay ahead at this moment when Lee had surrendered to Grant and Sherman was clearly moving to finish off the one remaining large Confederate force under Joseph E. Johnston.

Concerning "reconstruction, as the phrase goes," Lincoln called it a prospect "fraught with difficulty." Speaking defensively, he referred to a criticism "that my mind has not seemed definitely fixed on the question whether the seceded states, so called, are in the Union or out of it." Terming that question "a merely pernicious abstraction," he said, "We all agree that the seceded States, so called, are out of their proper practical relation to the Union, and that the sole object of the [federal] government, civil and military, is to again get them into that proper practical relation." Lincoln told the crowd, "We simply must begin with, and mould

from, disorganized and discordant elements." He added, using as an example the continuing efforts made during the war to form and maintain
a state government in Louisiana that was acceptable to the Union, that
in the coming period "no exclusive, and inflexible plan can be prescribed as to details." Again using the example of what might be accomplished in Louisiana, he spoke in favor of "giving the benefit of public
schools equally to black and white, and empowering the Legislature to
confer the franchise upon the colored man." As for who among the
blacks should have the right to vote, Lincoln was at the moment for being selective: "I would myself prefer that it now were conferred on the
very intelligent, and on those who serve our cause as soldiers."

The mist turned to rain; before Lincoln finished speaking, a considerable number of the crowd drifted away. The man who had led the nation
through its greatest crisis, the man who had held the border states in the
Union and nonetheless freed so many slaves was being honest and realistic enough to say of those who had supported the Union that even now
"we, the loyal people, differ among ourselves as to the mode, manner,
and means of reconstruction." This speech of Lincoln's on April 11, two
days after the surrender at Appomattox, was not the speech the crowd
wanted to hear.

The next day, in a cabinet meeting and in separate discussions with
several cabinet members, Lincoln discovered that his advisers had
learned about his recent dealings in Richmond with the Confederate
leader Campbell and his subsequent instruction to Union general
Weitzel to allow the Virginia legislature to assemble for the purpose of
withdrawing from the Confederacy. (Assistant Secretary of War Charles
A. Dana, who had been with Lincoln at City Point and knew about all
this, had sent reports to his superior, Secretary of War Stanton, about the
matter, and Stanton had passed on the information to Attorney General
James Speed and Postmaster General William Dennison.) Lincoln, who
took the position that he had "just seen" a letter from Campbell to
Weitzel in which Campbell spoke of reassembling the Virginia legislature in its official capacity, tried to argue that he was simply trying "to effect a reconciliation as soon as possible." He added that he had never
intended to treat the Virginia legislators as "a rightful body," but his cabinet resolutely opposed him on the issue. Finally admitting that he "had
perhaps made a mistake," Lincoln gave up on the initiative he had dis-

cussed with Campbell and later in the day wrote Weitzel, concerning the Virginia legislators, "Do not allow them to assemble, but if any have come, allow them safe-return to their homes."

While Sherman continued trying to bring things to an end in North Carolina, Julia Grant reported that in Washington "everyone was wild with delight" about the surrender. Grant returned to City Point from Appomattox shortly after four a.m. on Tuesday, April 13, and Julia noted, "About fifty generals of high rank and other officials breakfasted with us that morning." Later in the morning, she, Grant, and "a large number of other officers" went up to Washington aboard the *Mary Martin*, and as they came up the Potomac, "all the bells rang out in merry greetings, and the city was literally swathed in flags and bunting. The sun shone brightly, and the very winds seemed on a frolic." Julia was struck by the appearance of the American flag at the landing: the wind had come up strongly, and the Stars and Stripes was "broadly spread as though to show that not a single star was lost from that blue field . . . Our Union is safe." Grant drove in a carriage with Julia and the wife of John Rawlins to the fashionable Willard Hotel and then, Julia said, "as soon as he saw me comfortably located, went straight to the Executive Mansion" to meet with Lincoln.

The excitement in the city that Julia saw on arriving by boat was a foretaste of the evening to come: Washington was to have a "Grand Illumination" that night, with government buildings lit by flaming gas jets, some designed as stars and eagles, and others spelling out "Peace" and "Victory." Candles would shine brightly in every window of the government buildings and many of the residents' houses; there would be bonfires in the streets and in front of statues, and fireworks in the sky. Julia was looking forward to the events of the evening, which would include a reception honoring Grant at Stanton's house, to take place after the Grants and the Stantons rode around the city together in a carriage to see the city in its festival mood. Then, at a moment when Grant was out, a note from Mrs. Lincoln was delivered to the Grants at the Willard, and Julia opened it. Addressed only to Grant and not mentioning Julia, it said, "Mr. Lincoln is indisposed with quite a severe headache, yet would be very pleased to see you at the [White] house this evening about 8 o'clock & I want you to drive around with us to see the illumination."

When Grant returned—apparently Lincoln was feeling so bad that he held meetings with Grant, Secretary of War Stanton, and Navy Secretary Gideon Welles but did not venture out of the White House until the next day—Julia showed him this invitation. Grant's first reaction was that this would be fine; Julia would ride around the city with the Stantons and he would ride with Mrs. Lincoln and come by himself to the Stantons' after that, but he quickly learned that would not do. "To this plan," Julia recalled, "I protested and said I would not go at all unless he accompanied me." Grant retreated and said he would first ride around Washington with Julia and the Stantons, then "escort the wife of our President to see the illumination" as he felt it his duty to do, and then come to the party at the Stantons'. "This was all satisfactory to me," Julia said, "as it was the honor of being with him when he first viewed the illumination in honor of peace being restored to the nation, in which he had so great a share—it was this I coveted."

And so Julia and her "Ulys" drove through the brilliantly lit streets—the Capitol's marble dome and portico were gleaming, and many flags hung from the balconies and windows of the White House. There were thirty-five hundred candles in the windows of the Government Post Office, and six thousand illuminated the Patent Office. Lanterns glowed everywhere, and fireworks exploded high above them as crowds surged through the streets while bands playing patriotic tunes marched in every direction. Later, at the Stantons' house, Grant was the center of attention. Julia wrote, "All of the great men of the nation who were necessarily in Washington at that time were assembled that night. Such congratulations, such friendly, grateful grasps of the hand and speeches of gratitude!" The next day would mark the fourth anniversary of the surrender of Fort Sumter after the Confederate bombardment that began the fighting. On that day in 1861, Ulysses S. Grant, who had been forced to resign as a captain, a rank three grades below that of colonel, was not yet back in the army. Now he was the victorious commander of a force that had grown to a million men.

On the morning of that next day, Good Friday, April 14, 1865, Julia had plans to go to Burlington, New Jersey, where she and Grant had rented a house for her and their four children to use when they were not visiting him at his headquarters at City Point. She had their seven-year-old son Jesse with them in Washington, but now she wanted to get back to the other three children and asked Grant to come with her on the evening train to Philadelphia and then on to Burlington. Grant explained that he had to go to the White House at eleven that morning to meet Lincoln and his cabinet to discuss "the reduction of the army" and doubted he could get away at any point that day. Finally, when Julia said she could not wait until the next day, he said, "Well, I will see what I can do. I will certainly go if it is possible."

At noon, a man whose looks Julia did not like arrived at the Willard Hotel and said to her, "Mrs. Lincoln sends me, Madam, with her compliments, to say that she will call for you at exactly eight o'clock to go to the theater." On hearing what he had to say—there was no written message, and in any event the tone of it "seemed like a command"—she told the man to convey her regrets to Mrs. Lincoln and to say that she and Grant would be leaving the city that afternoon. The man persisted, saying, "Madam, the papers say that General Grant will be with the President tonight at the theater." Julia told him to leave.

> I dispatched a note to General Grant entreating him to take me home that evening; that I did not want to go to the theater; that he must take me home. I not only wrote to him, but sent three of his staff officers who called to pay their respects to me to urge the General to go home that night. I do not know what possessed me to take such a freak, but go home I felt I must.

Grant sent word that if he could possibly accompany them to Burlington, he would pick Julia and Jesse up at the hotel and they would go together to the station. Lincoln had indeed invited him and Julia to the theater—Mrs. Lincoln wanted to see *Our American Cousin,* a popular farce that was having its last performance at the Ford Theatre—but Grant was able to make their excuses. Julia was having a "late luncheon with Mrs. Rawlins and her little girl and my Jesse" at the hotel when four men came in and sat at the next table. Julia thought she recognized the man who had brought the message purporting to be from Mrs. Lincoln, and she was particularly struck by the behavior of "a dark, pale man" who "played with his soup spoon, sometimes filling it and holding it half-lifted to his mouth, but never tasting it. This occurred many times. He also seemed very intent on what we and the children were saying. I thought he was crazy." Quietly, Julia asked Mrs. Rawlins what she thought of the men at the next table, and when Mrs. Rawlins agreed that their behavior seemed "peculiar," Julia said, referring to a famous Confederate raider, "I believe they are part of Mosby's guerrillas and they have been listening to every word we have said. Do you know, I believe there will be an outbreak tonight or soon. I just feel it and am glad I am going away tonight."

The cabinet meeting that Lincoln had invited Grant to attend began at eleven and went on for three hours. The entire cabinet was present, except for Secretary of State Seward, who had been severely injured in a carriage accident nine days before and was in a weakened condition at home in bed. On some matters there was agreement: as soon as the victory was complete, unhindered commercial relations should be established with the states that had seceded, and the Departments of Treasury and the Interior would resume their normal functions in the South, along with the reestablishment of one national postal service under the postmaster general.

Then the substance and the mood of the meeting changed. Lincoln not only believed in magnanimity toward the defeated South, but was convinced, as he now told his cabinet, that as a practical matter, "We can't undertake to run State governments in all these Southern states. Their people must do that—though I reckon at first some of them may do it badly." This did not precipitate an argument, but it was generally agreed that an army of occupation would be needed, with military governors ruling the former Confederate states under martial law until some form of civilian rule was reestablished. (In this connection, the Commonwealth of Virginia would become, in administrative terminology, Military District No. 1.)

Lincoln was ready to give up on some of the specifics of what was to be done in the postwar situation, even deferring for the moment the matter of who in the South should be allowed to vote, but he persisted in putting forward his philosophy of what was needed. He did not want trials of the Southern leaders, or hangings. His solution: "Frighten them out of the country," he told his advisers, waving his hands as if shooing something away. "Let down the bars, scare them off." He expressed his fears about the "feelings of hate and vindictiveness" among many in Congress.

Unanimity in the room was reestablished when Lincoln asked Grant to tell the cabinet about the surrender at Appomattox. Lincoln was clearly pleased when Grant said of the Confederates, "I told them to go back to their homes and families, and they would not be molested, if they did nothing more." Some of Lincoln's advisers then wanted to know about Sherman's progress toward finishing things with Joseph E. Johnston in North Carolina, and Grant responded that he was expecting to hear more about that at any hour.

The reference to Sherman prompted Lincoln to tell the meeting about a dream he had the previous night. He was aboard a ship moving with "great rapidity" toward a shore that he described as "vast" and "indefinite." Lincoln said that it was the same dream he had before receiving the news of Gettysburg, Vicksburg, and other victories. He felt certain that this time the good news was coming from North Carolina. "I think it must be from Sherman," Lincoln said. "My thoughts are in that direction, as are most of yours."

The cabinet meeting had finished in time for Grant to make good on his promise to Julia that he would accompany her to Burlington, New Jersey, that evening. As Grant, Julia, and their young son Jesse rode to the station with the wife of General Daniel Rucker, whose carriage they were in, Julia recalled, speaking of the strange person she had seen at lunch, "This same dark, pale man rode past us at a sweeping gallop on a dark horse—black, I think. He rode twenty yards ahead of us, wheeled and returned, and as he passed us both going and returning, he thrust his face quite near the General's, and glared in a disagreeable manner." When Mrs. Rucker said, "General, everyone wants to see you," Grant replied, "Yes, but I do not care for such glances. They are not friendly."

In Philadelphia, before taking a ferryboat across the Delaware to get the train for Burlington at Camden, New Jersey, the Grants stopped to dine at Bloodgood's Hotel, near the ferry slip. Grant had not eaten since nine that morning and ordered some oysters. "Before they were ready for him," Julia said, "a telegram was handed to him, and almost before he could open this, another was handed him, and then a third." She described her husband's reaction. "The General looked very pale. 'Is there anything the matter?' I inquired. 'You look startled.' 'Yes,' he answered. 'Something very serious has happened. Do not exclaim. Be quiet and I will tell you. The president has been assassinated at the theater. I must go back at once. I will take you to Burlington . . . see the children, order a special train, and return as soon as it is ready.' "

When Lincoln's assassin John Wilkes Booth was tracked down and killed while resisting capture, Julia thought from seeing pictures of him that he was the "dark, pale man" who had watched her so carefully at lunch and then ridden so close to the carriage taking them to the station. An unsolved element of mystery was added to the shock and horror of the night that Lincoln was shot, dying the next morning. In her memoirs, Julia recalled that the next morning an unsigned letter arrived, saying, "General Grant, thank God, as I do, that you still live. It was your life that fell to my lot, and I followed you on the [railroad] cars. Your car door was locked, and thus you escaped me, thank God." (Years later, Grant confirmed that such a letter had come; soon after the assassination he said that he wished he had been in the presidential box at the theater when the attack occurred, because he might have been able to disarm Booth, or step in the way of the bullet intended for Lincoln.)

It had indeed been a wider plot, involving a total of nineteen conspirators. The first part was the attack on Lincoln. At about twenty minutes past ten on the evening of April 14, the famous actor John Wilkes Booth, who was not in the play Lincoln and his wife were watching, entered the presidential box and fired his small derringer pistol into the back of Lincoln's head. Shouting "*Sic semper tyrannis!*"—Thus always to tyrants—Booth leapt to the stage, brandishing a dagger. According to one shocked witness, he exclaimed, "The South shall be free!" before he ran through the wings of the stage and escaped.

At the same time, six blocks away near Lafayette Park, another conspirator, a tall, strong, well-dressed man who was using the name John Powell, entered the house of Secretary of State William H. Seward, who was bedridden from his recent carriage accident. Seward was lying in his sickbed on the third floor, with his broken arm in a cast and a metal brace immobilizing his head and neck as his fractured jaw healed. Powell claimed that he had some medicine that must be given to Seward; brushing past a servant, he had reached the second floor when Seward's son Frederick confronted him. The mysterious intruder pulled out a revolver. When it failed to fire, Powell repeatedly struck Frederick on the skull with a pistol butt, causing five fractures and leaving him lying in his blood. Pulling out a big bowie knife, the assailant dashed up the next flight of stairs and into Seward's bedroom, where he ran into Seward's daughter Fanny and a male army nurse tending to Seward, who was lying helpless in his bed. Powell hit Fanny so hard that she fell to the floor unconscious, and then he slashed the male nurse across the forehead with his knife before hitting him so that he too fell unconscious to the floor. Finally coming to the bed where the defenseless Seward lay, Powell stabbed Seward deep in the cheek, nearly cutting his cheek from his face, and then knifed him several times more, including three stabs into his neck. At this moment Major Augustus Seward, another of the secretary of state's sons, rushed into the room, and in the ensuing struggle, Powell stabbed him seven times. By this time the army nurse had recovered consciousness and came at Powell, who wrestled with him and stabbed him four more times. Then Powell fled down the stairs, stabbing in the chest a State Department messenger who happened to be entering the house. Having left blood splashed all over the stairs, walls, and front steps, Powell ran from the house, shouting, "I'm mad! I'm mad!" He leapt onto a horse and dashed into the darkness.

As the night passed, the city of Washington, so joyous and relieved about Lee's surrender at Appomattox less than a week before, filled with chaotic horror. Lincoln was mortally wounded and was unconscious and sinking but not yet dead. It was believed that Seward could not live. At the Kirkwood House, the hotel where Vice President Andrew Johnson was staying, a detective searched the room of a man named George A. Atzerodt, who had checked in that morning and was no longer there, and found a concealed loaded pistol and a bowie knife. There was reason to think that other government figures were targets for assassination and might be killed at any time. Orders were given to army sentries surrounding the city to let no one pass through their lines, and soldiers were placed on guard at the houses of the nation's leaders. Every train leaving the city for Baltimore had soldiers aboard, searching for suspects. Inside Ford's Theatre, all the actors and employees were detained for questioning. In the streets, when the crowds saw police escorting individuals who were in fact witnesses who wished to tell what they had seen in the hope of helping the investigation, they thought they were seeing some of the plotters, and cries arose for them to be hanged on the spot.

At the boardinghouse near the theater where Lincoln had been taken, Mary Todd Lincoln, who had witnessed the attack as she sat beside her husband and had cradled him in her arms until he was carried away, screamed at Assistant Treasury Secretary Maunsell B. Field, "Why didn't he shoot me! Why didn't he shoot me! Why didn't he shoot me!" In the confusion, Secretary of War Stanton first went to Seward's house, where there was blood all the way from the front door to the third-floor

bedroom where Seward had been attacked; he and the other victims
were being treated by hastily summoned doctors. (All would survive.)
Anyone would have been severely shaken by the nightmare scene of
blood splashed everywhere; Stanton never commented on whether the
shattering moment reminded him of the suicide of his brother, during
which "the blood spouted up to the ceiling." Stanton went next to the
house where Lincoln lay. When he saw Lincoln, Surgeon General
Joseph K. Barnes whispered to him, "Mr. Lincoln cannot recover." Stanton, a man who never displayed his emotions, began sobbing loudly, his
shoulders convulsively shaking for several minutes. The other members
of Lincoln's cabinet joined Stanton there by midnight; Stanton soon
composed himself and took control of the situation, using the little
house as a command post for sending out communications about the
governmental emergency. Standing beside Lincoln's bed when the president died at 7:22 the next morning, April 15, 1865, Stanton said quietly,
"Now he belongs to the ages."

The news of the attack on Lincoln spread from Washington across
the North like a thunderclap. *The New York Times*, appearing on the
streets the morning of April 15 with a report sent from Washington before Lincoln died, had the headline, "AWFUL EVENT," with the subheadlines, *"President Lincoln Shot by an Assassin," "The Act of a
Desperate Rebel," "No Hopes Entertained of His Recovery."* Word of his
death followed swiftly, flashed across the nation's telegraph lines. It was
almost impossible to believe: in the hour of victory, with the restoration
of the Union a step away, the man who had led the nation through its
darkest hours had been torn from the people he loved and served.

On that Saturday morning, the bells began tolling. Stores shut their
doors; in New York City an art gallery closed, leaving in its big glass window only one empty picture frame. Broad ribbons of black crepe began
appearing on houses, churches, and public buildings. In the camps of
the Army of the Potomac, there was anger at the South, but the main reaction was one of stunned grief. One officer said his men "seemed stupefied by the terrible news." A young soldier of the 148th Pennsylvania
burst into tears, sobbing, "He was our best friend. God bless him," and a
private wrote home, "What a hold Old Abe had on the hearts of the soldiers of the army could only be told by the way they showed their mourning for him." When Admiral David Dixon Porter heard the news upon

landing at Baltimore aboard the USS *Tristram Shandy*, he wrote his mother, "The United States has lost the greatest man she ever produced."

During the next days, the headlines were followed by editorials and tributes; speeches and sermons were given everywhere. On April 19, the voice of intellectual New England was heard at the Unitarian Church in Concord, Massachusetts. Ralph Waldo Emerson told the assembled mourners, "The President stood before us as a man of the people . . . His occupying of the chair of state was a triumph of the good sense of mankind, and of the public conscience . . . His powers were superior. This man grew according to the need." Speaking of Lincoln's wartime leadership, he said, "There, by his courage, his justice, his even temper, his fertile counsel, his humanity, he stood a heroic figure in the centre of a heroic epic. He is the true history of the American people in this time."

At the time Emerson was speaking, Walt Whitman, who so greatly admired Lincoln, had already begun pouring out his grief in a poem, "Hush'd Be the Camps To-day." He would later publish two other heartfelt tributes—"O Captain! My Captain!" and the sublime "When Lilacs Last in the Dooryard Bloom'd," the latter with its line, "O powerful western fallen star!"—but having cared for so many wounded Union soldiers, in his first shock Whitman clung to something of which he was certain. The average soldier believed, rightly, that Lincoln cared about him and admired him, and using the word "celebrate" to mean commemorate, honor, and solemnize, Whitman wrote of the bond that he knew existed between them and their murdered leader:

Hush'd be the camps to-day,
And soldiers let us drape our war-worn weapons,
And each with musing soul retire to celebrate,
Our dear commander's death.

No more for him life's stormy conflicts,
Nor victory, nor defeat—no more time's dark events,
Charging like ceaseless clouds across the sky.

But sing poet in our name,
Sing of the love we bore him—because you, dweller in camps, know it truly.

As they invault the coffin there,
Sing—as they close the doors of earth upon him—one verse,
For the heavy hearts of soldiers.

Once again, while all in the North learned of Lincoln's death swiftly, because of the need for communications sent to Sherman's army to go part of the way by ship, news traveled slowly from Washington to Sherman's headquarters in North Carolina. At eight o'clock on the morning of April 17, forty-eight hours after Lincoln's death, Sherman and several of his staff were boarding the special train of one locomotive and two passenger cars taking him north from Raleigh to meet Joseph E. Johnston near Durham to discuss the terms of surrender of Johnston's army. At that moment, Sherman said, "The telegraph operator, whose office was up-stairs in the depot-building, ran down to me and said that he was at that instant of time receiving a most important dispatch in cipher from Morehead City, which I ought to see. I held the train for nearly half an hour, when he returned with the message translated [decoded] and written out."

The telegram was from Stanton: it gave the facts of Lincoln's assassination and said there was reason to think that Grant, newly sworn-in President Andrew Johnson, and other political and military leaders might be targets themselves. (Sherman would also receive a warning from Stanton of a report that he was a specific target, but nothing came of that.)

Reading this telegram as the train waited for his order to pull out, Sherman immediately found himself "dreading the effects of such a message at that critical time." He knew that "Mr. Lincoln was particularly endeared to the soldiers, and I feared that some foolish woman or man in Raleigh might say something or do something that would madden our men, and that a fate worse than that of Columbia would befall the place." Telling the telegraph operator to let no one learn of Lincoln's death, Sherman said nothing of this to his staff officers and simply let it be known that he would return to Raleigh that afternoon.

About two and a half hours later, Sherman was greeted politely by Johnston, whom he had never met, at a place in the countryside a few miles from Durham, and the two of them went into a small frame house that belonged to a local farmer and his wife, James and Nancy Bennett. "As soon as we were alone together I showed him the dispatch announcing Mr. Lincoln's assassination, and watched him closely. The perspiration came out in large drops on his forehead, and he did not attempt to conceal his distress. He denounced the act as a disgrace to the age, and hoped I did not charge it to the Confederate Government. I told him I could not believe that he or General Lee, or the officers of the Confederate army, could possibly be privy to acts of assassination; but I would not say as much for Jeff. Davis."

Turning to the immediate military situation, Johnston readily agreed that "any further fighting would be '*murder*;' but he thought that, instead of surrendering piecemeal, we might arrange terms that would embrace *all* the Confederate armies." When Sherman asked Johnston if he had the authority to surrender all the Confederate forces still spread out in places like Louisiana, Texas, Alabama, and parts of Georgia, Johnston said that he did not but indicated that he thought that, "during the night," he could get Davis's agreement for what Sherman termed "a universal surrender." They agreed to meet at the same place the next day and ended what Sherman called an "extremely cordial" conversation, "satisfying me that it could have but one result . . . to end the war as quickly as possible."

Back in Raleigh, Sherman issued a Special Field Order to his army, informing the troops of Lincoln's assassination, denouncing the crime and at the same time saying he knew that "the great mass of the Confederate army would scorn to sanction such acts." He cautioned his officers "to watch the soldiers closely, to prevent any violent retaliation by them," and it was well that he did. A major wrote in his diary, "The army is crazy for vengeance." Most of the troops contented themselves with storming around their bivouacs, shouting angrily and bellowing out the song, "We'll Hang Jeff Davis to a Sour Apple Tree," but about two thousand men from one encampment headed toward Raleigh, and only the threat of being blasted by their own artillery—a battery was placed right in the road, the cannon aimed at them—stopped the potential violence. Sherman spent the night riding from one of his divisions to another, keeping the peace, and later said, "Had it not been for me Raleigh would have been destroyed." The next day, a large meeting of citizens of Raleigh convened at the Wake County Court House. In an action that helped calm the situation to some degree, the meeting quickly drafted, approved, and circulated a resolution "to express our utmost abhorrence of the atrocious deed." (When a Southern lady expressed to Sherman her pleasure that Lincoln had been killed, he replied, "Madam, the South has lost the best friend it had.")

Conferring with his generals about the overall situation, Sherman found that their greatest concern was that he sign something that would guarantee an end to the fighting. They had chased Johnston and his hardened troops all over the South. Johnston still had forty-five thousand men in the area, and Sherman said that his generals now told him that if these negotiations failed "they all dreaded the long and harassing march in pursuit of a dissolving and fleeing enemy—a march that might carry us back again over the thousand miles that we had just accomplished. We all knew that if we could bring Johnston's army to bay, we could destroy it in an hour, but that was simply impossible in the country in which we found ourselves." Sherman, remembering his conversation with Lincoln at City Point, asked his generals if he should let Jefferson Davis and his fleeing cabinet "escape from the country" if they fell into his army's hands. One of them, mentioning the capital of the British island chain of the Bahamas off the Florida coast, replied that, "If asked for, we should even provide a vessel to carry them to Nassau from Charleston."

On his way to meet Johnston again that afternoon, Sherman felt mounting pressure to get the entire situation settled. Today he found Johnston, while making statements that proved to be entirely honest, also to be as adroit in negotiations as he had been in his long series of retreats. Johnston had prepared himself for this second meeting as fully as a lawyer prepares for a negotiation or a trial. Sherman said that he "assured me that he had the authority to surrender all the Confederate armies, so that they would obey his orders to surrender on the same terms with his own, but he argued that, to obtain so cheaply this desirable result, I ought to give his men and officers some assurance of their political rights after the surrender."

This was taking Sherman right into the area of "civil policy" that Stanton and Grant insisted he avoid. Sherman pointed out that an 1863 proclamation of amnesty by Lincoln, still in force, enabled every Confederate soldier below the rank of colonel to regain citizenship "by simply laying down his arms, and taking the common oath of allegiance"; he added that Grant's terms to Lee had embodied the same principle, extending it, as Sherman told Johnston, "to *all* the officers, General Lee included." The procedure for regaining full citizenship, Sherman reassured Johnston, was established; this meeting, however, was not to determine everyone's postwar civil status, but to stop the fighting.

Johnston conceded that "the officers and men of the Confederate army were unnecessarily alarmed about this matter, as a sort of bugbear," but he insisted that there needed to be some guarantees about their postwar status committed to paper. He then told Sherman that John C. Breckinridge of Kentucky was nearby and available to join their discussion.

This gave Sherman pause. Breckinridge, the prewar vice president of the United States under Democratic president James Buchanan, had run for president against Lincoln under the Southern Democrat banner in the four-way 1860 presidential race, losing to Lincoln but winning more electoral votes than the other two contestants. Named by Kentucky to serve in the United States Senate, late in 1861 he had resigned to go with the South, serving for three years in the Confederate Army and rising to the rank of major general. Ten weeks before this meeting between Sherman and Johnston, Breckinridge had been appointed by Jefferson Davis as the Confederacy's secretary of war. "I objected," Sherman said of this proposal that Breckinridge sit in with them, "on the score that he was then in Davis's cabinet, and our negotiations should be confined strictly to [military] belligerents." Johnston countered with the thought that "Breckinridge was a major-general in the Confederate army, and might sink his character of [not act as] Secretary of War."

Sherman thought about all this. He had an army that had already shown it could explode with wrath about Lincoln's assassination, and he could understand that now, more than ever, the soldiers of the South might want to know what lay ahead for them if they surrendered. He had his generals urging him to get it all settled; they did not want another long march to hunt down an elusive foe. The night before, Sherman had written to one of his generals who was stationed at New Bern, "There is great danger that the Confederate armies will dissolve and fill the whole land with robbers and assassins," and added that he needed to restore order: "I don't want Johnston[']s army to break up in fragments."

Sherman also remembered his meeting with Lincoln and Grant aboard the *River Queen*, during which, as Admiral Porter remembered it, Lincoln expressed the thought that "he wanted peace on almost any terms." Perhaps Sherman thought that to have Breckinridge—a man who was both a Confederate general and the Confederate secretary of war—enter the discussions now might add further weight to this parlay's authority to create a widespread cease-fire. He agreed to have Breckinridge join them; as soon as he arrived, Breckinridge "confirmed what he [Johnston] had said as to the uneasiness of the Southern officers and soldiers about their political rights in case of surrender." At this point, a Confederate courier brought in a sheaf of papers for Johnston, who explained that they were from John Reagan, the South's postmaster general, who was traveling with Davis as the Confederate president moved through the South to avoid federal capture. After Johnston and Breckinridge looked over these papers and had "some side conversation," as Sherman put it, Johnston handed one of the documents to Sherman. It was the Confederate government's proposal for peace, apparently ready to sign if Sherman would just do that. Sherman looked at the document: "It was in Reagan's handwriting, and began with a long preamble and terms, so general and verbose, that I said they were inadmissible."

There the three men sat, Sherman perhaps a bit nettled that the Confederate government, its armies defeated on the battlefield, would try to set the terms for peace. He was not going to sign what had been put before him, but he felt he must do something.

> Then recalling the conversation of Mr. Lincoln, at City Point, I sat down at the table, and wrote off the terms, and explained that I was willing to submit these terms to the new President, Mr. Johnson, provided that both armies should remain *in statu quo* until the truce declared therein should expire. I had full faith that General Johnston would religiously respect the truce, which he did; and that I would be the gainer, for in the few days it would take to send the papers to Washington, and receive an answer, I could finish [repairing] the railroad up to Raleigh, and be the better prepared for a long chase.

With that, there was some conversation while the papers were copied and then signed by Sherman and Johnston. During this, Sherman, perhaps acting on his memory that Lincoln had observed at City Point that Jefferson Davis should for everyone's sake "escape the country" if he could, advised Breckinridge to do that himself. Breckinridge replied that he would "speedily leave," and Sherman soon took his special train back to Raleigh. At that moment, despite the acclaim Grant had received in Washington during the days after Lee's surrender at Appomattox, Sherman's reputation throughout the North was nearly as great as Grant's: a

grateful Union saw him as the remarkable leader who had served promi-
nently in one successful battle or campaign after another, all the way
from Shiloh and Vicksburg to planning and riding at the head of his epic
marches through Georgia and the Carolinas. No one, including Sherman, would have imagined that he would soon be in a predicament
from which he could be saved only by Grant.

17

SHERMAN IN TROUBLE

The morning after his second meeting with Joseph E. Johnston, Sherman sent Major Henry Hitchcock of his staff to Washington, carrying two copies of the surrender terms he had devised. One copy was to go to Grant, who since Appomattox was commanding the United States Army from an office in the War Department. The other was addressed to General Halleck, who, unknown to Sherman, Grant had shifted from his chief of staff position in the War Department to take command of forces in Virginia. Probably thinking that Halleck still had quicker access to President Andrew Johnson that Grant did, in his covering letter to Halleck, Sherman asked him to urge Johnson "not to vary the terms at all, for I have considered every thing, and believe that, the Confederate armies once dispersed, we can adjust all else fairly and well." In a second letter to both Grant and Halleck, he said that what he had worked out would, "if approved by the President of the United States, produce peace from the Potomac to the Rio Grande." To Ellen he wrote, "I can see no slip. The terms are all on our side . . . If I accomplish this I surely think I will be entitled to a month[']s leave to come and See you . . . I now expect a week of Comparative leisure till my messenger returns from Washington, and I will try to write more at length."

The terms that Major Hitchcock was carrying north, to a capital aflame with new vengeful feelings toward the Confederacy, included several conditions. Rather than surrendering their muskets and cannon directly to the federal forces in the field, as had been done at Appomattox, the enemy regiments, apparently still carrying their muskets, were "to be conducted to their several State capitals." There they would put all their weapons and equipment in each state arsenal, where they would remain available "to maintain peace and order"—in whose hands was not specified. Sherman also set forth procedures by which enemy soldiers would "file an agreement to cease from acts of war, and to abide the action of both State and Federal authority," and guaranteed to the citizens of the Confederate states "their political rights and franchises . . . as defined by the Constitution of the United States and of the States respectively."

All that language about the ongoing authority at this time of states, when they were the states that had seceded from the Union and started a war that cost 360,000 Northern lives, was bad enough, but Sherman reinforced this by the article in his agreement that pledged the president of the United States to leave the present Southern state governments in place, as soon as their "officers and Legislature" took an oath of allegiance. There was in addition a provision about restoring to Southerners "their rights of person and property" that could be construed as continuing the right to own slaves.

Whatever he thought he had written and signed, Sherman had in fact abandoned his statement to Grant and Stanton that he would "be careful not to complicate points of civil policy." His terms greatly exceeded the purely military surrender agreement that Grant and Lee had signed. In the last paragraph of his own agreement, Sherman stated that neither he nor Johnston was "fully empowered by our respective principals [the governments of the United States and the Confederacy] to fulfill these terms," but went right on to say that "we individually and officially pledge ourselves to promptly obtain the necessary authority and to carry out the above programme." Sherman and Johnston also agreed to give each other forty-eight hours' notice before resuming the fight, if either of their governments rejected the document they had signed, but they clearly thought they had brought the war in the South to an end.

At four o'clock on the afternoon of April 21, two days after Sherman sent Hitchcock north to Washington carrying the copies of the signed surrender document, the major delivered one copy to Grant in his office at the War Department. As soon as Grant read its terms, he saw that they were unacceptable and immediately wrote this note to Stanton.

I have rec'd, and just completed reading the dispatches brought by Special Messenger from Gen. Sherman. They are of such importance that I think immediate action should be taken on them, and that it should be done by the President, in council with his whole Cabinet.

I would respectfully suggest whether the President should not be notified, and all his Cabinet, and the meeting take place tonight?

At eight that evening, with Grant present, President Andrew Johnson sat down with his cabinet. The preceding days had been ones of immense drama and tension. Two days before, Lincoln's funeral services had been held in the overflowing East Room of the White House, with Grant standing alone at the head of the coffin as its chief military guard, tears rolling down his cheeks during the ceremony. The following day, thousands of grieving ordinary citizens had filed past Lincoln's body as it lay in state in the Capitol's rotunda. Twelve hours before this cabinet meeting, crowds had silently lined Washington's streets to watch the procession as the slain president's body was taken to the Baltimore and Ohio railroad station and placed aboard the black-wreathed train that would carry him to Philadelphia, New York, and other cities that would honor him during a twelve-day trip through the grieving North, before he was buried in his home state of Illinois.

It was in this highly charged atmosphere that President Johnson and his cabinet heard Stanton announce that Grant would read them the terms written and signed by Sherman. As soon as Grant finished, everyone immediately agreed that the document must be disavowed and rejected, but that was only the beginning. Grant characterized the meeting as being in a state of "the greatest consternation." Johnson, Stanton, and Attorney General James Speed denounced Sherman as a traitor. According to Secretary of the Navy Gideon Welles, Stanton, who was still fearful that he might be assassinated himself, "seemed frantic," and the attorney general voiced the fear that Sherman might march his "victorious legions" up to Washington and take over the government.

Certainly Stanton, who just a week before had walked through blood in Secretary of State Seward's house and been at Lincoln's side when he died, might in his present state have associated what he knew of Sherman's racial views, and these lenient terms Sherman had now committed himself to in dealing with Joseph E. Johnston, with an effort to undermine the victory. Stanton, like the other cabinet members present, was determined that the fighting should come to an end on terms acceptable to them and lead to a peace in which the victor's word was law. Some

Radicals felt that even the terms Grant gave Lee at Appomattox had been too generous and that Lee should now be in prison, awaiting trial and a possible death sentence for treason.

There were no notes taken at this tempestuous meeting; one second-hand account said that in the midst of this uproar, Grant, who had been the first to see that the terms could not stand, defended Sherman's motives, but others made no mention of that. For the moment, in the still-shocked atmosphere of these days after Lincoln's death, Sherman's enormous contribution to victory was forgotten amid this sudden suspicion that he might somehow be selling out the Union at the last hour, or undercutting the results of all that had been sacrificed and won.

The thrust of the meeting became to undo what Sherman had done, and to do it swiftly. Again, there is some question as to whether Grant volunteered to take the next step and was then given authorization to implement it, or whether the meeting simply turned to him to solve the problem, or whether he was ordered to act. In any event, before the evening was out, he received from Stanton instructions that read, referring to Sherman's agreement with Johnston, "You will give notice of disapproval to Gen Sherman and direct him to resume hostilities at the earliest moment." Stanton enclosed a copy of the message he had sent Grant after Lee's earlier peace overture in March, stating on Lincoln's instructions that, concerning a political settlement, "such questions the president holds in his own hands," and added that this also expressed "the views of President Andrew Johnson, and will be observed by Genl Sherman." Stanton's instructions closed with, "The President desires that you proceed immediately to the Hd Qtrs of Gen Sherman and direct operations against the enemy."

This was in its way an order as difficult to execute wisely as any Grant had ever received, and it had the possibility of cutting the oft-tested bond between Sherman and Grant. Stanton's language was clearly bellicose, indicating a perfect willingness to start shooting at the Confederates again and keep it up until they agreed to anything put before them. The order that Grant should "direct operations against the enemy"—words used in referring to battlefield movements, not to negotiations for a surrender—reinforced the authority that Grant possessed as general in chief to supersede Sherman and take active field command of his army.

Grant made plans to leave for North Carolina at midnight. At that moment, a telegram might have reached Sherman a few hours before Grant could get to Raleigh himself, but evidently not wishing any telegraph operator to see his words, and perhaps worried that he himself might be struck down by an assassin en route, Grant chose to write a letter stating what he wanted Sherman to know. He told Sherman of the "disapproval" of his agreement and instructed him to inform Johnston that, as matters stood, the truce would come to an end. With characteristic clarity, Grant set forth what was expected of the Confederates: "The rebels know well the terms upon which they can have peace and just where negociations [sic] can commence, namely: when they lay down their Arms and submit to the laws of the United States."

For whatever reasons, Grant did not use any means of communication to tell Sherman that he was on his way to join him in North Carolina. As events would show, he intended this to be a mission of peace, not of war; determined that the fighting should not resume, he also hoped to save his friend Sherman from the crisis he had created. He enclosed to Sherman the copy of Lincoln's distinction between military surrender and political settlement—Sherman was to say that he had never heard of that letter or its contents, and that may well have been true—and then wearily wrote Julia. Tired as he was, near the end of his letter he wrote words that indicated his sense of the international destiny that awaited the nation whose future unity he had done so much to ensure. "It is now nearly 11 O'Clock at night and I have received instructions from the Sec. of War, and the President, to start at once for Raleigh North Carolina. I start in an hour . . . I find my duties, anxieties, and the necessity for having all my wits about me, increasing instead of diminishing. I have a Herculean task to perform and shall endeavor to do it, not to please any one, but for the interest of our great country which is now begining [sic] to loom far above all other countries, modern or ancient."

Despite the hour and the pressures on him, Grant developed this thought concerning the country's future international role: "That Nation, united, will have a strength which will enable it to dictate to all others, *conform to justice and right*." Speaking of the limits of power, the good it could achieve if used wisely, and the dangers of using it in an immoral way, he added, "Power I think can go no further. The moment conscience leaves, physical strength will avail nothing, in the long run." Then Grant reverted to the purpose of his letter.

I only sat down to write you that I was suddenly required to leave on important duty, and not feeling willing to say what that duty is, you must await my return to know more.

Love and kisses for you and the children.

U. S. GRANT

In speaking of the "Herculean task" that lay before him, Grant almost certainly was thinking of more than what awaited him in North Carolina, pressing and crucial though the need to solve that problem was. In the days since Lincoln's death he had seen political polarization in Washington. Vice President Andrew Johnson, sworn in as president three and a half hours after Lincoln's death, was an unknown quantity; minutes after learning of the attack on Lincoln, Grant had told Julia, "I dread the change." The Radical faction of the Republicans had already wanted to enforce an iron peace in the Confederate states as soon as they surrendered, and Lincoln's murder greatly strengthened their hand. Before his death, Lincoln had agreed to the arrangement that the conquered South would for a time at least be divided into military districts, administered by the army and other federal officials. With every hour in his office at the War Department, Grant could see that, whatever else lay ahead, he had three major tasks before him. He had to stop the fighting throughout the South and Texas, prepare the postwar United States Army for some type of military occupation of the former Confederate states, and plan and execute across the next seven months of 1865 the demobilization of more than eight hundred thousand of his soldiers, almost all of whom were wildly eager to go home.

In addition to that, there was a problem involving Mexico, the land in which Grant first experienced combat. In December of 1861, while Washington was preoccupied with the war, European troops had landed in Mexico in a punitive response to President Benito Juarez's decision to suspend payments of foreign debt. There were forty thousand excellent French troops still there, including regiments of the Foreign Legion, serving the puppet regime of the Austrian Duke Maximilian, who had been installed by Emperor Napoleon III.

This situation was a dramatic affront to the Monroe Doctrine, but there was more to it than that. Although Maximilian had refused the Confederacy's overtures to create an alliance, some thousands of Confederate soldiers, among them former guerrillas fearing federal punishment, were now crossing into Mexico, where Maximilian was willing to have them join his forces and was ready to condone ownership of slaves. Combined with the fact that organized Confederate units in Texas still had not surrendered, there was a need for a strong United States Army force to resolve the situation within Texas and on its borders. (A few days after Appomattox, Grant had said lightly to an aide that the new slogan would soon be, "On to Mexico.") Ironically, considering that the Mexican War in which Grant, Lee, and other American soldiers fought resulted in the taking of a vast area of northern Mexico, there would now be both overt and secretive American efforts to help Juarez and his Mexican nationalists in their successful struggle to throw out Maximilian and these more recent foreign invaders.

To deal with all these matters, Grant brought to the challenge the prestige of a victorious general, the authority of his rank as lieutenant general and position as general in chief, a wealth of administrative as well as battlefield experience, and his legendary determination. At the same time, the famous soldier who wrote Julia at eleven that night, readying to leave Washington in an hour on a daunting mission, was a tired man who had for four years borne a steadily increasing load. A few days earlier, he had balanced these assets and liabilities in a letter to his old friend Charles W. Ford: "For myself I would enjoy a little respite from my cares and responsibilities more than you can concieve [sic]. But I have health, strength and endurance and as long as they are retained I am willing to devote all for the public good."

The next day, April 22, as Grant traveled south along the Virginia coast aboard the "special steamer" Alhambra accompanied by three of his officers, a servant, and Major Hitchcock of Sherman's staff, Secretary of War Stanton received a letter from General Halleck, writing from his headquarters in Virginia. "Old Brains," who had received the same explosive report from Sherman to Grant that precipitated the tempestuous cabinet meeting, knew that Sherman and Johnston had resumed surrender talks but did not know their status. Halleck told the secretary of war he was receiving intelligence that Davis and his fleeing Confederate governmental colleagues were carrying a large amount of gold with them and added, "They hope, it is said, to make terms with General Sherman or some other Southern [Union] commander by which they will be permitted, with their effects, including this gold plunder, to go to Mexico or Europe. Johnston's negotiations look to this end. Would it not be well to put Sherman and all other commanding generals on their guard in this respect?"

Stanton, who had "seemed frantic" the day before, was no more calm today. He added to Halleck's just-arrived letter his own misunderstanding of a cavalry movement that Sherman had recently ordered: the troopers would actually be placed directly across the route along which Davis was thought to be moving, but Stanton thought that Sherman had deliberately sent them in the opposite direction, to aid Davis in his effort to elude capture.

At that point, Stanton decided on his own initiative to disavow publicly anything Sherman had recently done or might do, and in the process distance himself and Johnson's cabinet from Sherman. Stanton prepared a signed statement for a number of newspapers, including *The New York Times* and the *Chicago Tribune*. Acidly referring to "a memorandum for what is called a basis for peace," he released to the public his account of the previous night's heretofore secret meeting of the cabinet. Stanton wrote that Sherman's agreement with Johnston had been disapproved, and that Sherman had been "ordered to resume hostilities immediately." After including the text of Lincoln's earlier instructions to Grant that, regarding the political aspects of a peace settlement, "such questions the President holds in his own hands," Stanton set forth the text of Sherman's agreement with Johnston and then wrote that "this proceeding of Gen. Sherman was unapproved for the following among other reasons." Putting the darkest construction upon every ambiguity, he said that "by the restoration of the rebel authority in their respective States, they would be enabled to reestablish slavery."

Stanton went on to say that the agreement might require the federal government to pay the debts that had been incurred by the Confederacy—a possibility nowhere mentioned in Sherman's terms—and indicated that the Confederates would be "relieved" of facing legal action for any kind of crimes they had committed. He included the idea that the captured Confederate weapons might be available to the men of the South "as soon as the armies of the United States were disbanded, and used to conquer and subdue the loyal States." In his ninth and final objection, Stanton concluded that this agreement left the Confederates "in condition to overthrow the United States Government." He also quoted selectively from Halleck's letter to him and added some words in Sherman's orders to his cavalry, leaving the impression that Sherman might have been bribed with Confederate gold to sign an easy peace and let Jefferson Davis escape the country.

As Grant continued his trip to Sherman's headquarters, neither he nor Sherman knew of Stanton's mixture of fact, speculation, and falsehood, nor did they know that Halleck, apparently on his own initiative, had told generals Meade and Horatio Wright "to disregard any truce or orders of General Sherman suspending hostilities" and had suggested to Stanton that General James H. Wilson, serving directly under Sherman, "obey no orders from Sherman." Halleck's letter about this would also make its way into *The New York Times*. (In a letter to his friend General George W. Cullum, Halleck said of Sherman that he feared "there is some screw loose again.")

On April 23, Sherman, who had billeted himself in the recently vacated (and much run-down, despite the name) Governor's Palace in Raleigh, received a telegram from his aide Major Hitchcock. Sent from Morehead City, on the coast, the major informed him of his arrival the next day but, on Grant's orders, did not mention that the general in chief was with him. Shortly after six the next morning, Sherman was up but not dressed when Grant walked in. Quietly taking Sherman aside, he told him that his surrender terms had not been approved but gave no indication of the "consternation" they had caused at the cabinet meeting. Grant left Sherman in no doubt as to what he was to do: get in touch with Joseph E. Johnston, tell him that the shooting would start again within forty-eight hours if he did not sign a new agreement based on the terms of the Appomattox surrender, and then go to Johnston and get such a document signed. Sherman accepted this—Grant later said, "like the true and loyal soldier that he was, he carried out the instructions I had given him"—but at that moment, when Grant realized that Sherman was expecting him to come along to the new parley, he told him that quite the opposite was true. He was going to stay quietly in Raleigh and evidently did not want Johnston to know that he was not still in Washington. Sherman's mission was to get this done as fast as he could, and give the papers to Grant, who would speedily take them back to Washington. With all that had been going on—Lee's surrender at Appomattox, victory celebrations, the North suddenly plunged into grief by Lincoln's assassination, and the massively attended funeral ceremonies as Lincoln's body proceeded home to Illinois—Grant hoped that a distracted public might not realize there was a delay in bringing to a close the long struggle between the forces of Sherman and Johnston.

Sherman promptly wrote two messages to Johnston, the first telling him that their truce would end within forty-eight hours, and the second saying that, pursuant to instructions he had just received, "I therefore demand the surrender of your army on the same terms as were given General Lee at Appomattox . . . purely and simply." (Some of Sherman's generals soon learned that Grant had slipped into Raleigh. Major General Henry Slocum noted that "Grant is here. He has come to save his friend Sherman from himself.")

On this same morning that Grant arrived and Sherman wrote Johnston that he was rescinding his previous terms—Sunday, April 24—*The New York Times* came out with Stanton's statement plastered all over the front page. Among a third of a column of subheadlines were: "Sherman's Action Promptly Repudiated," "The President and All His Cabinet Rebuke Him," "Gen. Grant Gone to North Carolina to Direct Our Armies," and "Possible Escape of Jeff. Davis with His Gold." In an editorial titled "An Extraordinary Operation," the *Times* said that "it looks very much as if this negotiation was a blind to cover the escape of Jeff. Davis and a few of his officials, with the millions of gold they have stolen from the Richmond banks." On the same day, the *Chicago Tribune* had this to say: "Sherman has been completely over-reached and outwitted by Joe Johnston . . . We cannot account for Sherman's signature on this astounding memorandum, except on the thesis of stark insanity . . ."—an eerie echo of the *Cincinnati Commercial*'s "General William T. Sherman Insane" headline of 1862, when Sherman had been so nervously overestimating the forces opposed to him in Kentucky. The *Tribune* added that it had information that Sherman intended to lead a proslavery political party composed of unrepentant Confederates and Northern conservatives.

No word of this sensational release of garbled information reached either Grant or Sherman in Raleigh, but, from talking to Grant, Sherman realized he needed to mend some fences. The next day, knowing only that Johnston had received his messages but still awaiting a response to the changed terms, and unaware of Stanton's disclosure to the press, Sherman wrote Stanton that "I admit my folly in embracing in a military convention any civil matters, but, unfortunately, such is the nature of our situation that they seem inextricably united." He added that he would carry out President Johnson's and the cabinet's wishes, as conveyed to him by Grant, but that was about as conciliatory as Sherman could bring himself to be: referring to Stanton's instructions that Grant should "proceed immediately to the headquarters of Major-General Sherman, and direct operations against the enemy," he said in closing that "I had flattered myself that, by four years of patient, unremitting, and successful labor, I deserved no such . . . [censure]."

At about the same time Sherman was writing this, Grant, probably sitting a few yards from Sherman in the mansion of the departed governor of North Carolina, penned a letter to Julia. He told her that Raleigh was virtually untouched by the war, but added, "The suffering that must exist in the South the next year, even with the war ending now, will be beyond conception. People who talk now of further retaliation and punishment, except of the political leaders, either do not conceive of the suffering endured already or they are heartless and unfeeling and wish to stay at home, out of danger, whilst the punishment is being inflicted."

The next day, Sherman and Johnston met again, in the same place. Johnston began trying for somewhat easier terms but quickly sensed that he could either sign or hear the Union cannon fire resume. He signed terms virtually identical to those Grant had offered Lee. As soon as he did, Sherman began a humane policy similar to the one Grant had begun at Appomattox: there would be ten days' rations for every surrendered Confederate soldier, and his army would even lend them enough horses and mules "to insure a crop." Sherman would go farther than that, having his quartermasters issue thousands of bushels of corn and tons of meal and flour to hungry civilians throughout the South. This would soon bring from Johnston a letter praising Sherman's "enlightened and humane policy," and stating, "The enlarged patriotism exhibited in your orders reconciles me to what I had previously regarded as the misfortune of my life, that of having to encounter you in the field."

When Sherman returned to Raleigh on the evening of April 26, Grant immediately read and approved the new document he brought with him. (Throughout the meeting with Sherman earlier in the day, Johnston had no idea that Grant was in North Carolina and was startled when his copy was quickly returned to him with Grant's written endorsement on it.) At ten that night, Grant sent Stanton a telegram saying that the surrender had been signed, "on the basis agreed upon between Lee and myself for the Army of Northern Virginia."

From a standing start five days before in Washington, Grant had called for a cabinet meeting, traveled 380 miles by combinations of ship and train, caused the resumption of disastrously conducted surrender arrangements, brought about a successful conclusion to the matter, and done it in a manner that, thus far, left his and Sherman's friendship intact. As Grant boarded the train for the coast on the morning of April 27, both men had reason to feel they had put a most unfortunate episode behind them. Back in Washington on the 29th, he wrote Julia that "I have just returned after a pleasant trip to Raleigh N.C. where Gn. Sherman succeeded in bringing Johnston to terms that are perfectly satisfactory to me and I hope will be well received by the country. I have not yet been able to look over the [news]papers and see what has transpired in my absence."

When Grant did "look over the papers," he saw that *The New York Times*, after its first criticism of Sherman, was now characterizing the original surrender agreement as "Sherman's surrender to Johnston." *The New York Herald* told its readers that "Sherman's splendid military career is ended, he will retire under a cloud . . . Sherman has fatally blundered." *The Washington Star* characterized his dealings with Johnston as "calamitous mischief." In New York, when Lincoln's body was halted there to receive the city's homage, the historian George Bancroft said in a funeral oration that Sherman had "unsurped more than the power of the executive, and has revived slavery and given security and political power to traitors from the Chesapeake to the Rio Grande." Radical Republicans, who had long worried that the sacrifices of war might be undercut by too generous a peace, leapt into the situation: Senator William Sprague of Rhode Island telegraphed Stanton that "loyal men deplore and are outraged by Sherman's action. He should be promptly removed." Two days before Grant read the newspapers from the week while he was gone and out of touch, the *New Haven Journal* carried a story suggesting that Sherman had played a part in the plot to kill Lincoln. (One historian concluded that, apart from major Union victories and Lincoln's assassination, Sherman's supposed treason received more newspaper attention throughout the North than any other event of the war.)

Grant described his reaction to the uproar. "I knew that Sherman must see these papers, and I fully realized what great indignation they would cause him, though I do not think his feelings could have been more excited than were my own." In this last, Grant was mistaken. As he had prepared to leave Washington for North Carolina, Grant burst out to Sherman's aide Hitchcock with this indictment of what he had heard said of Sherman's suspected treason: "It is infamous—infamous! After four years of such service as Sherman has done—that he should be used like this!" Indignant as Grant was, his anger did not match the fury of his friend. On the day after Grant left Raleigh, Sherman saw *The New York Times* of April 24 that carried Stanton's de facto indictment of him. An officer came upon Sherman in his headquarters, surrounded by a dozen generals, acting "like a caged lion, talking to the whole room with a furious invective which made us all stare. He lashed Stanton as a mean, scheming, vindictive politician who made it his business to rob military men of their credit earned by exposing their lives." As for his old enemy, the press, "the fellows that wielded too loose a pen" should be put in prison. (Sherman's rank and file had a similar view of what the press was doing to their "Uncle Billy"; when General Henry Slocum saw a crowd of soldiers standing around a blazing cart on a street in Raleigh and sent a staff officer to investigate, the man returned with the message, "Tell General Slocum that cart is loaded with New York papers for sale to the soldiers . . . We have followed Sherman through a score of battles and nearly two thousand miles of the enemy's country, and we don't intend to allow these slanders against him to be circulated among his men.")

That same day, April 28, Sherman wrote Grant an anguished letter. He began by saying that "I do think that my Rank, if not past services, entitled me at least to the respect of Keeping secret what was known to none but the Cabinet, until further inquiry could have been made," and went on to say, accurately, that Stanton was "in deep error" in his portrayal of Sherman's orders to his cavalry as aiding Jefferson Davis's continuing flight. He told Grant that the idea that he was insubordinate and "have brought discredit on our Government" would cause "pain and amazement" to his generals. He put it to Grant that he had "brought an army of seventy thousand men in magnificent condition across a country deemed impossible, and placed it just where it was wanted almost on the day appointed," and said he felt that alone "entitled me to the courtesy of being consulted before publishing to the world a proposition [Sherman's first agreement with Johnston] rightfully submitted to higher authority for proper adjudication." He inveighed against Stanton's "other state-

ments which invited the Press to be let loose upon me," and in his post-
script added, "As Mr. Stanton's singular paper has been published, I demand that this also be made public."

The following day, in Goldsboro on his way to Charleston to reposi-
tion his forces in the South for the postwar duty that would soon be
theirs, Sherman wrote Grant's chief of staff Rawlins at some length, en-
closing a copy of his previous day's letter to Grant and asking him to
"send a copy to Mr. Stanton, and say to him I want it published." He
characterized Stanton's criticism of him as "untrue, unfair, and unkind
to me, and I will say undeserved." Sherman went on to point out, cor-
rectly, that "there has been at no time any trouble about Joe Johnston's
army," and told Rawlins that "the South is broken and ruined, and deserves our pity. To ride the people down with persecutions and exactions
would be like slashing away at the crew of a sinking ship." On another
point, he commented that "the idea of Jeff. Davis running around the
country with tons of gold is ridiculous." (Sherman calculated that if the
fleeing Davis had with him as much as six million dollars in bars of gold
bullion—Stanton and Halleck were now saying he might be trying to
escape with more than twice that—it would take fifteen slow-moving
teams of six mules apiece to move it through the South. When Davis was
captured, disguised as a woman while wearing his wife's raincoat and
shawl, the figure was found to be half a million, all of which was speedily confiscated.)

In something of an undertone running through this letter to Rawlins,
Sherman revealed what else was on his mind, in addition to justifying
himself to the world. He had not seen Grant since they parted at Raleigh
two days before, and he was worried about him, about their friendship,
and about Grant's overall reaction to this avalanche of criticism.

> I doubt not efforts will be made to sow dissension between Grant and myself,
> on a false supposition that we have political aspirations, or, after Killing me
> off by libels, he will next be assailed. I can keep away from Washington, and
> I confide in his good sense to save him from the influences that will sur-
> round him there . . .
>
> If, however, Gen. Grant thinks that I have been outwitted by Joe John-
> ston, or that I have made undue concession to the rebels to save them from
> anarchy and us the *needless* expense of military occupation, I will take good
> care not to embarrass him.

In short, Sherman wanted Grant to know that he had learned his lesson, but that, while moving right along with the duties of commanding his army, he still had unfinished business with Stanton: using the word "resent" in its meaning of an aggressive reaction to an affront, he told Rawlins that "I have no hesitation in pronouncing Mr. Stanton's compilation of April 22 a gross outrage upon me, which I will resent in time."

By mistake, this letter to Rawlins, asking him to forward on to Stanton a copy of his relatively restrained letter to Grant, was sent on to Stanton. The letter to Grant had said nothing about repaying Stanton for "a gross outrage," or linking Stanton with an effort to drive a wedge between Grant and Sherman; now it was all there, in Sherman's handwriting, for Secretary of War Stanton to see.

Angry as Sherman was at Stanton, a different mixture of emotions swept over him a few days later. Stopping at Hilton Head, South Carolina, as he headed north after inspecting forces at Savannah, on May 2 he read in *The New York Times* the text of Halleck's letter to Stanton, stating that Halleck had recently ordered Union generals not to obey Sherman's orders. As would remain true for another two weeks, there would be delays and crossed communications between Grant and Sherman. Back in Washington, Grant had already made these orders "not to obey" Sherman inoperative, but Sherman did not know that, and the fact that Halleck had countermanded his orders struck Sherman as a betrayal that he would soon publicly characterize as "an act of Perfidy." Despite Ellen Sherman's warnings to him earlier in the war about Halleck's "lawyerly ambiguities," Sherman saw him as the man who had saved his career at a time when many judged him to be insane. That had been in good part true, although Halleck, aware of Sherman's political connections in Washington, had wished to curry favor with men like Sherman's father-in-law, Thomas Ewing, and his brother Senator John Sherman, while at the same time writing cautionary internal reports that would protect him from any consequences of Sherman's actions. Again, after Shiloh, Sherman had seen Halleck sideline Grant to the point that only Sherman's intervention persuaded Grant not to leave the army and go home, but his own gratitude to Halleck, who to this moment he had thought of as a friend, had led him for a time to see Halleck as being a better commander than Grant.

Sherman went on the offensive. On May 6, sitting aboard the steamer *Russia* while she rode out a storm in the harbor at Beaufort, North Carolina, during his trip north to rejoin his main army, he issued his blister-ing Special Field Orders 69, an astonishing military document. It began: "The General commanding announces to the Armies under his com-mand that a most foul attempt has been made on his fair fame." Suggest-ing that there were some mysterious figures behind "this base attempt," who "used the Press[,] the common resort of libellers," he promised his soldiers that they "will be discovered and properly punished." Then, after saying that these shadowy people "made use of the gossiping official Bul-letins of our secretary of war, with their garbled statements and false con-texts," he stopped using the image of nebulous conspirators and took aim at the man he had thought was his friend. It was bold language to de-scribe a man who through date of rank was still nominally senior to him. "Maj. Gen. Halleck, who as long as our Enemy stood in bold & armed array sat in full security in his Easy chair at Washington, was suddenly seized with a Newborn Zeal & Energy, when that Enemy has become (by no agency of his,) defeated, disheartened & submissive. He publicly disregarded [a] Truce of which he was properly advised." Sherman criti-cized the countermanding of orders, saying that the instruction to his generals to do that had been withheld from him "but paraded before the Northern Public in direct violation of the Army regulations, of the orders of the War Department . . . as well as Common decency itself."

As Sherman portrayed it, in contrast to the villains Stanton and Hal-leck, there was a hero. "But thanks to our noble and honest commanding officer, Lt. Genl. Grant, [who] after coming in person to Raleigh, and seeing and hearing for himself was enabled to return to the North" with, as Sherman assured his troops, the proper resolution to confound these calumnies against "one of the most successful results of the war."

This was, of course, a selective picture, absolving Sherman of any blame for his own actions, but Halleck immediately ran up the white flag. On May 9, when Sherman, reunited with his northward-moving army, was nearing Richmond, he received an obsequious letter from Halleck, who as the post-Appomattox commander of the Army of the James [River] had his headquarters in the former Confederate capital. "You have not had during this war nor have you now a warmer admirer than myself," Halleck told Sherman. "If in carrying out what I knew to be the wishes of the War Department in regard to your armistice I used language which has given you offense it was unintentional, and I deeply regret it. If fully aware of the circumstances under which I acted I am certain you would not attribute to me any improper motives. It is my wish to continue to regard and receive you as a personal friend." Knowing that Sherman was marching his army north to be demobilized and would be passing through Richmond, Halleck invited Sherman to stay at his headquarters and apparently expected to review Sherman's army as it passed through, standing in a place of honor and receiving the salutes of his regiments as they marched by.

Sherman answered this the next day. Contrasting these professions of friendship with the published letter in *The New York Times* in which Halleck had assured Stanton that he was countermanding Sherman's orders, he told Halleck that "I cannot possibly reconcile the friendly expressions of the former with the deadly malignity of the latter, and cannot consent to the renewal of a friendship I had prized so highly." As for the idea that Halleck would be receiving the salutes of his men at a review, Sherman had this to say: "I will march my Army through Richmond quietly and in good order without attracting attention, and I beg you to keep slightly perdu [lost], for if noticed by some of my old command I cannot undertake to maintain a model behavior, for their feelings have become aroused by what the world adjudges an insult at least to an honest commander. If loss of life or violence result from this you must attribute it to the true cause, a public insult to a Brother officer when he was far away on public service, perfectly innocent of the malignant purpose and design."

On the same day, May 10, Sherman wrote a letter to Grant, marked "Private & Confidential," that showed him not only to be still enraged against Halleck and Stanton but also puzzled and troubled about the state of their own relationship. Until a brief telegram had come that morning ordering him to march his army to Washington and encamp at Alexandria, Virginia, about three miles down the Potomac from the capital, he had not heard from Grant since they had parted at Raleigh thirteen days before. There had been no answer to his letters to both Grant and Rawlins attacking Stanton and asking that his sentiments be passed on to Stanton and published. At midnight two days before, Sherman had penned a plaintive note to Grant, saying that he had immediate reason to issue some orders to General James H. Wilson, one of the officers who had been told "not to obey" him, and asked, "Does the Secretary of War's news-paper order take General Wilson from my command, or shall I continue to order him—If I have proven incompetent to manage my own command, let me know it." The following day, still not hearing from Grant, he communicated with him again, saying that Wilson needed some instructions, but that because of "secretary Stanton's newspaper order taking Wilson substantially from my command I wish you would give the orders necessary."

Although Sherman had not yet received it, these messages had produced a brief, businesslike telegram from Grant sent on May 9, saying, "I know of no order which changed your command in any particular." Referring to the fact that Sherman had been all over the South since they parted, Grant added, "Gen. Wilson is in telegraphic communication with Washington whilst you have not been[,] consequently instructions have been sent to him direct." The unspoken message: you have been wronged, but I consider you to be in full command, just as you were before you got yourself into trouble and I did my best to get you out of it, and we need to get on with our duties, and you need to spend less time feuding.

Now, still having heard nothing from Grant except a brief order to bring his army on to encamp across the river opposite Washington, Sherman poured out his emotions to Grant in his letter. "I do think a great outrage has been enacted against me," he wrote. "Your orders and wishes shall be to me the Law, but I ask you to vindicate my name . . . If you do not I will . . . No man shall insult me with impunity . . . No amount of retraction or pusillanimous excusing will do. Mr. Stanton must publicly confess himself a Common libeller or—but I won[']t threaten. I will not enter Washington except on yours or the Presidents emphatic orders, but I do wish to stay with my army, till it ceases to exist, or till it is broken up and scattered to other duty."

Having expressed his own feelings, he told Grant that Stanton "seeks your life and reputation as well as mine . . . Whoever stands in his way must die." About Halleck, a copy of whose recent sycophantic letter he enclosed, Sherman said, "Read Halleck's letter and see how pitiful he has become. Keep *above* such influences, or you also will be a victim— See in my case how soon all past services are ignored & forgotten. Excuse this letter. Burn it, but heed my friendly counsel. The lust for Power in political minds is the strongest passion of Life, and impels Ambitious Men (Richard III) to deeds of Infamy."

With that, Sherman subsided for the moment. In fact, not only had Grant written Sherman the recent telegram saying "I know of no order which changed your command," with its implicit signal that Grant considered him to still have all the authority he held prior to the political explosion, but Grant had sent him a longer letter on May 6 that in effect dealt with much of what Sherman now expressed. In his earlier letter, Grant had explained that Sherman's long letters to Rawlins and himself about Halleck and Stanton had been delayed and had only just arrived, but what he went on to say about the controversy was much less than what Sherman wanted to hear. Grant wrote that he did "not know how to answer" Sherman's concerns about what he agreed was an "insult"— Halleck's countermanding of Sherman's orders and publishing the fact he had done that. Evidently trying to deflect some of Sherman's anger against Halleck, Grant had written, "I question whether it was not an answer, in Halleck[']s style, to directions from the Sec. of War giving him instructions to do as he did." Then, as if keeping the targets difficult to hit, he added, "I do not know this to be the case although I have spoken to Mr. Stanton on the subject." Without comment on Sherman's wisdom in doing so, Grant stated that, having received Sherman's denunciation of Stanton and his request to have it published, "I requested its publication. It is promised for tomorrow."

Tentative as his language on political matters seemed to be in this response to Sherman, on another subject—Sherman as a soldier—Grant spoke with his usual directness. Telling him that there was room for disagreement between Sherman and himself concerning Sherman's original negotiations with Johnston, he said that what had happened then "made no change in my estimate of the services you have rendered or of the services you can still render, and will, on all proper occasions."

Sherman made no immediate answer to this. The next day, as he marched his main army through Richmond on its way north, Halleck stood for a time on the portico of the house he was using as his headquarters, apparently believing that Sherman's divisions would feel compelled to salute him as they passed by a few yards away. Every officer and man of the leading column of Sherman's army—fifty-three thousand of his sixty-five thousand soldiers—marched past that headquarters with eyes straight ahead as if Halleck were not in Richmond. Not a single officer's sword lifted in salute, and one of Sherman's ragged riflemen stepped out of the passing column and spat a stream of tobacco juice all over the polished boots of the "very spick and span" sentry standing guard there. (As soon as Ellen learned of this monumental snub, she wrote Sherman that the Ewing clan was "truly charmed" that he had "so good an opportunity of returning the insult of that base man Halleck . . . I would rather have seen that defiant parade through Richmond than anything else since the war began.")

As Sherman's men, many of them barefoot, left Richmond behind and began the last marches and bivouacs on their way to Washington, throughout the North other voices began to be heard. The *Cincinnati Commercial*, which four years before had carried the headline, "General William T. Sherman Insane," now said, "As to the charge of insanity being made . . . We wish there were a few more such insane men in the Army." *The Louisville Journal*, published in the Kentucky city where Sherman had made panicky claims of an impending advance by overwhelming Confederate forces, deplored "the most cruel attacks . . . upon the integrity and patriotism of the illustrious soldier."

Among the Shermans and Ewings, there had been some dismay about Sherman's first agreement with Johnston. On reading the earliest newspaper accounts, Ellen immediately wrote him that "I think you have made a great mistake" in giving such lenient terms to what she called "perjured traitors [and] deserters," but added, "I know your motives are pure . . . I honor and respect you for the heart that could prompt such terms." Then the family closed ranks and began one more of their campaigns to help him in Washington. Sherman's brother John told him candidly that "for a time you lost all popularity gained by your achievements" but added that public opinion was turning against Stanton and Halleck for their "gross and damnable perversions" of what Sherman had done, and his brother Charles advised him that if Sherman would "act prudently," it might all turn out well for him.

By no means placated, Sherman sent ahead a letter to a friend in Washington that said, "It is amusing how brave and firm some men become when all danger is past," and made an unmistakable comparison between Halleck and Shakespeare's Falstaff. At Sherman's request, that letter too would soon be published in the Washington papers and, in a demonstration that Sherman was not the only member of the family who knew how to strike fear into the enemy's heart, his brother the senator wrote anonymously in the *Washington Chronicle* of May 15 that Stanton "must expect open defiance and insult, and neither his person nor his rank can shield him." What the public did not hear, fortunately, was the sentiment expressed in a letter from Sherman to Ellen in which he told her that the men against him in Washington were "a set of sneaks who were hid away as long as danger was rampant" and that he would "take a regiment of my old Division & clear them out."

As it was, there were those in Washington who really believed that Sherman was going to march his army straight into the capital. (Their opinion of Sherman was not improved by the fact that Ellen's brother Tom, a lawyer who had risen to brigadier general before resigning from the army in 1864, had since May 12 been the attorney defending three men accused as lesser conspirators in Lincoln's assassination.) Sherman had not the remotest idea of participating in or permitting some sort of coup—he had already written Grant that he would enter Washington only upon his express wish, or that of the president, and in addition he would soon go out of his way to reject any suggestions that he would make a good future presidential candidate for the Democratic Party. Nonetheless, his army was still angry about what they considered to be the slurs upon him: Theodore Upson, the combat-hardened sergeant of the 100th Indiana who had described the celebrations among Sherman's men when they heard of Appomattox, complete with a general marching through camp banging on a bass drum in the midst of drunken revelry, wrote in his diary that newspapermen had better "look out . . . or they would have General Sherman's army to reckon with the first thing they know."

So there matters stood. Since Grant and Sherman had mapped out their grand strategy in a hotel room in Cincinnati fourteen months before when Grant was taking command of the entire Union Army, they had met twice, both times within the last seven weeks. On the first occasion, they had conferred with President Lincoln at City Point. Since then, Lee had surrendered, Lincoln had been assassinated, and Grant had rushed to North Carolina to correct Sherman's dealings with Johnston, his worst mistake of the war. Grant was deeply involved with the myriad issues and details of moving the army from a wartime footing to a peacetime status in a politically tense postwar situation. Sherman was coming to Washington, angry and determined to clear his name. How all this would affect their friendship remained to be seen.

18

GRANT, SHERMAN, AND THE RADICALS

As Sherman's army marched north through Virginia, now only days away from Washington, at his office in the War Department Grant was trying to juggle a number of matters in the middle of a volatile atmosphere. The congressional Joint Committee on the Conduct of the War, a powerful body formed to give the Senate and the House authority to investigate, and thus influence, war policies and the conduct of military affairs, was holding hearings concerning the end of the war and what might lie ahead in dealing with the South. With the sudden vacuum in power caused by Lincoln's death, and the succession to office of the untried, little-known Andrew Johnson, both political parties were maneuvering vigorously to further their interests. As the Radicals saw it, the white South must be punished for seceding, and the freed blacks be given the vote, or else the war had been fought for nothing. At the opposite end of the spectrum, the former "Peace Democrats" (sometimes called "Copperheads") believed in gentler treatment for the defeated Confederacy and no vote for the blacks for the time being. Millions of Northern voters found themselves between the two positions.

Grant had been summoned to appear at a session of the committee on May 18, and when Sherman arrived in the Washington area on May 19 he would discover that he had been similarly requested to testify within the next few days. The senior committee chairman, Senator Benjamin Wade of Ohio, was a Radical Republican who was Stanton's political colleague and friend. (In an interesting example of how political positions could harden, and perceptions change, earlier in the war, after Shiloh, Wade had teamed with his fellow senator from Ohio, John Sherman, to endorse Halleck's recommendation that William Tecumseh Sherman be promoted to major general of Volunteers.) Lincoln's death had further hardened some positions: more than before, some Radicals in Congress viewed more moderate Republicans, and the great majority of Democrats, as their enemies.

On one issue, Grant soon confronted the position of those who wanted revenge against the Confederacy. Jefferson Davis, captured in Georgia on May 10, was on his way to confinement at Fortress Monroe on the Virginia coast as a political prisoner, but the Radicals also wanted to arrest Robert E. Lee, now at his house in Richmond, and try him for treason, a crime punishable by death. This would lead to an angry exchange between Grant and President Andrew Johnson. When Grant pointed out that Lee, along with all his army, had left Appomattox with a valid parole that guaranteed he could not be arrested as long as he lived a law-abiding life, the new president demanded to know on what grounds "a military commander interferes to protect an arch-traitor from the laws."

Grant had an answer to that. Referring to the document he and Lee signed at Appomattox, he said, "My terms of surrender were according to military law, and as long as General Lee observes his parole, I will never consent to his arrest. I will resign the command of the army rather than execute any order to arrest Lee or any of his commanders so long as they obey the law." Johnson knew better than to precipitate the resignation of the victorious commanding general of the United States Army, and for the time being the matter was put aside.

The paperwork piling up on Grant's desk also dealt with matters having important implications for the future: the continuing settlement of the Great Plains and the Indian wars. On May 17, Grant wrote to General John Pope, commander of army frontier outposts well to the northwest of any remaining Confederate holdouts. Responding to Pope's request that he be allowed to keep the people of the Sioux tribe "placed in that relation to the military forces which ensures their protection both against white and red rascals and enemies," Grant expressed an opinion seldom heard on this subject. After approving Pope's request, Grant added, "It may be the Indians require as much protection from the whites as the whites do from the Indians. My own experience has been that but little trouble would ever have been had from them but for the encroachments & influence of bad whites."

On the same day, Grant wrote lengthy instructions to his cavalry leader Philip Sheridan, "assigning you to command West of the Miss. [Mississippi River] . . . Your duty is to restore Texas, and that part of Louisiana held by the enemy, to the Union in the shortest practicable time, in a way most effectual for securing peace." Having future trouble with Mexico also in mind, he added, "To be clear . . . I think the Rio Grande should be strongly held whether the forces in Texas surrender or not and that no time should be lost in getting them there." This set in motion movements to deploy what would become a total of fifty thousand Union soldiers, including twenty thousand black troops, who would end Confederate resistance and begin the pressure that in time caused Maximilian's downfall and the restoration of the Juarez government in Mexico. (Sheridan would quietly supply Juarez with sixty thousand rifles for his followers.)

Meanwhile, Grant still felt that Sherman had been badly treated, but to perform his varied duties as general in chief during this transition from war to an uneasy peace, Grant felt he needed to keep a clear and effective line of communication open to Secretary of War Stanton, whose office was near his in the War Department. This soon moved beyond simple courtesies. When Grant fell ill at just this time, with Julia finding it difficult to care for him while they both stayed at a hotel, Stanton suggested to Halleck by telegram that Halleck might offer Grant and Julia the house in the Georgetown section of Washington that he had occupied before Appomattox. Halleck invited them to do so, and the Grants promptly moved in, with Grant thanking Halleck for his "very kind" act.

There was no record made of private talks between Grant and Stanton at the War Department, but certain memoranda he wrote to Stanton suggest that previous discussions occurred. Considering the Radical animus against Confederate soldiers, it is unlikely that, without Stanton's approval, Grant would have suggested a policy of making recently surrendered rebel soldiers eligible to be recruits for the United States Army. Grant's wording was that he "would respectfully recommend" that, in addition to opening recruiting stations in the North to enlist men to replace some of the masses of soldiers about to be mustered out, "Citizens of the Southern States, as well as persons who have served in the rebel Armies, be accepted as recruits, but all persons who have been engaged in the rebellion against the U.S. before being received will be required to qualify as loyal Citizens, in addition to taking the prescribed enlistment oath." Stanton agreed.

Obviously, some backroom compromises were being reached, and Grant felt the need to keep the United States Army, which he commanded, above the partisan political fray. This became clear when he made his appearance before the Committee on the Conduct of the War. Grant might well have been a lawyer who had the army for his client. When the committee's chairman, Stanton's friend Benjamin Wade, questioned Grant about Stanton, whose treatment of Sherman Grant had described as "infamous," the queries and responses went this way.

> *Question.* I wish to place upon our record your answer to the following question. In what manner has Mr. Stanton, Secretary of War, performed his duties in the supply of the armies and the support of the military operations under your charge?
>
> *Answer.* Admirably, I think. There has been no complaint in that respect— that is, no general complaint. So far as he is concerned I do not think there has been any ground of complaint in that respect.
>
> *Question.* Has there ever been any misunderstanding with respect to the conduct of the war, in any particular, between you and the Secretary of War since you have been in command?
>
> *Answer.* Never any expressed to me. I never had any reason to suppose that any fault was found with anything I had done. So far as the Secretary of War and myself are concerned, he has never interfered with my duties, never thrown any obstacle in the way of any supplies I have called for. He has never dictated a course of campaign to me, and never inquired what I was going to do. He has always seemed satisfied with what I did, and has heartily cooperated with me.

So there it was. Ulysses S. Grant intended to get along with the men who represented the traditional American civilian control of the military. As occurrences within the next few days would prove, he was thinking about his friend Sherman, and would help him, but he had extraordinarily complicated tasks on his hands. In this moment of moving the army around, of garrisoning areas of the South, of preparing to demobilize more than eight hundred thousand men, and with problems ranging from the situations in Texas and Mexico to the question of how to handle the Sioux in Minnesota and the Dakota Territory, Grant was ready to protect Robert E. Lee from arrest as a matter of principle, but he wanted all his generals, including Sherman, to say and do nothing that would cause unnecessary political repercussions.

There were at this moment two military realities. In Washington, Grant was actively winding down the army's wartime commitments, and the soldiers of the Army of the Potomac who had faced and defeated Lee's Army of Northern Virginia were in comfortable encampments near the capital, still disciplined and equipped for battle but fully realizing that the war was over and that most of them would soon be going home. In Sherman's army, the situation was different. Marching northward for hours every day, sleeping in makeshift conditions every night, his men still had the mentality of soldiers whose campaign has not ended. There were of course thoughts of home—an Irishman in the Twenty-first Wisconsin said that on his return he would hire a fifer and a drummer to come to his door in the morning to play reveille, and every day he would roll over and say, "To hell with your reveille"—but the troops continued to think as the highly professional soldiers they had become. As they passed the recently defended Confederate entrenchments between Richmond and Washington, they studied them and decided that they had faced far more difficult defensive positions at Vicksburg, at Missionary Ridge and Seminary Ridge, and at Kennesaw Mountain. When they encountered defeated Confederates, they treated them kindly, but when they saw Union troops of the Eastern army, their esprit de corps led them to make taunting remarks about the "white glove" soldiers, and this led to frequent fistfights.

Among the units of Sherman's Western army were some from the East that, before being sent west, had begun the war on battlefields south of Washington such as Bull Run. Of Sherman's 218 regiments, many of them literally decimated, thirty-three were from the East—sixteen from New York, ten from Pennsylvania, three from New Jersey, and two each from Connecticut and Massachusetts. A war correspondent had noted how these Easterners had changed, not only in switching from the forward-sloping visored caps of the Eastern army to the broad-brimmed slouch hats of the Western divisions, but also in the way they marched:

they moved with the long, relaxed strides of men who had campaigned across not hundreds, but thousands, of miles.

At the head of all these men, Sherman also studied the battlefields they passed. At Chancellorsville, the scene of Lee's greatest victory, Sherman's soldiers saw him walking with his hands clasped behind him, deep in thought as he passed along the ridge near the key position of Chancellorsville House. When he came to Bull Run, Sherman was at the place where he had fought in his first battle of the war. For him, and some of the regiments now following him, that had begun an enormous, bloody, roughly circular route, 2,500 miles in all—out to the Mississippi River and down to Vicksburg, over to Chattanooga, down to Atlanta and east to Savannah and the sea, followed by the final marches north through the swamps of the Carolinas, with this last movement up through Virginia done only to the sound of feet and hooves and wagon wheels on the road, in the unfamiliar silence after the guns stopped firing.

Sherman had some definite ideas of what might lie ahead of his army, if not for himself, when they got to Washington. As far back as his time in Raleigh, he had written Ellen that his final destination, as told to him by Grant during Grant's visit there to revise Johnston's surrender terms, would be Alexandria, across the Potomac from Washington, "where I will move my Head Qrs. in anticipation of mustering out the Army." He told her then that "if I could take all the family to Alexandria to witness the final success attending 'Shermans' army it would be a prize in the memory to our children," and on May 16 he had wired Ellen to come on to Washington. Now, reaching Alexandria on Saturday, May 19, he wrote John Rawlins a letter that announced his arrival and said, "All my army should be in camp near by to-day." In a slightly offended tone, he indicated that he knew that a great two-day final parade of the Union Army was now planned for May 23 and May 24, but said that "I have seen the orders for the review in the papers, but . . . it is not here in official form. I am old-fashioned, and prefer to see orders through some other channels, but if that be the new fashion, so be it. I will be all ready . . . though in the rough. Troops have not been paid for eight or ten months, and clothing may be bad, but a better set of arms and legs cannot be displayed on this continent." In a mixture of defiance and a desire for acceptance, Sherman closed with, "Send me all orders and letters you may have for me, and let some one newspaper know that the vandal Sherman is encamped near the canal bridge, halfway between Long Bridge and Alexandria, to the west of the road, where his friends, if any, can find him. Though in disgrace, he is untamed and unconquered." He signed it, "As ever, your friend."

This brought an almost instant reply, not from Rawlins but from Grant, who tried to soothe Sherman's feelings about having to learn of his army's part in the forthcoming "Grand Review" by reading the newspapers. "I am just in receipt of yours of this date," Grant began. "The orders for review was only published yesterday, or rather was only ready for circulation at that time, and was sent to you this morning."

Then Grant got down to more serious business.

> I will be glad to see you as soon as you can come to the City, can you not come in this evening or in the morning? I want to talk to you upon matters about which you feel sore, I think justly so, but which bear some explanation in behalf of those who you feel have inflicted the injury.
>
> Yours truly,
> U.S. Grant Lt. Gen.

By the next morning, Sherman had not yet gone to see Grant, and members of Sherman's family, across the river from the huge encampment of Sherman's army at and around Alexandria, started worrying about him. Washington was abuzz with talk that Sherman and his men wanted to cross the Potomac and physically punish and perhaps remove Stanton and other officials. In his diary entry for that day, Ellen's brother Major General Hugh Ewing—three of her four brothers had become generals—said that "the threats of Gen. Sherman against the authorities, that we heard on the streets this morning, made it necessary that he be counseled, and I found John Sherman in Willard's barbershop in the chair, took him to Charles Sherman's room, where the Shermans and Gen. Tom [Ewing] and myself held a consultation over his condition, and had John go to his camp and quiet him."

Clearly, the Sherman-Ewing family feared that they might be hearing more than just rumors based on vindictive talk coming out of Sherman's encampment. The reference to Sherman's "condition" may have indicated a concern that they might be seeing another form of the breakdown he had suffered in Kentucky in 1861, this episode brought on by a different form of pressure and anxiety. His brother John made no mention of what happened when he visited the general at his headquarters near Alexandria, but later in the day Sherman came in to see Grant at the War Department. Apparently Grant enabled Sherman to understand to some degree just how terrified and bewildered so many people, in and out of government, had been after Lincoln was killed in the middle of their city and how easy it had been to read a traitorous intent into a flawed, lenient truce agreement.

Talking to Sherman in a government building that had until recently been hung with long black streamers in mourning for the slain president, Grant assured him that, whatever might have happened in North Carolina a month before, Sherman now had the grateful support of virtually all the government leaders. To prove his point, Grant led Sherman from the War Department, which was at Seventeenth and F Street, to a house at the corner of Fifteenth and H occupied by Johnson, who had not yet moved into the White House. Sherman found the president and his cabinet, with the exception of Stanton, waiting to greet him. In a marvelous example of political hypocrisy, President Johnson, who had joined Stanton and others of his cabinet in calling Sherman a traitor when Grant had read them Sherman's first agreement with Joseph E. Johnston, stretched out his arms and said, "General Sherman, I am *very* glad to see you—*very* glad to see you—*and I mean what I say*." He then told Sherman that he had known nothing about Stanton's letters to the newspapers until they were published, and that Stanton had shown them to no one before sending them out on his own initiative. Almost all the other cabinet members told Sherman the same thing, which was evidently true.

As Grant and Sherman parted at the end of the afternoon, they were coming to the end of a remarkable time in their lives and in the life of their nation. On this evening in May, Sherman was forty-five and Grant was forty-three. Much lay ahead for both of them—the presidency for Grant, succession in peacetime to Grant's position as general in chief for Sherman—but it was as wartime soldiers that they had made their mark in history. Forty-seven months before, the men of the Twenty-first Illinois had hooted at Grant, this shabby figure of a colonel who had come to command them, and on this date in 1861, Sherman, who had held four jobs in the four years before that, was just reentering the army.

As they went their separate ways that evening, there was one thing they had to do, something that was equally important to these two West

Pointers. They had to say good-bye to the Union Army, and do it in a way that made the troops take justified pride in themselves and gave the public the chance to show its feelings for those who had served and sacrificed. The Grand Review, a two-day parade through Washington, would begin in three days. Between now and then, Sherman had to appear before the Committee on the Conduct of the War, Grant had to deal with the continuous flow of paperwork involved in the shift from war to peace, and each of them had to prepare for his part in a tremendous spectacle. When they saw each other again, it would be as two immensely famous figures appearing in front of enormous crowds, but as they parted now in the hour before sunset, it was two men shaking hands on a quiet street.

19

A PARADE FOR EVERYONE,
AND A HEARING FOR SHERMAN

During the morning of the next day, May 21, the streets of Washington filled with preparations for the great two-day victory parade, now only forty-eight hours away. Although many black wreaths mourning Lincoln remained in place throughout the city, both government buildings and private dwellings began to be decorated with red-white-and-blue streamers, banners, and bunting of every design and size. Along Pennsylvania Avenue near the White House, scores of workmen erected grandstands. In front of the White House itself, a special covered pavilion with a capacity for two hundred of the most prominent spectators went up, a few yards from where the troops would pass. Hung with blue-and-white drapes that had big white stars on them, with baskets of flowers hanging from the pavilion's supports, the decorations facing the avenue included signs that proclaimed, "Donelson," "Shiloh," "Vicksburg," "Gettysburg," and the names of other Union victories. In the pavilion would sit the president and cabinet, Grant and his staff, some of the leading officers not in the parade, the ambassadors of other nations, and the families of all these dignitaries. On the other side of the avenue, festooned with the state flags of the Union, was the grandstand in which would be the justices of the Supreme Court, senators and congressmen, and governors of several states. On either side of the presidential reviewing stand were stands, paid for by private contributions, that would each hold five hundred wounded or sick Union soldiers, including withered men recently released from Southern prison camps.

Tens of thousands of visitors poured into the city, many on special trains. Under the headline "THE GRAND REVIEW," two subheadlines in *The New York Times* said, "Great Rush of Visitors to See the Boys in Blue" and "Washington Crowded Beyond All Precedent." The hotels filled to the point that cots had to be placed in the hallways. Some enterprising people, finding no accommodations available anywhere, rented streetcars in which to spend the night. The newspapers joined in the enthusiasm, offering interesting examples of the size of the forces that would march down Pennsylvania Avenue: they told their readers that they would see more men marching than had been in the combined armies of Napoleon, Cromwell, and Augustus Caesar, and that the legions with which Caesar conquered his empire had fewer men than were in just one wing of the Army of the Potomac.

At the War Department, Grant kept sending out orders, balancing out the smaller number of troops that were still needed against those that were to be sent to their home states and mustered out as soon as possible. Also awaiting his attention were matters as varied as many men going out of service wanting to buy their muskets to keep—Grant ruled that they could—and the fact that, as Grant would soon write Halleck in Richmond, "I am informed that a great many bodies have been left unburied [*sic*] at Appomattox C. H."—a situation Grant solved by instructing Halleck to send cavalry units there and to Sayler's Creek to bury the dead.

At his tent headquarters across the river, Sherman was pondering what to say the next day at the hearing of the Committee on the Conduct of the War. He had by now come to understand that the Radicals wanted to portray Lincoln, posthumously, as a man who would have extended to the South peace terms so soft that they would have undercut the war aims for which 360,000 Northern men died. Therein lay a defense for Sherman, if he wanted to use it. He could portray his first agreement with Joseph E. Johnston as having been an implementation of what Lincoln had told him to do during his meeting with Lincoln and Grant and Admiral Porter at City Point seven weeks before, and justify his actions: he had been a soldier, carrying out his commander in chief's unwise instructions.

At least one man ready to strengthen that claim came to see Sherman at his headquarters. Colonel Absolom H. Markland had served under both Grant and Sherman, in an unusual capacity. As a military postmaster general, in the Western theater he had implemented Grant's idea of sorting mail for the troops as it was being carried on special railroad cars. Grant thought so highly of his services to the army throughout the war that, two days before this, he had sent him the saddle that he had used in all the battles and campaigns from Fort Henry in February of 1862 through to the final day at Appomattox. "I present this saddle to you not for any intrinsic value it possesses," Grant had written him, "but as a mark of friendship and esteem."

Coming now to Sherman, Markland reminded him that, soon after Sherman's conference with Lincoln and Grant at City Point, Sherman had told him about it in some detail. Markland now told Sherman that the moment he read of Sherman's first truce terms with Johnston, he had recognized them as being what Sherman had said that Lincoln wanted. Markland urged Sherman to use that information and promised to make written statements or testify on Sherman's behalf in any way that he could. To Markland's surprise, "General Sherman was in no mood to take up the subject and very clearly intimated to me that I should be silent concerning it."

No one could ever know all that was in Sherman's mind, but as the hours passed on the day before the hearing, he wrote to Major General Stewart van Vliet, who had been one of his two closest friends at West Point. On April 27, when the news of Sherman's dealings with Johnston were first in the newspapers, van Vliet had written him that he thought Sherman's original set of terms were better suited to produce a just and lasting peace, and that he believed time would prove that true. "Dear Van," Sherman wrote him now, "Stanton and Halleck . . . thought they had me down, and when I was far away on public business under their own orders, they sought the opportunity to ruin me by means of the excitement naturally arising from the assassination of the President, who stood in the way of fulfillment of their projects, and whose views and policy I was strictly, literally following." He went right on to demonstrate just how far his thinking differed from that of the Radicals. "I prefer to give votes to rebel whites, now humbled, subdued and obedient, rather than to the ignorant blacks that are not yet capable of self-government." This was a distillation of something he had recently written to Ellen, saying that "Stanton wants to kill me because I do not favor the scheme of declaring the negro of the South *Now Free*, to be loyal voters whereby Politicians may manufacture just so much more pliable Electionary material. The Negros dont want to vote. They want to work & enjoy property, and they are *no* friends of the negro who want to complicate him with new prejudices."

Sherman's place in history at that moment was ironic. If he had not captured Atlanta just when he did in the autumn of 1864, the Republicans, including the Radicals whom he would be facing at the hearing the following day, might well have lost the election and the political power they now wielded, including the power they were bringing to bear in pressing for immediate voting rights for blacks. It was Sherman's bold and brilliant marches through the South that had freed hundreds of thousands of blacks, but he continued to view them as inferior beings. In an exchange of letters with Chief Justice Salmon P. Chase after seeing him during Chase's inspection trip of newly freed Southern areas the past January, he had written, "Of course I have nothing to do with the Status of the Negro after [the] war. That is for the law making power, but if my opinion were consulted I would say that the negro should be a free race, but not put on an equality with the whites."

As night fell on Washington, no one, and seemingly not Sherman himself, could say just how he intended to handle himself at the hearing the next day. His letter to van Vliet, with its statement that in his first agreement with Joseph E. Johnston he was "strictly, literally following" Lincoln's "views and policy," implied that he had documents or knowledge in his possession that "not only justify but made imperative" his action in offering those terms. To van Vliet he said, "I am to go before the war investigating committee, when for the first time, I will be at liberty to tell my story in public," adding that "thus far I have violated no rule of official secrecy, though severely tempted, but so much the worse for [Stanton and Halleck] when all is revealed."

As the morning of May 22 dawned, Washington was far more focused on the next day's parade than on what would happen in one governmental hearing room. Visitors continued to arrive by the thousands, and the soldiers of the encamped armies began appearing in the streets during their off-duty hours. Walt Whitman was out and about and taking it all in. An ardent Union patriot, he had seen terrible suffering among the men being brought from the battlefields and dying in the hospitals where he helped care for them, and was deeply relieved by the end of the fighting. His notes for that day included: "Have been taking a walk along Pennsylvania Avenue and Seventh street north. The city is full of soldiers, running around loose. Officers everywhere, of all grades. All have the weather-beaten look of practical [active] service. It is a sight I never tire of. All the armies are now here (or portions of them,) for tomorrow's review. You see them swarming like bees everywhere."

In his office at the War Department, Grant may well have been thinking of what the day ahead might hold for Sherman, but nothing he put on paper that day referred to it. In a telegram to Meade, the commander of the Eastern troops who were to parade down Pennsylvania Avenue the next day, he said, "Please direct your Eng. [engineering] officer to place in the review a pontoon train say of four boats and two Chess wagons," the wagons used to transport the planks that formed the bridge supported by the boats. Other than suggesting this touch, sure to interest the crowds along the parade route, his only other communications that day involved assignments of commanders in the West.

And so Major General William Tecumseh Sherman made his way to the hearing of the Joint Committee on the Conduct of the War. This was to be its last day in existence. Every Union general knew of its activities. Formed in December of 1861 to investigate the Union military disaster at Bull Run and a lesser fiasco at Ball's Bluff and given "the power to send for persons and papers," its evolving Radical faction had pressed for more aggressive Union Army campaigns and had pushed Lincoln for an earlier emancipation of slaves than he had felt it politically wise to grant. The committee looked at everything, not only defeats and such matters as the Confederate treatment of Union prisoners, but also whether victories might have been better exploited, and the status of government contracts, naval shipbuilding, and matters such as the cotton trade in the occupied South. Some of its inquiries were legitimate and possibly helpful, but generals who appeared before the committee felt like defendants rather than witnesses.

Among the committee's clear prejudices was that against West Pointers. Its members felt that even academy graduates from the North were likely to be Southern sympathizers at heart, men who had no interest in the issue of slavery and retained friendly feelings for their brother officers of the prewar army who had chosen to fight for the Confederacy. They also saw the career army men as having little interest in democratic institutions and as possessing little faith in the American tradition of civilian control over the military. Some of the committee's targets were reasonable ones: their sometimes-devious investigation into George McClellan's inertia had been part of what pushed Lincoln to relieve him of command of the Army of the Potomac.

On the other hand, the committee sometimes acted in the role of thought police. An example of the unprincipled ferocity of this committee that Sherman would now appear before was its shameful and groundless arrest and six-month imprisonment of General Charles Stone, who was both a West Pointer and a Democrat. Stone, although not actually present, was technically in command at the defeat of Ball's Bluff, Virginia, on October 21, 1861, a battle in which the federal losses of all categories totaled 921, compared with 149 Confederate casualties. This in itself justified an investigation, but the committee's motives were political to the core. The inept general who was directly in charge at Ball's Bluff and was killed there was Edward Baker, a major general of Volunteers who had been a Republican senator and a strong Radical. On the other hand, Stone's wife had relatives in the South, and, prior to army orders not to do so, Stone had returned fugitive slaves to their owners in accordance with then-existing laws. There were also rumors that rebel couriers were able to cross the Potomac River in his area, and it was true that he was one of the favorite generals of McClellan, whose Democratic Party loyalties were well-known.

The committee, in good part because it saw a chance to discredit McClellan, decided that Stone was a traitor who had deliberately planned an action in which Baker and his men would be killed. A month after a committee hearing began in which Stone was in effect treated as a defendant in a criminal case who was never told the charges against him and given no chance to prepare an informed defense, he was, with Secretary of War Stanton's tacit compliance, arrested at midnight and placed in solitary confinement at Fort Lafayette in New York Harbor. Released by an act of Congress after six months, he failed in his repeated attempts to have a formal military court of inquiry convened to hear his case. Finally, on February 27, 1863, seven months after his unjustified and illegal imprisonment ended, Stone had the chance to confront his original accusers in a committee hearing, during which he categorically disproved every charge that had been brought against him. Going on to serve bravely later in 1863 at the battle of Port Hudson and in the Red River campaign, he was nonetheless a marked man, never able to escape the personal tragedy of the unfair indictment against him. Returning east to the Army of the Potomac in which he had been serving at the time of Ball's Bluff, he found that gossip pursued him and that his opportunities for advancement were closed. In April of 1864, Stanton arbitrarily mustered him out of his Volunteer commission as brigadier general, and as a Regular Army colonel this late in the war, Stone waited for orders that never came and resigned in September of 1864. (Ironically, nineteen years later, Stone, in his capacity as a civil engineer, was in charge of constructing the massive base for the Statue of Liberty.)

Stone was by no means the committee's only victim. General George Gordon Meade, who was summoned to appear after the Fredericksburg defeat in May of 1863, two months before his great victory at Gettysburg, said in a letter that "I sometimes feel very nervous about my position, [the committee is] knocking over generals at such a rate."

And here was Sherman, who had commanded Louisiana's state military academy before the war, was known to have a low opinion of the black population's abilities, and had extended such generous peace terms to his fellow West Point alumnus Joseph E. Johnston. He was indeed the witness, and one the committee saw as a questionable and possibly dangerous figure, but he was not the real target. Senator Benjamin Wade, the prominent Radical Republican and a former trial lawyer who was the committee chair, wanted to use Sherman to get at and discredit what he considered to be the misguided policies of Abraham Lincoln. Among other things, Wade had never forgotten what had happened to the Wade-Davis Bill, legislation that he had put forward in 1864 with Congressman Henry Winter Davis of Maryland. Motivated by the Radicals' belief that only Congress had the right, through legislation, to determine the terms of Reconstruction, this had been a challenge to Lincoln's assumption of wartime emergency powers to be enacted by executive orders. When Lincoln killed the bill with a pocket veto, it infuriated the legislation's sponsors and further widened the gap between Lincoln and the more ardent Radicals.

As the committee's session opened, Wade went to work, questioning Sherman himself. He soon found that Sherman, who had aggressively torn apart the South, was employing a defensive strategy. Sherman loved his Army of the West, to which he would say good-bye the day after tomorrow after leading it down Pennsylvania Avenue on the second day of the Grand Review; he loved the United States Army and wanted to continue serving in it; he hated Stanton. He wanted Wade's jaws to snap shut on thin air, and he wanted to clear his name.

In his testimony—the transcript ran to thirty-two printed pages— Sherman kept faith with the officers and men who trusted him and did not hide behind Lincoln, in whom he had seen more goodness and greatness than in any other man he had met. He rebutted any effort to smear him in such a way that there could be grounds for his dismissal from the army, and he caused no new trouble for Grant and the postwar United States Army. And Sherman had his day in court, where Stanton was concerned.

Senator Wade tried and tried. Thinking that Sherman would, for his own defense, agree with the suggestion that in his first dealings with Johnston he was carrying out explicit orders from Lincoln, he gave Sherman the chance to clear himself and portray Lincoln just as the Radicals wanted him to be seen: a man who wanted a soft peace for the South. "Did you have," Wade asked him, "near Fortress Monroe, a conference with President Lincoln, and if so, about what time?"

Sherman said that he had, and gave the dates of his stay at Grant's headquarters at City Point. Wade moved in. "In those conferences was any arrangement made with you and General Grant, or either of you, in regard to the manner of arranging business with the Confederacy in regard to the terms of peace?"

"Nothing definite. It was simply a matter of general conversation, nothing specific and definite."

Here was the mystery. Either Sherman did not possess any proof of Lincoln's instructions to him, or he had it and chose not to produce it, or, more likely in the light of his family's worries about him, he had indulged in feverish wishful thinking the day before and was now facing the reality of this hearing. Thwarted, the best Wade could get out of Sherman was the statement, "Had President Lincoln lived, I know he would have sustained me."

Turning to Sherman's first agreement with Johnston, Wade asked why he had not dealt with the subject of slavery in that document and had left open an interpretation that slavery could continue in the former Confederate states. Sherman answered that Lincoln's Emancipation Proclamation had settled that subject once and for all, long before his meeting with Johnston, and that "for me to have renewed the question when that decision was [already] made would have involved the absurdity of an inferior undertaking to qualify the work of his superior." He went on to point out that the agreement he had reached with Johnston specifically stated that it had to be approved in Washington, and that he had immediately sent the document there by special messenger. In what was clearly a misrepresentation of what had happened—Sherman had never expected his terms to be rejected by his superiors—he now portrayed his initial terms as being "glittering generalities," hastily penned to stop the fighting and to have something to send to Washington for improvement. On this last point, that of stopping the fighting, Sherman stressed to the committee that both his own generals and Johnston had wanted an immediate end to the war, and that he in particular had feared that if there were not a general surrender, Johnston's experienced men would slip away and continue fighting throughout the South indefinitely as guerrilla bands.

Finally, Wade gave Sherman his opportunity to talk about Halleck and Stanton. Sherman told the committee, with no letters or documents to prove it, about a conversation that he had with Stanton, when the secretary had come to Savannah in January of that year. Sherman said that Stanton had encouraged him to include civil matters in any opportunity to end the war. For Stanton to turn on him after that, Sherman said, as he had of Halleck, was "an act of perfidy." This description of his conversation with Stanton in Savannah was the only new piece of information that Sherman had to offer, but for a moment the committee saw the man who had slashed apart the South. "I did feel indignant—I do feel indignant. As to my own honor, I can protect it."

When Sherman left the hearing room, the Joint Committee on the Conduct of the War had taken testimony from its last, and most controversial, witness. It would still receive final written reports from the generals of all the major commands, including Sherman, but the war and the work of the committee had ended. Sherman had held his own and, like everyone else in Washington, was looking ahead to the next day's beginning of the Grand Review, but in this hour after he was excused by the committee he was in for an experience that balanced out the Radicals' view of him. An Associated Press dispatch reported: "Gen. Sherman and his brother Senator Sherman passed down Pennsylvania-avenue this evening. His appearance caused the gathering of crowds, who repeatedly cheered him, while ladies waved their handkerchiefs. A large number of persons followed him, and the press soon became so general that he was compelled to call a carriage to escape the labor of a severe hand-shaking, which had already commenced."

That evening presented an interesting contrast between life in the encampments of Meade's Eastern army, both north and south of the Potomac, and Sherman's Western army, bivouacked to the south of the river in and around Alexandria. Meade's Fifth Corps of the Army of the Potomac was in a comfortable permanent camp on Arlington Heights above the southern bank of the river, near Arlington House, the Custis mansion occupied by Robert E. Lee up to the moment of secession. On this last night of that army's existence, there were many parties, and the officers of the First Division of the Fifth Corps, a hard-fighting unit whose flag displayed a red Maltese cross against a white background, were gathering to honor the corps commander, Major General Charles Griffin. Four enormous hospital tents had been put together to accommodate the officers and their guests.

A special gift for General Griffin had been arranged for the occasion, paid for by a collection taken up among the officers. Made by Tiffany's in New York, it was a gold pin whose enameled white surface bore the division's red Maltese cross, with the cross outlined in diamonds, and with a center diamond that cost a thousand dollars. The man who designed it, and pinned it on Griffin's uniform during the party to great applause, was the commander of the First Division, Major General Joshua Chamberlain, one of the war's most interesting figures. When the war started, he had been a professor of religion and Romance languages at Bowdoin College in Maine. Given a leave of absence to study in Europe during 1862 and 1863, he had instead entered the army and first risen to be colonel of the legendary Twentieth Maine, winning the Congressional Medal of Honor for his defense of Little Round Top at Gettysburg. Wounded several times—he had been hit twice at Hatcher's Run during the final Appomattox campaign, seven weeks before, and during the war the horse that he would ride in the next day's parade had been shot from under him three times and had healed well enough to go into action again—Chamberlain had been assigned to a special duty at Appomattox Court House. After Grant, Lee, and other commanders had left the area, Chamberlain was designated to be the senior officer present when the surrendered remnants of Lee's Army of Northern Virginia marched up to lay down their arms. On his own initiative, he had ordered the Union regiments lined up guarding the area to come to "present arms" in salute to their gallant foes—a gesture that, climaxed by a spontaneous three cheers given by Chamberlain's men as the last Confederates marched up to surrender, caused one of Lee's soldiers to write, "Many grizzled veterans wept like women, and my own eyes were as blind as my voice was dumb."

No one in the Fifth Corps went to bed that night. At two in the morning they began lining up to come marching down from Arlington Heights. The Long Bridge could handle normal traffic, but it took two hours for all these men and horses and cannon to cross the river and get to their assembly place near the Capitol, where they would have to wait until the Eastern army's cavalry and many other units preceded them in the parade.

In Sherman's camps around Alexandria that evening, columns of quartermaster wagons had arrived, filled with new uniforms and boots for Sherman's men to wear when their turn came to parade, two days hence. To the surprise of the well-meaning Eastern supply officers who brought them, most of the men refused to take a single item of clothing. Let them see us the way they are, they said. Clean weapons and bare feet. Let them look at us in our rags. They'll know who we are. We're Uncle Billy's men. They don't have to cheer. It's our last march, and we're going to do it our way.

Most of the men of the Western army had no idea of the intense curiosity about them felt by the crowds gathering in Washington. Among the visitors who had come to see the Grand Review was a small group of young ladies, friends who were members of prominent Boston families. The only place they had found to stay was in an attic room of a house near the Willard Hotel. On the day before the parade, they hired a carriage and drove out to Georgetown, finding themselves on a street where companies of Sherman's men were passing. Letting their friendly curiosity get the better of their New England reserve, they called out, "What regiment are you?" Back came the shouted answer, a regiment's number, and "Michigan!" As if on an expedition into unknown territory, the delighted girls asked the next unit where they were from and heard, "Wisconsin!," and another exchange produced what was to them the undoubtedly exotic, "Iowa!"

What the girls from Boston could not be expected to see, what few Americans of the day really understood, was the significance of those fine soldiers, farm boys from Michigan and Wisconsin and Iowa, being on the streets of Washington. It was the old Northwest, the West as it was then called, the Midwest as it later came to be known, that had given the nation such men as Lincoln and Grant from Illinois, and Sherman from Ohio. The soldiers of the Eastern states had done their full share in winning the war, but it was this new dimension, the political and military power of the West, that had welded itself to the older Eastern states in the great national crisis, and together these regions and their forces had won the war. The next two days would be a dramatic demonstration of that reality.

THE PAST AND FUTURE
MARCH UP PENNSYLVANIA AVENUE

The Grand Review turned out to be something so vast, so moving, that its effect overwhelmed even the generals who organized it. The parade was scheduled to step off from the Capitol Building at nine, but the crowds packed the sidewalks soon after sunrise, listening to the whisper in the air of drums beating miles away, as regiments kept marching into the city from their encampments. Everywhere, people positioned themselves at windows and on rooftops to watch, and boys climbed into trees along the route.

It was a beautiful May morning. For the first time since Lincoln's death five weeks earlier, the flag at the White House was raised from half-mast to the top of the pole. In the crowd stood many thousands who had lost members of their families in the war—mothers and fathers grieving for their sons, sisters remembering their brothers, young widows bringing their children to see the army that their father had joined and from which he had not returned. Along the route stood choirs of school-girls dressed in white, who would sing patriotic songs as the soldiers passed. Other young women dressed in white carried woven garlands to present to generals and colonels as they rode at the head of their divisions and brigades and regiments, while thousands of others in the crowd held bouquets that they intended to throw at the soldiers' feet in tribute as they passed. Many of the spectators carried little American flags, and women had white handkerchiefs in their hands, ready to wave.

Sherman came to the presidential reviewing stand early, along with Ellen, their eight-year-old son Tommy, and Ellen's father. Thomas Ewing was old and frail now, but he was happy to be back in the city where he had been a major figure and immensely proud of the son-in-law who had never seemed able to make a career for himself. After a time, Julia Grant joined them, bringing her son Jesse, who was seven. Julia and the Shermans sat together, talking as more dignitaries were escorted into this pavilion and the congressional reviewing stand across the avenue. Sherman was his usual animated self but seemed relatively at ease. It was his day to be a spectator as the nation gave thanks and said good-bye to the Army of the Potomac; his moment to ride up to this stand at the head of sixty-five thousand men would come tomorrow. Various members of the cabinet arrived. Presumably Secretary of War Stanton was somewhere among these powerful figures, but there is no account of his encountering Sherman on this first day.

At nine o'clock, the time for the parade to begin, neither President Johnson nor Grant had appeared, but, just as ordered, the signal gun fired from the Capitol. General George Gordon Meade, his horse decorated with chains of flowers placed there by admirers, came riding around the side of the Capitol, which had on its western portico a huge banner that read, "THE ONLY NATIONAL DEBT WE CAN NEVER REPAY IS THE DEBT WE OWE TO THE VICTORIOUS UNION SOLDIERS." He turned up Pennsylvania Avenue, followed by the officers of his staff, also on horseback and with their sabers drawn. A band marched behind them, playing. Soon after that came the first mounted troopers of a column of cavalry *seven miles long*. As the crowd caught a glimpse of the first of eighty thousand soldiers from the Army of the Potomac coming toward them, a roar of "Gettysburg! Gettysburg!" swept along the avenue, accompanying Meade and his men as they neared the presidential reviewing stand. When Meade came directly in front of that pavilion, with the president and Grant still not there, everyone in the stands on both sides of the avenue rose and cheered: the justices of the Supreme Court, senators, congressmen, governors, and their families. Meade raised his sword in salute and rode his horse into the White House grounds. As he dismounted to come and sit in a place of honor in the presidential pavilion, President Johnson arrived in a carriage, and Grant and some of his staff appeared, briskly walking through the White House grounds as they came from the War Department. (Neither man ever offered an explanation of this astonishing failure to be on time.) Johnson took his place in the stands, with Grant and Sherman sitting within a few feet of him.

As the cavalrymen clattered past, thousands upon thousands of them,

Grant and Sherman began watching the parade intently—two professional officers, less concerned with the history of the occasion than with the identity of the regiments and the performance of the troops. As the generals at the head of each division came up, Grant rose and saluted. The horsemen rode by briskly; there was much that the public probably did not grasp about what passed before them. Here came the Second Massachusetts Cavalry, but men from California had also fought in its ranks. Early in the war, starting with a hundred volunteers known as "the California Hundred," young men had come east to fight in a cavalry unit that was incorporated into this Massachusetts regiment. In all, 504 men from California had served this regiment, and their survivors were passing Grant and Sherman now, perhaps triggering memories for Grant of the loneliness of his early tour of duty on the West Coast that led to his drinking and forced resignation from the army, and also reminding Sherman and Ellen of Sherman's time there as an officer and, later, as a banker.

Someone that many in the crowd did indeed recognize was the dashing cavalryman General George Armstrong Custer, who eleven years later would die with the Seventh Cavalry at the hands of the Sioux during the Battle of the Little Big Horn, at a time when Grant was president and Sherman had Grant's old job as commander of the United States Army. As Custer rode up to the reviewing stand to salute, his foot-long golden locks streaming from under his wide-brimmed officer's hat, a wreath thrown from the stands landed in front of his horse. The mount wheeled, bolted so swiftly that Custer's hat blew off, and galloped away with him, causing Sherman to say later, "he was not reviewed at all." The unintended drama of that moment was followed, during the *two hours* that it took for the Cavalry Corps to ride past, with thrilling periodic displays by different units of the Horse Artillery. A long interval was allowed to develop in the line of march, and then artillery batteries came speeding down the avenue, the six-horse teams cantering past in a rumble of caissons and cannon that conjured up the scene when the guns had been brought swiftly onto battlefields, to be unlimbered and fired in situations where their timely arrival often saved the day.

As the last of the horsemen passed the reviewing stand and rode into history, some soldiers and civilians alike saw not only them but also the ghosts of those who had once ridden with them. In the 1864 Shenandoah campaign led by Philip Sheridan, who had desperately wanted to lead the Cavalry Corps today but who on Grant's orders was taking command in Texas, they had suffered losses of 3,917 men.

No sooner had the sound of hooves died away than the infantry came into view. Grant's aide Horace Porter said that when the long column marched toward the reviewing stand, "Their muskets shone like a wall of steel." The men of Sherman's army might be right in thinking that they had no equal as a combined striking force, but coming up Pennsylvania Avenue now were regiments second to none. The Sixty-ninth Pennsylvania, which also had a component of California volunteers, had asked to have that regimental number in tribute to the distinguished fighting record of New York's "Fighting Sixty-ninth." There were forty-five battle streamers weighing down its regimental flagstaff as it came up the avenue, signifying that they had fought in that many separate engagements during the past four years, including the Seven Days', Antietam, Fredericksburg, Chancellorsville, Gettysburg, the Wilderness, Spottsylvania Court House, Cold Harbor, and Petersburg, right through to Lee's surrender at Appomattox. As was the case with so many regiments, they had lost nearly as many men from disease as from enemy action; in one unlucky unit, the Fifteenth Maine, assignments to militarily quiet malarial areas in Mississippi, Louisiana, and Florida resulted in a loss of only five men in combat, but 343 died from illness. The Fighting Sixty-ninth itself, marching up Pennsylvania Avenue behind its green banner that had a gold harp and shamrocks embroidered on it, every man with a sprig of green in his cap, was filled with Irishmen from New York City; it had fought in many of the same battles and campaigns as its Pennsylvania namesake, also ending the war at Appomattox. The Scots of the Seventy-ninth New York, "the Highlanders," who had fought under Sherman's command at Bull Run and had remained in the Eastern theater, marched past behind the only bagpipe band in the Grand Review.

The sight of these regiments, some shrunken to small numbers and marching behind flags shot through so often that some were only shreds dancing in the breeze, moved many in the crowd to tears. Every few yards, someone, perhaps from the family of a man who died serving in those ranks, would suddenly run out from the sidewalk and kiss what was left of a flag. The bands marched past, playing the great songs—"The Battle Hymn of the Republic," "When Johnny Comes Marching Home Again," "Tramp, Tramp, Tramp, the Boys Are Marching"—and the crowds sang the words, some smiling as they did and some sobbing. Many a soldier found a woman coming out of the crowd to hand him a bouquet, and carried it swinging along in one hand while he kept his musket on his shoulder with the other.

On the reviewing stand, Sherman was making mental notes of what he saw. As he later recalled, when many of Meade's troops marched past the presidential pavilion, they "turned their eyes around like country gawks to look at the big people on the stand." The regiments marched well when their own bands were playing, but when they came up to the reviewing stand those bands fell silent, and the music as they passed was provided by two orchestras that had no sense of military cadence — Sherman called them "pampered and well-fed bands that are taught to play the very latest operas." He decided to dispense with the services of those two orchestras when his army marched the next day. Turning to Meade at one point, Sherman said, "I'm afraid my poor tatterdemalion corps will make a poor appearance tomorrow when contrasted with yours." When Meade replied that the spectators would make allowances for the appearance of Sherman's men, Sherman swore to himself that his men would show Washington some marching the public would never forget.

As the hours went by, with three hundred men moving past the reviewing stand every sixty seconds — a spectator called it "a Niagara of men" — here came General Joshua Chamberlain and his division, including the Twentieth Maine. As he rode up the avenue at the head of his troops, he had an ecstatic moment.

Now a girlish form, robed white as her spirit, presses close; modest, yet resolute, fixed on her purpose. She reaches up to me a wreath of rare flowers, close-braided, fit for viking's arm-ring, or victor's crown. How could I take it? Sword at the "carry" and left hand tasked, trying to curb my excited horse . . . He had been thrice shot down under me; he had seen the great surrender. But this unaccustomed vision — he had never seen a woman coming so near before — moved him strangely. Was this the soft death-angel — did he think? — calling us again, as in other days? For as often as she lifted the garland to the level of my hand, he sprang clear . . . I managed to bring his forefeet close beside her, and dropped my sword-point almost to her feet, with a bow so low I could have touched her cheek. Was it the garland's breath or hers that floated to my lips? My horse trembled.

Later in the presidential reviewing stand, watching his regiments receive Grant's salute as they marched by, Chamberlain, like many another soldier that day, saw the living men but felt the ghosts as well. "These were my men . . . They belonged to me by bonds birth cannot create nor death sever. More were passing here than the personages on the stand could see. But to me so seeing, what a review—how great, how far, how near! It was as the morning of the resurrection . . . the ten iron-hearted regiments that made that terrible charge down the north spur of Little Round Top into the seething furies at its base, and brought back not one-half of its deathless offering . . . I see in this passing pageant— worn, thin hostages of the mortal."

The future was there at the review, as well as the past. In the congressional stands sat James A. Garfield, who had fought under Grant at Shiloh and gone on to fight at Chickamauga, before retiring from the army as a brigadier general to run for Congress. Near him was another future president, four-times-wounded Major General Rutherford B. Hayes, who was still in uniform. Waiting to march with their men in Sherman's army the next day were two more men who would occupy the White House, Brigadier General Benjamin Harrison of Indiana, who had campaigned with Sherman to Atlanta, and William B. McKinley, who had enlisted as an eighteen-year-old private in 1861 and four years later was a major.

The spectators sometimes moved almost as one being. War correspondent Charles A. Page of the *New York Tribune*, who at Bull Run noted that the same stretchers used to carry wounded men back from the front were quickly loaded with boxes of cartridges and sent forward again, wrote that "when the crowd would surge up to the stand, at any brief interval in the procession, and demand a sight of their favorites, the President would rise, and bow repeatedly, but say never a word. Grant when called for would but rise for an instant, with lifted hat, and if his face told any story at all it was one of shyness and surprise." As for those the crowd rushed forward to see, "There never was so perfectly happy a set of men as those in the main pavilion—the President and Cabinet, General Grant, and the score or two of distinguished officers. It wasn't self-complacency, but a sort of calm quiet; a settled peace and gratitude seemed to pervade them all."

At three-thirty in the afternoon, the last of Meade's eighty thousand men passed the reviewing stand where Grant and Sherman had been since morning. The spectators poured from the sidewalks into the center of Pennsylvania Avenue, a sea of little American flags and white handkerchiefs waving good-bye to the backs of the last ranks of the Army of the Potomac as they marched into history.

Grant and Sherman parted, and Grant went into the White House grounds and mounted the horse that had been brought there for him. On an impulse, he told an enlisted orderly to mount another horse and come with him, and rode out into the crowd that was slowly dispersing along Pennsylvania Avenue, all talking about what they had seen. Grant on a horse always was an imposing figure, one of the great horsemen of his day. Startled to see him right there, a few feet away, the people cheered as he rode quietly among them, occasionally lifting his hat. Perhaps he too did not want the day to end.

During the first day of the Grand Review, Sherman's army had moved to new bivouac areas just south of the Long Bridge across the Potomac, and that night Sherman held a meeting with his generals and their adjutants. Sharing his observations of what he had seen from the reviewing stand, he said, "Be careful about your intervals and your tactics. I will give [the troops] plenty of time to go to the Capitol and see everything afterward, but let them keep their eyes fifteen feet to the front and march by in the old customary way."

These were the kinds of instructions Sherman was giving, with his generals noting what they were to do, but *The New York Times* had a different idea of what might happen in the morning. Citing the *Washington Tribune* as its source, the *Times* said that "it is mentioned in political circles that an influence is organizing among the superior officers of Sherman's army *to demand* of President Johnson the removal of Secretary Stanton, for his warfare upon their Commander . . . There is a public expectation throughout the city of a demonstration of the feeling of the rank and file of Sherman's army toward the Secretary of War when it shall march past the official stand in front of the White House." In an example of praise being so faint as to be inaudible, the *Times* added that, while the actions of Stanton and Halleck might have been "somewhat hasty and ill-advised," it was to be hoped that Sherman would not "forfeit the respect in which he is held by the great body of the people, and add another to the many proofs already existing, that one may be a great commander without being a wise man."

Whether Sherman even saw this article is unknown; he and his generals were concentrating on the last march they would make together, and nothing else, and his army finally got to sleep. At first light, a correspondent from *The New York World* heard bugles blowing, and he described Sherman's men forming up and following their regimental colors across the river on the Long Bridge: "Directly all sorts of colors, over a wild monotony of columns, began to sway to and fro, up and down, and like the uncoiling of a tremendous python, the Army of Sherman winds into Washington." The column was fifteen miles long.

Around the Capitol, young women were everywhere, chatting flirtatiously with Sherman's weather-bronzed soldiers as they pushed roses into the lapels of the men's uniforms and stuck flowers into the muzzles of their muskets. Numbers of girls had set up tubs of water with blocks of ice in them on street corners and brought cups of cold water for the men to drink as they waited for the parade to start. Everything was fair game for decoration: garlands were attached to the tops of the staffs of regimental battle flags, and horses were draped in flowers, as were cannon. The weather was even better than it had been the day before.

Sherman rode through all this, wearing a clean uniform and with his wiry red hair freshly cut. His men smiled as they saw their Uncle Billy "dressed up after dingy carelessness for years." His splendid horse, a "shining bay," was not only perfectly groomed for the occasion but was already covered in flowers put there by young women. For the rest of his life, female admirers would fuss over Sherman—and he loved the attention—but this morning he had just one thing on his mind: he wanted his army to make a good showing. He knew how well his men could fight, but he did not know if they could march well in a massive parade like this, and he was not sure that they cared what Washington thought of them. His officers cared: they were passing along their ranks, saying, "Boys, remember it's 'Sherman' against the 'Potomac'—the west against the east today."

Sherman's orders to his men instructed that the first units of his army were to form "opposite the northern entrance to the Capitol grounds, prepared to wheel into Pennsylvania Avenue at precisely 9 A.M.," and at the sound of the signal gun Sherman turned up Pennsylvania Avenue on his flower-decked horse. At his side rode Major General O. O. Howard, who had lost his right arm in combat three years before. In a sense, Howard symbolized both the sacrifices of war and the hopes of the peace for which it had been fought: still holding his military rank, just eleven days earlier he had been named to head the Freedmen's Bureau, the new federal agency formed to protect the interests of the former slaves.

From the outset, Sherman's march up the avenue conveyed the reality of his army when it was campaigning: a contemporary account said that behind the mounted officers of his staff "was a group of orderlies, mounted servants, pack mules, &c., and behind these a body of cavalry, known as the headquarters guard and escort." Only after that came a band, playing "The Star-Spangled Banner." One observer's first impression of Sherman's troops, muskets with fixed bayonets on their shoulders, was that these leathery young men, their faces made old by war, were marching along sullenly, but their expressions changed as the successive columns "wheeled" into Pennsylvania Avenue and saw what awaited them. The crowd was larger than the day before; a *New York Times* reporter wrote, "The enthusiasm to-day far exceeded that of yesterday." Many thousands of men from the Army of the Potomac who had marched the day before had come back today on their own, ready to cheer the Army of the West. Two new banners had gone up overnight, stretching across the avenue: the first read in part, "HAIL TO THE HEROES OF THE WEST!" and the second said, "HAIL CHAMPIONS OF BELMONT, DONELSON, SHILOH, VICKSBURG, CHATTANOOGA, ATLANTA, SAVANNAH, BENTONVILLE—PRIDE OF THE NATION." People in the crowd were holding up babies, so that the infants could one day be told they had seen Sherman's men. Some of Sherman's regiments were marching behind bare flagstaffs, because their battle flags had been literally shot away, while other banners were darkened and stained from powder burns and weather. Yesterday the crowds had shouted, "Gettysburg!"; today, the cry of "Vicksburg!" rang along the avenue.

Sherman and his men, so many of them barefoot and in rags, were showing the North the "sea of bayonets" that had recently convinced the residents of Raleigh, North Carolina, that for them the war was over. At the end of large units, ambulances came along, the horses drawing them well groomed and the ambulances clean; the crowds hushed as they saw the rolled-up canvas stretchers on the sides of the ambulances, which had been washed but still had deep brown bloodstains from the wounded men they had carried. Sometimes wild cries came from the crowd, giving voice to feelings that perhaps none understood. Other people "raised their hands to heaven in prayer." So many flowers were thrown from the sidewalks that barefoot men marched through petals that lay ankle-deep.

Sherman's progress up Pennsylvania Avenue was literally triumphal: the *Times* said, "He was vociferously cheered all along the line," and added, "The greeting of this hero was in the highest degree enthusiastic." When he rode by, raising his hat in answer to the shouts of admiration and welcome, people jumped up and down to get a better look at him, waving flags and handkerchiefs as they did. One spectator caught up in the crowd felt that "there was something almost fierce in the fever of enthusiasm." A woman journalist observed "in his eye . . . the proud, conscious glare of the conqueror, while his features, relieved of the nervous anxious expression of war times, assumed an air of repose which well became him." Sherman's soldiers began responding to what Sergeant Upson of the 100th Indiana called "one constant roar," and Upson noted that his troops marched better and better: "on the faces of the men was what one might call a *glory look*."

A young private from Wisconsin was thinking about Abraham Lincoln: if he had been there, this veteran of many battles decided, the units would simply have broken ranks to crowd around him, and the parade would have stopped right then. Even in the midst of the bands playing and the crowds cheering, some houses still bore wreaths of mourning; it was as if Lincoln, the man with the solemn face and the sudden sweet smile, still hovered over the soldiers of whom he asked so much and loved so well.

At the head of his sixty-five thousand men, Sherman had a great deal happening in only a few minutes. As he rode up the incline to the corner of Pennsylvania Avenue where he and the leading division of his army would come to the first two turns before passing the reviewing stand in front of the White House, Sherman gave in to an impulse to look back and see if his soldiers were marching as well as he fervently hoped they were. Underneath the roar of the crowd he had been hearing behind him what a soldier from Minnesota called "one footfall"—a good indication that the men were marching in step—but now, at the top of this little slope, he turned on his horse to see the column of thousands of men that stretched more than a mile behind him down Pennsylvania Avenue to the Capitol. "When I reached the Treasury-building, and looked back, the sight was simply magnificent. The column was compact, and the glittering muskets looked like a solid mass of steel, moving with the regularity of a pendulum."

Sherman was to say in later years that "I believe it was the happiest and most satisfactory moment of my life," and it may well have been: he would serve as both commander of the United States Army and as secretary of war, honors would descend upon him, he would remain nationally and internationally famous, and be lionized during the final years of his life in his adopted New York City, but he would always be happiest when he was among his groups of veterans at their frequent reunions in many states, occasions he unfailingly attended.

Coming into Lafayette Square, Sherman rode over to the side of the street toward the front of the house that served as the army's headquarters for the defense of Washington. Secretary of State Seward, who was still recovering from the knife wounds he received during the attempt to assassinate him, had been brought there to watch the parade. Sherman brought his horse to a halt and "took off my hat to Mr. Seward, who sat at an upper window. He recognized the salute, and returned it." Finally, as Sherman came to the presidential pavilion and the other grandstands, with the bands striking up "Marching Through Georgia," the New York World said, "The acclamation given Sherman was without precedent . . . The whole assemblage raised and waved and shouted as if he had been the personal friend of each of them . . . Sherman was the idol of the day." When he entered the pavilion after dismounting from his horse in the White House grounds, everyone was still on his feet to welcome him. Witnesses would differ on whether Secretary of War Stanton extended his hand to Sherman or simply nodded in greeting but, in a historically memorable instance of one person "cutting dead" another, Sherman walked past Stanton as if he were not there. He shook hands with President Johnson and every other member of the cabinet. Then, to loud applause, he and Grant greeted each other warmly.

There it was: the apotheosis of the friendship and military partnership that had brought the Union and its armies to this day. They were the men, the two generals, who more than any other soldiers had made this moment happen, and everyone there knew it. Sitting on either side of Johnson as the Army of the West continued to pass, Grant and Sherman rose and returned salutes whenever it was appropriate, but they seemed to become lost in thought, occasionally saying a few words to each other, and it was others who studied and recorded what the celebrities and the crowd now saw. Journalists remarked on how the Western soldiers were bigger men and marched with longer strides. One reporter reacted this way: " 'Veteran' was written all over their dark faces, browned by the ardent Southern sun, and health almost spoke from their elastic step and erect figures . . . They seemed almost like figures from another planet." Walt Whitman saw them as "largely animal, and handsomely so." Two *New York Times* stories vied with each other in praise, one describing the Westerners as "tall, erect, broad-shouldered, stalwart men," and the other calling them "the most superb material ever molded into soldiers."

Sherman's army kept passing, like a torrent controlled only by itself. Someone in the crowd noted that so many garlands were draped on the musicians that the bands as they marched past appeared to be "moving floral gardens." There were not only the muskets and bayonets and some highly polished brass cannon gleaming in the sun, but also heavy supply wagons and the components for pontoon bridges like those in Meade's column the day before. Signalmen carried slender staffs sixteen feet high, at the top of which were little emblazoned flags that a *New York World* reporter likened to "talismanic banners" that might be found in some medieval pageant.

And there was this: marching at the head of each brigade of the Fifteenth Corps, and at some other places in the parade, was "a battalion of black pioneers [engineering troops] . . . in the garments he wore on the plantation, with shovel and axe on the shoulder, marching with even front, sturdy step and lofty air." Sherman had not brought these freed slaves into his army as combat soldiers, but he had come to appreciate their strength and skill as they laid the plank roads through the Carolina swamps that Joseph E. Johnston thought Sherman's army could never pass. In other similar units, a reporter saw "the implements being carried on the shoulders of both white and black soldiers."

At the end of each of several brigades came some of Sherman's "bummers," the independent operating foragers, "first in an advance and last in a retreat"—with examples of their foraging and of the newly freed slave families that had attached themselves to the army as it moved through the South. The crowd reacted to this as if watching a circus parade, and a *Times* reporter described it this way.

It was a most nonchalant, grotesque spectacle—two very diminutive white donkeys bestrode by two diminutive black contrabands. If that is not paradox, a dozen patient pack mules, mounted with Mexican pack saddles, camp equipage on one side and boxes of hardtack the other; half a dozen contraband females on foot; a dozen contraband males leading the mules; a white soldier or two on horseback, to see that everything was all right; the servants of the mess, and the mess kit, and scattered about on the panniers [cargo baskets] of the mules, reclining very domestically, half a dozen game cocks, a brace of young coons, and a sure-footed goat, all presenting such a scene that brought laughter and cheers from end to end of the avenue.

Here was complex irony again: these black Americans, being treated as figures of fun by the crowd, were no longer slaves because of the sacrifices made by the white men in blue uniforms marching ahead of them, as well as by those made by black regiments. (As many as 180,000 black soldiers served in the Union Army at one time or another; no uniformed contingent of these United States Colored Troops, as they were designated, was included in the parade.)

Julia Grant was greatly enjoying the Grand Review. She was to remember thinking, "How magnificent the marching! What shouts rent the air!" when suddenly she saw Mrs. Herman Canfield. She was the "tall handsome" woman "clad in deepest mourning" who had come to call after Shiloh, to tell of Grant's kindness to her when she came to see her wounded husband, the colonel of an Ohio regiment, who died before she could reach his hospital bed. The last thing she had said to Julia that day three years before was, "I have determined to devote my time to the wounded soldiers during the war." Julia saw that she had. "I saw Mrs. Canfield, the soldier's widow, the soldiers' nurse, when all this was passing. She, yes, she had grown older in these three long, weary years, for her dark hair showed threads of silver, her fair face and brow were furrowed and browned by exposure, her mourning robes looked worn and faded, as did the flag of her husband's old regiment as it passed on that glorious day up Pennsylvania Avenue."

Grant and Sherman continued to return salutes and to greet division commanders who would dismount and sit with them as their regiments passed. This parade was truly a good-bye: most of these tens of thousands of men were marching together for the last time. Their units would be disbanded, some within a few days, and they would return home, honored as veterans but taking up their future lives as individual civilians.

In the midst of this day's fame and excitement, the future was indeed waiting for Grant, and for Julia, and for Sherman and Ellen. Forty-one months after this parade, Grant would be elected president and serve two terms marred by political scandals caused by men who betrayed the governmental trust he reposed in them. Historians would differ as to what degree it was a failed presidency, but it had its moments. Soon after he entered the White House, Grant invited Robert E. Lee to call on him there. At eleven in the morning of May 10, 1869, six years to the hour after the first shots were fired in his great victory at Chancellorsville and forty-nine months after Appomattox, Lee stepped out of a carriage and walked into the Executive Mansion to be greeted by Grant. The visit was brief and formal, and not without its political repercussions. Many Republicans who had voted Grant into office were aghast at what he had done, but both Grant and Lee understood the meaning of the occasion: Grant was inviting the South back to the White House, and Lee was accepting the invitation.

Grant and Sherman's friendship would to some extent survive, but it had some exceedingly difficult times. Sherman's life after the war had in it a mixture of national and even international fame, along with professional frustration and disappointment in his friend Grant. Less than two months after the Grand Review, with Grant remaining the army's commander, Sherman was assigned to command what was then designated the Military Division of the Mississippi, a territory which, with the exception of Texas, included all the land from the great river to the Rocky Mountains. Named a lieutenant general the following year, he found himself holding a key command in an army whose size the Congress was steadily reducing, at a time when the Indian Wars were under way and federal forces were required for the military occupation of the South during the early Reconstruction years. Fourteen months after the surrender at Appomattox, at which time the Union Army had numbered a million men, Congress reduced the size of the peacetime Regular Army to forty-four thousand, with further reductions coming despite a continuing responsibility to garrison 225 posts ranging from coast artillery installations to wooden forts deep in Indian territory.

When the cuts in military appropriations also reduced the pay for generals, Sherman pointed out that he felt all the Union generals had been underpaid during the war and certainly should not be treated this way. He wrote to a friend, "What money will pay Meade for Gettysburg? What Sheridan for Winchester and the Five Forks & what Thomas for Chickamauga, Chattanooga or Nashville?" Referring to what the taxpayers would save by these reductions in military pay, Sherman added, "Few Americans would tear these pages from our national history for the few dollars saved from their pay during their short lives."

Never politically adept and always disliking the press, Sherman's views caused more controversy than he wanted. Convinced by the concept of Manifest Destiny, which held it self-evidently right that the American white population should spread across the plains and settle all of the Western lands, he regarded the Indians as being an inferior people, who, he told a graduating class at West Point, had an "inherited prejudice . . . against labor." (Grant, who had written in a letter to Julia of the future international power of the United States that was an extension of the idea of Manifest Destiny, was for a gentler treatment of the Indians, but his efforts were hampered by both the inefficiency and occasional corruption of the Indian Agency, and the enormous thrust of post–Civil War Western expansion.)

At times Sherman tried to ensure that white settlers treated the Indians fairly, but as he became increasingly convinced that the Indians would not change and become the kind of domesticated, productive citizens he wanted them to be, he came to something like his wartime policy toward the South. It would be more realistic, and better in the long run for everyone concerned, to crush the Indians' resistance to white settlement sooner than later, by applying harsh force. The Indians simply had to be gotten out of the way of the inevitable settlement of the Western lands that they persisted in thinking were theirs. His underlying attitude toward the Indians was not a desire to wipe them out, but at least once, voicing his belief in the need to take strong military measures against the Indians, he admitted that he "believed in the doctrine" expressed in the saying, "The only good Indian is a dead Indian."

During the immediate postwar years, before Grant became president, Sherman's resentment of what he considered to be the dictatorial attitude toward him taken by President Andrew Johnson and Secretary of War Stanton increased the tension between him and his two civilian superiors. Nonetheless, his wartime fame ensured that he would be sounded out by one group or another as a possible presidential candidate during every election from 1868 to the one to be held in 1892, overtures that he most memorably finally dismissed with, "If nominated I will not run; if elected I will not serve."

When Grant was elected president in the autumn of 1868, he named Sherman to replace him as the army's commander. Taking that position of general of the army in March of 1869, Sherman felt that he and his old friend Grant would work well together, as they had during the war. At that moment, the secretary of war was General John M. Schofield, who had served under Sherman during the war. Schofield had been a compromise appointment made by Andrew Johnson after Johnson failed in his efforts to remove Stanton during their dispute about Reconstruction policies, with Stanton resigning on his own initiative after Johnson's impeachment. Sherman, who on Grant's recommendation had just been promoted to the new rank of full general, had reason to assume that, among himself, Grant, and Schofield, three greatly experienced West Pointers, army matters would be dealt with in a harmonious and effective fashion.

No sooner had Sherman taken command of the army than Grant replaced Schofield as secretary of war with his old wartime chief of staff and confidant John A. Rawlins, who was now gravely ill. In an action that substantially reduced Sherman's authority, Rawlins immediately issued orders that in effect rescinded Grant's own recent order setting forth the wide scope of the general of the army's powers. Hurrying to the White House, Sherman tried to get this reversed, only to encounter Grant's statement, about Rawlins and his illness, "I don't like to give him pain now; so, Sherman, you'll have to publish the rescinding order." When Sherman still protested, with the two men still addressing each other as "Grant" and "Sherman" in the manner of their wartime meetings, Grant said, "Well, if it's my own order, I can rescind it, can't I?"

Sherman, who had written his brother during the war that Grant "has an almost childlike love for me," stood up and said, "Yes, Mister President, you have the power to revoke your own order; you shall be obeyed. Good morning, sir."

Things were never quite the same between them after that. When Rawlins died five months later, Grant named Sherman as his interim secretary of war, but when the post was filled by W. W. Belknap, a civilian who was another of Sherman's former subordinate generals, Belknap soon exceeded Rawlin's actions in restricting the powers of the general of the army. Sherman began to see that Grant was in the hands of politicians. Nonetheless, Sherman remained in command of the army for a total of fourteen years, serving not only under Grant through his two terms as president but also under President Rutherford B. Hayes, another Civil War general, as well as James A. Garfield, who had fought as a brigadier general at Shiloh, and Chester A. Arthur. On his sixty-fourth birthday, February 8, 1884, nearly forty-eight years after he was sworn in as a cadet at West Point, William Tecumseh Sherman resigned from the army.

During the last year of Grant's life, it would be the newly retired Sherman's turn to lift his friend's spirits. In 1884, seven years after he left the White House, Grant, then living in New York, lost all his money because of the fraudulent machinations of a Wall Street figure to whom he had entrusted everything he had. To try to regain the loss and provide for his family, he began to write his accurate, powerful, historically valuable memoirs, a work that brings the reader to the close of the war. Lavishly praised across the next century by figures as diverse as Mark Twain, Gertrude Stein, and Edmund Wilson, the book was destined to succeed with the public immediately and guarantee Julia a comfortable income, but soon after he began to write, Grant started to suffer from the throat cancer finally brought on by thousands of cigars. Racing against his illness as he wrote, Grant welcomed Sherman's repeated visits to him: on December 24, 1884, Sherman wrote Ellen, "Grant says my visits have done him more good than all the doctors."

In a gallant final effort, Grant finished his classic work on July 19, 1885, and died four days later. A crowd of one and a half million—the largest to assemble in the United States to that time—lined the streets of Manhattan to watch his funeral cortege pass. In the solemn parade were not only Union Army veterans but also a contingent of Confederates who had served in the Stonewall Brigade. Sherman was a pallbearer at the funeral ceremony and bowed his head and wept as a bugler played Taps. Two months later, Sherman said of his friend, "It will be a thousand years before his character is fully appreciated." He became the defender of Grant's military reputation: when an argument was made that Lee was the greater general, Sherman countered that it was Grant who had seen the Civil War as a strategic seamless web, and added, of Lee, "His Virginia was to him the world . . . [He] stood at the front porch battling with the flames whilst the kitchen and whole house were burning, sure in the end to consume the whole."

In that future, still far distant on the day of the Grand Review, Ellen would be the next to go. To the end, she and Sherman were very different and disagreed on many things as they always had—twelve years after the war, they lived apart for close to two years—yet deep in Ellen was the little five-year-old girl who "peeped with great interest" as her father brought the red-headed nine-year-old boy from next door home to live with them. Two years after Sherman retired from the army in 1884, they went to live in New York City, where she remained at home in the evenings while he consorted with the millionaires of the age—the Astors, the Vanderbilts, the Carnegies—as well as with actresses, artists, and assorted celebrities. His flirtatious friendships with women, including those with a sculptress and a Philadelphia socialite, were well-known, but in the middle of all that, while she was away on a trip, Ellen wrote him a letter that rang the truth for both of them. "You are the only man in the world I ever could have loved," she said, and then told him that, whether he knew it or not, "You are true to me in heart and soul." They would never reconcile their views on Catholicism—when their son Tom, eight years old at the time of the Grand Review, decided at the age of twenty-three to become a Jesuit priest, his decision broke Sherman's heart and pleased Ellen—but their unending love for their dead son Willy was one of the bonds that held them together like steel.

The end for Ellen came on November 28, 1888, when she was sixty-four. She had been sick for some weeks, lying in bed upstairs in the house on Manhattan's West Seventy-first Street into which they had recently moved. Sherman wishfully thought she was exaggerating the gravity of her illness but had installed a nurse to take care of her. He was reading in his office when the nurse suddenly called down to him that Ellen was failing. Sherman raced up the stairs, crying out, "Wait for me Ellen, no one ever loved you as I love you!" When he reached her bed, she was gone.

After a time of mourning so deep as to worry those who remembered the mental states of his wartime years, Sherman resumed his New York social life. Cared for by his unmarried daughter Lizzie and with frequent visits from his married daughters Ellie and Minnie, he resumed his combination of sophisticated New York life and reunions of his veterans—he attended several hundred of those gatherings during the first fifteen years after the war, always as the guest of honor—but his fabled energy was deserting him. On February 8, 1890, his seventieth birthday brought forth greetings and tributes from all over the nation, and he was surrounded by a family that now included seven grandchildren, but a year later, he was stricken by an illness that appeared to be related to the asthma from which he had suffered earlier in his life. This soon became pneumonia. Sherman lay in bed, steadily growing weaker, with his mind sometimes wandering. On February 11, 1891, he asked his daughter Minnie to make certain that the words "Faithful and Honorable" be carved on his headstone. Three days after that, he died.

Sherman's body was to be taken west for burial in St. Louis, but first there would be a massive funeral service for him in New York, attended by President Benjamin Harrison, who had served in his ranks, former president Rutherford B. Hayes, another of his veterans, former president Grover Cleveland, and five of the surviving major generals of the Army of the West. The funeral procession from Sherman's house to the church would have thirty thousand men marching, in organizations ranging from the entire West Point Corps of Cadets to regiments of the Regular Army and National Guard, as well as thousands of Civil War veterans from the association known as the Grand Army of the Republic.

As the funeral procession was being organized, an erect, frail, eighty-four-year-old man got off a train in New York, bringing with him a valise and the honor of the South. From the time he had met Sherman in North Carolina for the purpose of surrendering his army to him, General Joseph E. Johnston had remained his admiring friend. They had corresponded and had frequently dined together in Washington during the postwar years during which Johnston had, among other things, served the reunited nation as a congressman from Virginia from 1879 to 1881, and was appointed United States commissioner of railroads in 1885. Learning that Sherman had named him an honorary pallbearer, Johnston had headed north from his home in Washington, despite the concern of his family and friends that he was not used to the winter weather he would find in New York in mid-February. As he stood at attention bareheaded outside Sherman's house while the casket containing his friend's body was brought down the steps to be placed in a hearse, someone behind him leaned forward and said, "General, please put on your hat, you might get sick." Johnston replied, "If I were in his place and he were standing here in mine, he would not put on his hat." By the end of the day, Johnston had a severe chill, which caused complications for his weakened heart. Back in the South, Johnston died a month later.

The trip of Sherman's body west to be buried in St. Louis was symbolic in itself—a man from Ohio, a soldier of the West, the commander of the Army of the West, returning to be buried near the Mississippi River that he and Grant had used as the strategic avenue that led them to the victory at Vicksburg and all that followed from that—but there was something more. Along the route of his funeral train, his veterans waited for him in daylight and darkness, ready to salute him a final time. Many of them had gathered in squads, small groups of survivors wearing their broad-brimmed slouch hats, saved from the days when they fought and marched beside their "Uncle Billy." Some even had their old muskets, loaded with a blank charge of powder, to raise and fire at the sky in the manner of military funerals. What they saw when the train came was a big picture of Sherman, fixed to the headlight, and his sword swinging beneath it as he went west.

The last survivor of the two couples, the Grants and the Shermans, was Julia Grant. She loved her eight years in the White House as first lady and greatly enjoyed the two and a half years that she and Grant spent traveling around the world after he left office. Nation after nation hailed him with twenty-one-gun salutes, and United States Navy vessels were placed at their disposal whenever they wanted to travel aboard them. When the Grants dined with Queen Victoria at Windsor Palace, before Grant took the queen into dinner on his arm, Julia and Victoria had a chance to talk and told each other about their children. "The Paris I remember," Julia wrote in her own wise and charming memoirs, "is all sunshine, the people all happy," and she also noted that "the President of France, that grand old soldier Marshal MacMahon and Madam MacMahon were unceasing in their attentions." At Heidelberg, Richard Wagner "performed some of his own delightful pieces of music for us." Arriving at Cairo by train from the port of Alexandria, the Grants were met by officials representing the khedive of Egypt, including former Confederate general William W. Loring, a West Pointer who had lost an arm in the Mexican War and had commanded troops fighting Grant in the Vicksburg campaign, and was employed by the khedive in modernizing and training his army. (Grant promptly asked Loring to ride with him in his carriage, and the two talked of Civil War campaigns.)

After Grant's death, Julia moved from New York to Washington, where she lived in a large, comfortable house on Massachusetts Avenue, receiving prominent visitors of every sort and often dining at the White House. In good health, lively as always, interested in everything, she began to outlive many of her contemporaries; when the Spanish-American War started in 1898, she became the active head of the Women's National War Relief Association, which sent its first shipload of supplies to Manila. (During that conflict, which began thirty-three years after the Civil War ended, former Confederate cavalry general Joseph Wheeler served as a major general of United States Volunteers at the age of sixty-one, and Robert E. Lee's nephew Fitzhugh Lee, who had been another young Confederate cavalry general, also became a major general and was a corps commander in Cuba, being present at the Battle of San Juan Hill when he was sixty-two. The Grants' son Frederick, who as a boy of thirteen was with his father at the surrender of Vicksburg, later graduated from West Point, and after time out of the service, including acting along with his brother Ulysses Jr. as confidential secretary to the president while Grant was in office, came back into the army when he was forty-seven to serve as a major general in the Philippines.)

In 1902, at the age of seventy-six, Julia Dent Grant had an attack of bronchitis, combined with heart and kidney failures. To the end, she remembered and cherished all of Ulysses S. Grant, not just the general and the president but also the young lieutenant who had ridden up to her house at White Haven so long ago. The squint-eyed girl from Missouri had gone with him every loving, supportive step of the way, from the morning rides across bright pastures along the shining Gravois Creek, through hard times when Grant was peddling firewood on the streets of St. Louis, to battlefield areas, to the White House, to Windsor Castle, to caring for him in his last illness as he wrote his account of the battles he fought and the campaigns he commanded. They had done it together, they had lived one of the great American love stories, and now in her last days she still yearned for her "Ulys," who signed all his letters with, "Kisses to you and the children." In the last lines of her memoirs she wrote, "I, his wife, rested in and was warmed in the sunlight of his loyal love and great fame, and now, even though his beautiful life has gone out, it is as when some far-off planet disappears from the heavens; the light of his glorious fame still reaches out to me, falls upon me, and warms me."

L'ENVOI

And so the living echoes of the friendship between Ulysses S. Grant and William Tecumseh Sherman came to an end. Both of them failures before the war, the two men, alike in some ways and so different in others, discovered their talents and strengths in the crucible of the great national crisis. They formed a partnership in which, often after significant differences of opinion, each resolutely and successfully supported the decisions and movements of the other. Both were formidable leaders, but it was their combined abilities and coordinated campaigns that proved literally irresistible and played such a major part in winning the Civil War.

Sherman's role appears at first to be the more dramatic and visionary half of the Grant-and-Sherman story. Coming into his own at Shiloh after earlier failure and headlines saying "General William T. Sherman Insane," his strategic vision not only led to his epic marches through the South but also ushered in a new era of modern warfare in which the destruction of the enemy population's will to resist can be as important as the defeat of its armies in the field. What is less obvious but exciting to watch is Grant's movement up the chain of command, proving himself equal to one extraordinary challenge after another. Grant was intuitively aggressive and instinctively tenacious. At Vicksburg, in saving the situation at Chattanooga, and in his risky clandestine change of front against Lee after Cold Harbor, Grant demonstrated that he too could be imaginative and bold. Beyond this, while Sherman had only his own armies to think of, as commander of the entire Union Army Grant proved to be a superb administrator who saw the overall picture of the war with a vision at least as clear as Sherman's.

What drew the two men together as friends? Each needed a military colleague whom he could admire and trust, and each wanted the other's warm approval. Grant had the love of Julia but yearned for affection from a flinty father who tried to turn a financial profit from his son's military position. Sherman had what a later generation would call a "support system"—his wife, Ellen, his brother Senator John Sherman, his prominent father-in-law, Thomas Ewing, who became increasingly appreciative of his achievements as the war progressed—but underneath there was the once-insecure boy whose father had died when he was nine, at which time he and his ten brothers and sisters had been split up among a number of households.

Sherman was right when he said of himself and Grant, "We were as brothers." They did the things that devoted brothers do: back each other up, help each other out, sacrifice for the other. It was Sherman, standing to gain if Grant resigned from the army, who talked him out of going home when Halleck sidelined him after Shiloh; it was Sherman who told Grant to go ahead and send him into action at Hayne's Bluff above Vicksburg, a move likely to hurt Sherman's reputation but one that might help the Vickbsurg campaign as a whole; it was Grant's steadfast support that led Sherman to say after Vicksburg, "I knew wherever I was that you thought of me, and that if I got in a tight place you would come if alive."

Grant worked to ensure that Sherman received his richly deserved promotions. When Sherman's brilliant victories in the South caused a bill to be introduced in Congress to give him a promotion making him eligible to rise to the supreme command above the beleaguered Grant, he wrote a letter to his senator brother urging that the effort be stopped. As the war came to a close, after Sherman had concluded the surrender agreement with Joseph E. Johnston that some in Washington found so generous as to be the act of a traitor, Grant arrived in North Carolina and swiftly and tactfully rectified a situation that could have ended Sherman's career at the height of his fame.

Grant thought that Sherman was entertaining, and thoroughly enjoyed his company; when they were together, Grant, usually reserved in manner, relaxed as he did with no other officer. Sherman was often baffled by the depths from which Grant pulled forth his successful military movements, but he came to see that Grant was a master of what he did. Each saw in the other a friendly, trusted partner who quickly grasped the other's ideas and made it possible to implement them for their mutual benefit and for the success of the cause to which they were dedicated. Each saw a man who wanted victory far more than he wanted promotion or fame; each saw a soldier's soldier. Whether they were campaigning together, or communicating by letter and telegraph at times when their headquarters were several hundred miles apart, each knew that the other made him more than what he was before they met.

NOTES

In citing works in the notes, short titles have generally been used. Works frequently cited have been identified by the following abbreviations. The full citation appears in the bibliography, under the name of the author or editor.

GMS   Ulysses S. Grant, *Memoirs and Selected Letters*
JDG   John Y. Simon, ed , *The Personal Memoirs of Julia Dent Grant*
LL   Lloyd Lewis, *Sherman: Fighting Prophet*
M   John F. Marszalek, *Sherman: A Soldier's Passion for Order*
PUSG   John Y. Simon, ed., *The Papers of Ulysses S. Grant*
SCW   Brooks D. Simpson and Jean V. Berlin, eds., *Sherman's Civil War: Selected Correspondence of William T. Sherman, 1860–1865*
SG   Jean Edward Smith, *Grant.*
SM   William Tecumseh Sherman, *Memoirs of General W. T. Sherman*

PROLOGUE
3 "put the river"   SG, 200.
3 "Well, Grant"   Ibid., 201.
4 "The South never smiled"   Brooks, *Grant,* 144.
4 "plain as an old stove"   Garland, *Grant,* 229.
4 "He is never quiet"   Kennett, *Sherman,* 99.
4 "as if he had determined"   SG, 300.
4 "When Grant once"   Porter, *Campaigning,* 223.
5 "In a moment"   Ibid., 417.

1. TWO FAILED MEN WITH GREAT POTENTIAL
7 "In my new employment"   PUSG, I: 359. Grant's letter of resignation, April 11, 1854, PUSG, I: 329–32. Lewis, *Captain Sam Grant,* 329, sets forth circumstances of the resignation. Also see SG, 87.
8 "Every day I like"   PUSG, I: 334.
8 Grant's financial situation   SG, 91.

8 "Why, Grant" Lewis, *Captain Sam Grant*, 346.

8 "Great God, Grant" SG, 91.

8 The pawn ticket PUSG, I: 339. The ticket also appears among this book's illustrations.

8 Crop freeze SG, 92.

8 "Julia and I" PUSG, I: 343.

9 "shabbily dressed" SG, 95.

9 "Grant was" Lewis, *Captain Sam Grant*, 377.

9 "I rarely read over" GMS, 39.

9 Jack Lindsay incident Fleming, *West Point*, 102.

10 "His hair was" SG, 26.

10 "the class, still mounted" Garland, *Grant*, 52.

11 "The farm of White Haven" Casey, "When Grant Went . . ." All quotations from Emma Dent Casey are from this recollection.

12 "That young man" Lewis, *Captain Sam Grant*, 110.

13 "a darling little lieutenant" Ulysses S. Grant Homepage, citing www.mscomm .com/~ulysses/page181.html.

13 "he was kind enough" JDG, 48.

13 "Saturday came" Ibid., 49.

14 "serious the matter with me" GMS, 37.

14 "On this occasion" Ibid., 38.

14 "We all enjoyed" Casey.

15 "I noticed, too" Ross, *The General's Wife*, 25, citing *Ladies' Home Journal*, October 1890.

15 "Before I returned" GMS, 39.

15 "In the thickest" PUSG, I: 86.

16 "I crossed at such" GMS, 81.

16 "I never went" Lewis, *Captain Sam Grant*, 249.

16 "You could not keep him" Garland, *Grant*, 100.

16 "the first two persons" Lewis, *Captain Sam Grant*, 251.

17 "found a church" GMS, 106.

17 "The shots from" Ibid., 109.

17 "every shot was" Ibid.

17 "I could not tell" Ibid.

17 "astonishing victories . . . frightful" PUSG, I: 146.

17 "one of the most" GMS, 41.

18 "the very best soldier" Freeman, *Lee*, I: 284.

18 "There's no danger!" Lewis, *Captain Sam Grant*, 245.

18 "was more bronzed" Casey.

19 "one of those beautiful . . . hand to glove" Ulysses S. Grant Homepage, interview, Julia Dent Grant, www.mscomm.com/~ulysses/page181.html.

19 "I enjoyed sitting" JDG, 56.

20 "a man of iron" Garland, *Grant*, 122.

20 "He seemed always to be sad" Ibid.

20 "a mail came in" PUSG, I: 320.

20 "He was in the habit" Lewis, *Captain Sam Grant*, 324.

21 "sprees" Ibid., 319.

21 Grant's resignation   PUSG, I: 329–32; also see Lewis, *Captain Sam Grant*, 329, and SG, 87.

22 "I peeped at him"   Bleser, *Intimate Strategies*, 138.

23 "I remember seeing"   SG, 25.

24 "These brilliant scenes"   Howe, *Home Letters*, 107.

24 "I have felt tempted"   Ibid., 116.

24 "What is that?"   M, 68.

25 "peculiarly bad luck"   Ibid., 79.

25 "firmly in the main"   Howe, *Home Letters*, 20.

26 "a terrible Civil War"   M, 78.

27 "protector"   Bleser, *Intimate Strategies*, 141.

27 "This is too bad"   Kennett, *Sherman*, 57.

28 "covered with sand"   SM, 120.

29 "a cry about Minnie"   Kennett, *Sherman*, 72.

29 "I would rather live" and "I would rather be"   Bleser, *Intimate Strategies*, 144.

29 "For the past seven months"   Kennett, *Sherman*, 74.

30 "Cump rubbed me"   Ibid.

30 "Cump & I"   Ibid.

30 "I have bet"   SCW, 563.

30 "sent jellycake," "Archbishop called," and "Prayed for the conversion"   Kennett, *Sherman*, 72.

31 "no symptoms of dishonesty"   Clarke, *Sherman*, 69.

32 "In giving his instructions"   Merrill, *Sherman*, 103.

33 "depression"   Clarke, *Sherman*, 66.

33 "Knowing insanity"   Ellen Sherman to John Sherman, November 10, 1861, *William T. Sherman Papers*, Library of Congress. Various printed sources give different versions of the words between "Cump" and "once in California." I believe that a photostat of the original reads as, "in the verge of it." See also SCW, 155–56, which renders this as, "in the seize of it."

33 "No doubt you are glad"   Bleser, *Intimate Strategies*, 145.

33 "that West Point"   M, 114.

34 "I look upon myself"   Ibid., 119.

34 Sherman's experience in this post in Louisiana is treated in Walter Fleming, *Sherman as College President*.

35 "I have heard men of good sense"   Howe, *Home Letters*, 163.

36 "You mistake, too"   LL, 138.

36 "It is hard to realize"   PUSG, I: 359.

37 "You are driving me"   M, 137.

37 "You are all in here"   Ibid., 139.

37 "whom I remember"   Fellman, *Citizen Sherman*, 88.

2. GRANT AWAKENS

39 "take command of the army to be brought into the field"   Freeman, *Lee*, I: 633–36. I interpret this to mean that, with the aged and infirm Winfield Scott, who was soon to retire, being in no condition to lead the Union Army, Lee would take field command and become general in chief upon Scott's retirement.

39 "I can anticipate" Ibid., 420.

39 "I could take no part" Ibid., 437.

39 "Civil War has only horror" Heidler, *Encyclopedia*, II: 568.

40 "I never went into" SG, 89.

40 "having been educated" PUSG, II: 6.

40 "Julia takes" Ibid., 22.

41 "Oh! how intensely" JDG, 87.

41 "I remember now" Ibid.

41 "fell in behind" Garland, *Grant*, 160.

42 "I might have got" PUSG, II: 21.

42 "at a little square table" and "one suit" McFeely, *Grant*, 74.

43 "I thought he was the man" and "McClellan never" SG, 107.

43 "got on one of his little sprees" Ibid., 83.

43 "I've tried" Garland, *Grant*, 168.

44 "[We] saw that" SG, 105.

44 Simon S. Goode Garland, *Grant*, 165–66.

44 "there wasn't a chicken" Ibid., 108.

45 "preferably Captain Grant" SG, 107.

45 "was dressed very clumsily" Ibid., 108.

45 "What a colonel!" Lewis, *Captain Sam Grant*, 427.

45 "What do they mean by" and "Rustic jokes" Garland, *Grant*, 173.

46 "Mexico" incident Woodward, *Grant*, 54.

46 "Howdy, Colonel?" Fuller, *Grant and Lee*, 71.

46 Orders No. 8 PUSG, II: 46.

46 "unostentatious" through "manner" SG, 110.

46 Orders No. 14 PUSG, II: 48.

47 "Alexander was not older" JDG, 92.

47 "Your Dodo" letter PUSG, II: 50.

47 "They entered" GMS, 246.

48 "My own opinion" PUSG, II: 21.

48 "This is an infantry" SG, 111.

49 "Fred enjoys it" PUSG, II: 59.

3. SHERMAN GOES IN

50 Meeting with Lincoln SM, 185–86.

51 "I shall, to the extent" Lincoln, *Speeches and Writings*, 231.

51 "so as to be independent" SCW, 88.

52 "I am convinced" Kennett, *Sherman*, 114.

52 "Of course I could no longer defer" SM, 192.

52 "tall gaunt form" and descriptions of Sherman's face and hat M, 147.

52 "volunteers called by courtesy" SCW, 127.

53 Letter of July 16 Ibid., 117–18.

53 "The march" SM, 198.

53 "As soon as real war" SCW, 98.

53 "On to Richmond!" Trefousse, *Radical Republicans*, 174.

54 Bettie Duvall and intelligence sources Leech, *Reveille*, 95–96.

54 "for the first time" SCW, 124.

55 "Up to that time" SM, 202.

55 "there stands Jackson" Roland, *Iliad*, 52.

55 "After I had" SM, 205.

55 "We could see" Johnston, *Him on the One Side*, 34.

56 "There was no positive" SM, 203.

56 "Shameless flight" and "seen the confusion" Ibid., 124–25.

57 "Though the North" SM, 199.

57 "were so mutinous" Ibid., 207.

57 "one of the . . . best" Ibid.

57 "We were all trembling" Ibid., 209.

57 "Some young officer" Ibid.

58 "offered the command" Ibid., 210.

58 "In this interview" Ibid.

59 "nearly all unfriendly" Kennett, *Sherman*, 132.

59 Figures of opposing forces LL, 122.

59 "I'm afraid" SCW, 143.

59 "I don't think" Ibid., 145.

59 "said he could not" SM, 216.

59 "to meddle as little" SCW, 127.

59 "My own belief" Ibid., 146.

60 "I am sorry" Ibid., 150.

60 "Do write me" and "How any body" SCW, 148n1, 147.

60 "our Gun Boat Fleet" PUSG, III: 36.

61 "You ask if" Ibid., II: 67.

61 "some of Washburne's work" SG, 113.

61 "I could not discover" SCW, 138.

62 "He usually wore" SG, 119.

63 "almost untenable" Woodward, *Grant*, 190.

63 General Orders No. 5 PUSG, II: 207.

63 "I have nothing to do with" Ibid., 194.

64 "Steamers . . . prizes" Ibid., 262.

64 "required here" Ibid., 218.

64 "Remember me" Ibid., 148–49.

65 "Woods should not" SCW, 145.

65 "old Baron Steinberger" and "had drawn to St. Louis" SM, 214.

66 "Now we'll have news" Davis, *Sherman's March*, 140.

66 description of the meeting in Louisville SM, 218–20.

67 "absolutely crazy" Merrill, *Sherman*, 176.

67 "promptly replied" Kennett, *Sherman*, 140.

67 "riding a whirlwind" and "the idea" SCW, 154.

68 "Sherman's gone in the head" M, 163.

69 "Send Mrs. Sherman" SCW, 156n.

69 "in a great, barnlike" Kennett, *Sherman*, 141.

70 "of such nervousness" M, 164.

70 "General Halleck is satisfied" Fellman, *Sherman*, 100.

71 "completely 'stampeded'" M, 164.

71 "acted insane" Ibid., 167.

71 "I would like" PUSG, II: 300.

72 "Veterans could not"   GMS, 179.

72 "The alarm 'surrounded' was given"   Ibid., 180.

72 "I saw a body"   Ibid., 183.

72 "There is a Yankee"   Ibid., 185.

73 "I was the only man"   Ibid., 184.

73 Confederate musket ball   Woodward, *Grant*, 211.

73 "the enemies [*sic*] loss"   PUSG, III: 129. This report was written by Captain William S. Hillyer of Grant's staff.

73 "The General Comdg."   Ibid., 130.

73 Quotations from *New York Herald* and *New York Times*   SG, 131.

74 "whose disorders"   Hirshson, *White Tecumseh*, 103.

74 "it seemed to affect him   Kennett, *Sherman*, 144.

74 "then came telegraphic"   Hirshson, *White Tecumseh*, 103.

74 "well convinced" and "the President evinced"   Ibid.

74 GENERAL WILLIAM T. SHERMAN INSANE   LL, 201.

75 "Nature will paint"   Hirshson, *White Tecumseh*, 201.

75 "Sir"   SCW, 161.

75 "distressed almost to death"   Kennett, *Sherman*, 141.

76 "I feel desolate" and "So now my dearest"   Hirshson, *White Tecumseh*, 105.

76 "true lawyer-like ambiguity"   LL, 205.

76 "Mr. Lincoln, Dear Sir"   Ibid.

77 "seemed very anxious"   Hirshson, *White Tecumseh*, 109.

77 "Dearest Ellen"   SCW, 173.

78 "I am so sensible"   Ibid., 174.

78 "Do you know who I am?"   M, 168.

78 President's General War Order No. 1   SG, 139.

79 "was cut short"   GMS, 190.

79 "Make your preparations"   Ibid., 140.

4. GRANT MOVES FORWARD, WITH SHERMAN IN A SUPPORTING ROLE

80 "If you can reinforce"   SG, 154.

81 "The sight of our camp fires"   PUSG, IV: 153.

81 "It must be victory or death"   SG, 146.

81 "a modest, amiable"   Ibid., 147–48.

81 "[A Union] officer came in"   Ibid.

82 "a few more"   Ibid.

82 "I felt that"   Ibid., 153.

82 "You have no conception"   PUSG, IV: 180.

82 "I will let him"   Ibid., 188.

83 "did not approve"   GMS, 197.

84 "I had no idea" and "I met Captain Hillyer"   GMS, 204.

84 Grant's quotations from his initial appearance on the battlefield   Ibid., 204–206.

85 "No flinching now"   Ibid., 159–60.

85 "General Smith"   Ibid.

86 Nathan Bedford Forrest   Brooks, *Grant*, 118.

86 "the appointment of Commissioners"   PUSG, IV: 218.

86 "No terms except"   Ibid.

86 "the largest capture"   Ibid.

87 "ungenerous and unchivalrous"   Ibid., 218.

87 "There will be nothing"   SG, 164.

87 "You are separated"   Ibid., 165.

87 "do everything in my power"   PUSG, IV: 215n.

88 "I feel anxious"   Ibid., 216.

88 "At that time"   GMS, 213.

88 "Send all reinforcements"   PUSG, IV: 248.

88 "Some of your wounded"   Ibid., 261n.

89 "Make Buell"   SG, 164.

89 "If the Southerners think"   SG, 164.

5. THE BOND FORGED AT SHILOH

90 "the vertebrae of the Confederacy"   Nevin, *Shiloh*, 157.

91 "the eyes and hopes"   SG, 184.

92 wired Halleck's headquarters   PUSG, IV: 245.

92 Halleck to McClellan   SG, 168.

93 Halleck's exchange with Stanton   Ibid., 168n, 169.

93 "It is my impression"   PUSG, IV: 257.

93 "I am disgusted"   SG, 169.

93 "Learning some days past" and "they carried off"   SCW, 195.

95 "It is hard to censure" and "The future success"   Ibid., 172.

95 "A rumor has just"   PUSG, IV: 320.

95 "working himself into a passion"   SM, 245.

96 "Forces going" and "I am not aware"   PUSG, IV: 317–18.

96 "Dearest Ellen"   SCW, 196.

97 "instead of relieving you"   PUSG, IV: 354–55.

98 "You have done" and "No one has sympathized"   Nevin, *Shiloh*, 157.

99 "magnificent plain"   SM, 252.

99 "not to advance"   PUSG, IV: 367.

99 "we must strike no blow"   Ibid., 367n.

99 "an engagement"   Ibid., 367. See also p. 392.

99 Halleck rebukes Grant   Ibid., 404n.

99 "I am clearly"   Ibid., 411.

99 "had no expectation"   GMS, 223.

99 "When you will hear"   PUSG, IV: 389.

99 "Diaoreah"   Ibid., 443.

100 "Soon I hope"   Ibid., V: 7.

100 "Oh, they'd call me crazy again"   LL, 214.

100 "We are constantly"   SCW, 199.

100 "Now is the moment"   Nevin, *Shiloh*, 107.

101 "until an hour"   GMS, 224.

101 "The night was"   Ibid.

102 "to-morrow"   PUSG, V: 16.

102 "I have scarsely"   Ibid., 14.

102 "the enemy is saucy"   Ibid.

102 "This is puerile!" Nevin, *Shiloh*, 108.

103 "a line of men" Ibid., 111.

103 "General Sherman says" LL, 219.

104 "Now they will be entrenched" Nevin, *Shiloh*, 110.

104 "remarked that this" Ibid.

104 "Gentlemen" and "I would fight them" Ibid.

105 "The battle has opened" through "This is no place for us!" Ibid., 114.

106 "a beautiful sorrel" SCW, 201.

106 "General, look to your right!" and "Appler" Nevin, *Shiloh*, 114.

106 "I saw the rebel lines" SM, 250.

106 "Tonight we will" Nevin, *Shiloh*, 113.

106 "a very early breakfast" and "heavy firing" GMS, 228.

107 "Gentlemen, the ball is in motion" SG, 190.

107 "bringing on this engagement" Nevin, *Shiloh*, 113.

107 "Tell Grant" Ibid., 114.

108 "desperately engaged" and "This gave him" SM, 266.

108 "pretty squally" and "Well, not so bad" Nevin, *Shiloh*, 120.

108 "this point was the key" GMS, 210.

108 "During the whole of Sunday" Ibid., 231.

109 "All around him" LL, 222.

109 "trouble keeping his cigar lit" and "I was looking for that" Ibid., 223.

109 "It's a hornet's nest" Nevin, *Shiloh*, 123.

110 "Then I will help you" through "Governor, they came near" Ibid., 128–29.

111 "We shall all be dead" and "I guess that's so" Catton, *Grant Moves South*, 232, citing *Chicago Tribune*, January 27, 1869.

113 "Whichever side takes the initiative" LL, 230.

113 "our troops were exposed" and "I made my headquarters" GMS, 234.

114 "to put the river" through "Lick 'em tomorrow" SG, 200–201.

114 "A COMPLETE VICTORY" and "I thought I had" Nevin, *Shiloh*, 147.

115 "heavy lines of skirmishers" GMS, 234.

115 "Move out" and "I leave that" SG, 202.

115 "If he had studied" Brooks, *Grant*, 142.

115 "At daybreak" SG, 202.

115 "the rebels fall back" Ibid., 203.

115 "along the northern edge" GMS, 237.

116 "The fire and animation" through "I intend to withdraw" Nevin, *Shiloh*, 151.

116 "wanted to pursue" GMS, 237.

117 Here was a long line Nevin, *Shiloh*, 152.

117 "Charge!" and "I and my staff" Ibid.

118 "rolling down the line" through "Boys, you have won" M, 181.

118 "Dear Julia" PUSG, V: 27.

118 "Dearest Ellen" SCW, 201.

119 "they were a disgrace" Kennett, *Sherman*, 169.

120 "The South never smiled" Brooks, *Grant*, 144.

120 *New York Times* and *New York Herald* press reports of Shiloh SG, 204–205; *New York Tribune*, Woodward, *Grant*, 255.

121 *Tribune* editorial Ibid.

121 "No, I can't do it" Ibid., 256.

121 "the blundering stupidity" LL, 234. For this controversy, see also SCW, 226n, 237–45, 245n.

121 "The accusatory part" SCW, 241–43.

121 Some details on the Sherman family defense are in LL, 235.

122 "so shockingly abused" PUSG, V: 116.

122 "Is *success* a crime?" Ibid., 79n.

122 "should never have occurred" Ibid., 116.

122 "Shame on such a Demagogue" Ibid., 83.

122 "not an enemy" Ibid., VI: 62.

122 Julia Grant's encounter with Mrs. Canfield JDG, 99–100, 116n.

124 "constructed seven distinct" Marszalek, *Commander*, 124.

124 "Halleck crept forward" Ibid.

124 a siege on the move GMS, 250.

124 "*should relate to one matter*" SG, 207.

125 Sherman describes Grant's situation and the details of their meeting SM, 275–76.

125 "Necessity however" PUSG, V: 246.

126 Sherman to Grant, June 6, 1862 SM, 276.

126 Grant to Ellen Sherman PUSG, V: 200.

127 "I feel it a duty" Ibid., 34.

127 "In Gen. Sherman" Ibid., 111.

127 "Although Gen. Sherman" Ibid., 140.

127 "Grant's victory" SCW, 193.

127 "you obtained" Ibid., 233.

127 "he is as brave" Ibid., 236.

127 "I cannot express" Ibid., 255.

129 "the People are as bitter" Ibid., 231.

129 "one more fight" PUSG, V: 47.

129 "it is possible" GMS, 244.

6. POLITICAL PROBLEMS, MILITARY CHALLENGES: THE VICKSBURG CAMPAIGN DEVELOPS

130 "scattered" SM, 275.

130 "write freely" SCW, 278. The words are from a letter to Grant from Sherman that says, "A letter from you of Aug. 4 asking me to write more freely and fully on all matters of public interest did not reach me till yesterday." SCW, 208n, states that this letter from Grant to Sherman has not been found, and it does not appear in PUSG.

131 Incident in Memphis church LL, 243–44.

131 "the Military for the time being" M, 191.

131 "Sherman never utters" LL, 252.

131 "felt loving towards us" Ibid., 244.

133 "Your orders about property" LL, 246.

133 "I have no hobby" PUSG, V: 264.

133 "Their *institution*" Ibid., 310.

133 "such men as are not fit" and "It will be the duty" Ibid. VI: 316–17.

134 "the commanding General directs" Kennett, *Sherman*, 178.

134 "leaving one house" and "the regiment has returned" Sherman to Rawlins, September 26, 1862, SCW, 306.

134 "The Boats coming down"   Ibid., 305.

134 "excites a smile" and "without uniform"   Ibid., 317.

135 "fire on any boat" and "You initiate the game"   Ibid.

135 "They cannot be made to love us"   M, 196.

135 "it is about time"   SCW, 301.

135 "I hope this"   M, 194.

135 "We found"   JDG, 102–203.

136 "in a handsome" through "Each day"   Ibid., 105.

137 "very little respect"   Dana, *Recollections*, 74.

137 letter from Rawlins to Grant regarding Grant's drinking   LL, 283.

137 "We all knew"   http://www.mscomm.com/~Ulysses/page47.html. This cites the 1932 edition of Lewis, *Sherman*, 614. I cannot find this quotation on that page, or any other page of the 1932 and 1958 editions. With that caveat, it is offered here because it so closely matches Dana's description of the same situation (see Ulysses S. Grant Homepage, 47). Other references to Grant's drinking or abstemiousness are to be found in PUSG IV: 111–14n, 115n, 116n–19n, 227n, 296n, 320n, 344n; ibid. VI: 87n, 242n; ibid. VIII: 322n–25n.

138 "Who is this strange"   JDG, 103.

138 "thin & worn" through "cheerful & well"   M, 200.

138 "My Dear Children"   SCW, 340–41.

138 "Audacity, more audacity"   SG, 216.

139 "dispose of "   Ibid., 217.

140 "an ugly place"   Dana, *Recollections*, 54.

140 "Heretofore I have"   PUSG, VII: 480.

141 "A ship without Marines"   Lyman, *Quotations*, 1151, citing an 1863 letter from Porter to John Harris.

141 "Admiral Porter"   Glatthaar, *Partners*, 165.

142 "determination"   Ibid., 164.

142 Porter meets Sherman   Kennett, *Sherman*, 174.

142 "may change our plans" through "Come over and we will talk"   PUSG VI: 404.

144 "Commerce must follow"   LL, 247.

144 "full of Jews"   Hirshson, *White Tecumseh*, 129.

144 "If the policy"   Kennett, *Sherman*, 175.

144 "I cannot take an active part"   PUSG, III: 226.

145 "The Jews"   Ibid., VII: 50.

145 "And so the children of Israel"   SG, 226n.

145 "immediately revoked" and "the President"   Ibid., 227n.

146 "the spirit of the medieval age"   Ibid., 226.

146 "incompetent"   Ibid., 223.

147 "reserved for some special"   PUSG, VI: 288.

148 "The mysterious"   Ibid., 310.

148 "You are hereby authorized"   SG, 227.

149 "Admirable for defense"   GMS, 359.

150 "Well we have been to Vicksburg"   SCW, 349.

150 "I assume responsibility"   Kennett, *Sherman*, 192.

150 "unaccountable"   LL, 264.

150 "Of course, General Sherman"   Ibid.

151 "Sherman managed his men"   M, 210.

151 "General Sherman is" and "Come with a sword"   LL, 266–69.

152 Grant to Sherman, April 27, 1863   PUSG, VIII: 130.

152 Sherman to Grant   Ibid., 131*n*.

152 "foolish, drunken, stupid . . . ass"   Woodward, *Grant*, 292.

152 "Suppress the entire press"   PUSG, VIII: 38.

153 "I make these suggestions"   SCW, 444.

154 "the pleasant impression"   Dana, *Recollections*, 36–37.

155 "I'm glad you've come" through "fully endorsed by Grant"   Ulysses S. Grant Homepage, 47, citing interview with Wilson in Hamlin Garland Papers, USC.

155 "Grant wound up"   Dana, "General Grant's Occasional Intoxication," New York *Sun*, April 28, 1891, ibid.

156 Dana's decision to stand by Grant is confirmed in Wilson, *Dana*, 232, although Wilson was wrong in saying that Dana never spoke of Grant's drinking.

156 "we dined on board"   JDG, 112.

157 "The great essential"   GMS, 307.

157 "perilous trip"   Ibid.

157 "manned them with soldiers"   SM, 343.

157 "their summer songs"   JDG, 112.

157 "Just before ten o'clock"   Dana, *Recollections*, 54.

157 "All was going well"   JDG, 112.

157 "were immediately under"   Dana, *Recollections*, 55.

158 "As soon as"   SM, 343.

158 "The air was"   JDG, 112.

158 "had a few words" through "to the shore"   SM, 344.

158 "ordered that"   Dana, *Recollections*, 550.

159 "Thus General Grant's army"   SM, 344.

7. THE SIEGE OF VICKSBURG

161 "I know Hooker well"   SCW, 452.

161 "affectionate regards" and "He has lost"   Freeman, *Lee*, II: 560.

161 "My God! What will the country say!"   Donald, *Lincoln*, 436.

161 "It's unnecessary for me"   PUSG, VIII: 151.

162 "stop all troops"   SG, 244. See also SCW, 470*n*.

162 "what rations of hard bread"   SG, 244.

162 "I knew well"   SG, 245.

162 "You may not hear"   PUSG, VIII: 196.

162 "the operatives were told"   GMS, 338.

163 "mounted on two"   Catton, *Grant Moves South*, 438. See also Wilson, *Dana*, 219–20.

164 "is not in good plight"   SG, 249.

164 "I was close enough"   Ibid.

165 "A pontoon-bridge"   SM, 349.

166 "These were still"   GMS, 326.

167 "Until this moment"   LL, 277.

167 "was in my rear"   GMS, 355.

167 "resulted in securing"   Ibid., 354.

167 "The heads of Colums" through "a dirty dog"   SCW, 472.

168 "I want this planted" LL, 279.

168 "The attack was gallant" GMS, 335.

168 "This last attack" Ibid., 356.

168 "I now determined" Ibid., 357.

169 "a dozen or two of poultry" through "But the intention was good" Ibid., 364.

169 "Among the earliest arrivals" Ibid.

170 "I am too weak" Symonds, *Johnston*, 212.

170 "Not a day passed" Hoehling, *Vicksburg*, 147.

170 "I was just within" Ibid., 75–76.

171 "When I was driving stakes" Ibid., 146.

171 "Say! You old bastard" Ibid., 93.

172 "Dear General" SCW, 474.

173 "I would add" PUSG, VIII: 395.

173 "A force of some two thousand" Dana, *Recollections*, 93.

174 "I am anxious to get" PUSG, IX: 23.

174 "The negro troops" Ibid., 110.

174 "I would prefer" LL, 303.

174 "the great solicitude" SG, 232.

174 "delivered that admirable communication" Ulysses S. Grant Homepage, 47, citing "General Grant's Occasional Intoxication," New York *Sun*, April 28, 1891.

175 "only served to increase" GMS, 356.

175 "did great injustice" Ibid., 367.

176 "we lost, needlessly" SCW, 487.

176 "I should have relieved him" PUSG, VIII: 385*n*.

176 "most pernicious consequences" Ibid., 386*n*.

176 "not an officer" SCW, 501.

176 "You will go" PUSG, VIII: 408.

176 "I have given" Ibid., 402.

176 "I did hope" SCW, 500.

176 "with him I am" Ibid., 580.

177 Grant to Sherman, June 23, 1863 PUSG, VIII: 411.

177 "very often oxen" Hoehling, *Vicksburg*, 165.

177 "Hotel de Vicksburg" and menu Ibid., 163.

177 "How the other troops" Ibid., 169.

177 "What's become of Fido?" Ibid., 162.

177 "I'm going down" and "I want to see" LL, 287.

178 "I am personally acquainted" PUSG, VIII: 414.

179 The effort to capture Frederick Grant is from Casey, "When Grant Went A-Courtin'."

180 "Many Soldiers" letter Hoehling, *Vicksburg*, 241; LL, 290.

180 "Fred. Has returned" through "Kiss the children" PUSG, VIII: 445.

181 Details of the surrender negotiations GMS, 374–79; SG, 254–56; PUSG, VIII: 455–59.

181 "Pemberton was much excited" Dana, *Recollections*, 101.

182 "sitting on my little cot" through "a general rejoicing" Hoehling, *Vicksburg*, 272.

183 "Not a cheer went up" LL, 290.

183 "it was good to see" Ibid., 291.

183 "At Vicksburg" GMS, 384.

183 "We met"   Hoehling, *Vicksburg*, 276.

183 "What a contrast"   SG, 256.

184 "I rode into"   Hoehling, *Vicksburg*, 280.

184 "No one"   SG, 256.

184 "I judge"   PUSG, VIII: 460.

184 "When we go in"   Ibid.

185 "I want Johnston"   Ibid., 461.

185 "If you are"   Ibid., 461n.

185 "There is but little"   Ibid.

185 "The news is so good"   Ibid., 463n.

185 "I can hardly" through "sling the knapsack for new fields"   LL, 291–92.

185 Grant to Sherman, July 4, 1863   PUSG, VIII: 479.

186 "Never mind, General"   Freeman, *Lee*, III: 130.

186 "I had been a most bitter"   Vandiver, *Civil War Battlefields*, 79.

187 "stating that Meade"   PUSG, IX: 18.

187 "the news from the Potomac"   SCW, 503.

187 "Victory! Waterloo Eclipsed!"   Wagner, *Civil War Desk Reference*, 31.

188 "The Father of Waters"   SG, 258.

188 "envelop the insurgent states"   Wagner, *Civil War Desk Reference*, 334.

188 "Grant is my man"   SG, 259.

188 "My Dear General"   Ibid., 257.

8. PAIN AND PLEASURE ON THE LONG ROAD TO CHATTANOOGA AND MISSIONARY RIDGE

190 "The dirt road" LL, 294.

191 "If Johnston is pursued"   This exchange between Grant and Sherman is in PUSG, IX: 66–68.

191 "a large, white"   JDG, 119.

192 "It combines"   SCW, 521.

192 "Victor"   Ross, *The General's Wife*, 153.

193 "I may wish to use"   SG, 261.

194 "The people of these states"   Ibid., 377.

194 "Rude Barbarians"   SCW, 448.

194 "I doubt if History affords"   Ibid., 492.

194 "a Civil Government now"   Ibid., 546–48.

195 "boned"   Glatthaar, *Partners*, 143.

195 "He is not"   SCW, 236.

195 "we have in Grant"   Ibid., 500–501.

195 "To me he is a mystery"   M, 385.

195 "As we sat in Oxford"   SCW, 506.

196 "stunned and confused"   SG, 263.

198 "Willy then told me"   Hirshson, *White Tecumseh*, 165–66.

198 "Mrs. Sherman, Minnie, Lizzie, and Tom"   SM, 374.

198 "this is the only death"   PUSG, IX: 274.

198 "private letter"   Ibid., 272.

199 "My Dear Friend"   SM, 374–75.

199 "I have got up early"   SCW, 552.

200 "The moment I begin to think"   Ibid., 556.

200 "My heart is now"   M, 238.

200 "He knew & felt"   SCW, 565.

200 "We must all now"   Ibid., 537.

201 "Hold Chattanooga"   PUSG, IX: 302.

201 "was seated entirely alone"   SG, 265.

202 "a horse-back ride"   PUSG, IX: 317.

202 Porter's account of his first experience with Grant   Porter, *Campaigning*, 1–5.

203 "Please approve"   PUSG, IX: 308.

203 "He had scarcely begun"   Porter, *Campaigning*, 6.

203 "bluntly but politely" through "material correction"   Ibid., 6–7.

205 "a special train"   SM, 376.

205 "some shallow rifle-trenches"   Ibid., 377.

205 "I am coming"   LL, 310.

205 "as though" and "I was somewhat"   M, 239.

205 "The enemy"   SM, 378.

206 "As soon as"   Porter, *Campaigning*, 9.

207 "During the fight"   Ibid., 9–10.

207 Howard's description of the meeting between Grant and Sherman   McFeely, *Grant*, 118–19.

9. CONFUSION AT CHATTANOOGA

210 "I am convinced"   Downey, *Storming*, 132.

210 "it was considered"   LL, 319.

210 "the most sensible"   Ibid., 321.

210 "I need not express"   Kennett, *Sherman*, 213. Charles A. Dana wrote that "Grant says the error is his," but it was clear that Grant had expected Sherman to leave his wagons and arrive sooner (ibid.). Also see LL, 317–18.

211 "It isn't possible"   Downey, *Storming*, 162.

212 "Up and up they went"   Ibid., 165.

212 "Here come fresh troops"   Ibid., 164.

212 "General Sherman carried" and "impracticable"   PUSG, IX: 443.

213 "A full moon made" and "no report"   Dana, *Recollections*, 140.

213 "Hail to the Chief"   LL, 320.

213 "When General Grant"   Williams, *McClellan, Sherman and Grant*, 100.

214 "that General Thomas"   SM, 402.

214 "I had watched"   Ibid., 404.

215 "vast masses"   Ibid.

215 "Where is Thomas?" through "All servants, cooks, clerks"   The account of this part of the action at Missionary Ridge is from LL, 320–22.

217 "A crash like a thousand thunderclaps"   SG, 278.

218 "Thomas, who ordered those men"   Downey, *Storming*, 179.

218 "On, Wisconsin!"   SG, 280.

218 "My God, come and see 'em" and "It was the sight of our lives"   LL, 323.

218 "You'll all be court-martialed!"   LL, 324.

219 "Almost up" and "A fellow of the Twenty-second Indiana"   Ibid.

219 "drawn vast masses"   SM, 404.

219 "to march at once"   Ibid.
219 "prompt pursuit"   Morris, *Sheridan*, 147.
220 "Glory to God"   Dana to Stanton, November 24, 1863, Dana, *Recollections*, 141.
220 "The storming of the ridge"   Wilson, *Dana*, 293.
220 "Damn the battle!"   Morris, *Sheridan*, 64.
220 "The whole philosophy"   SCW, 576.
220 "the whole plan"   SM, 396.
220 "Discovering that the enemy"   Thomas, *General Thomas*, 447.
220 Regarding his ability to see the entire battlefield clearly, on December 2, 1863, Grant wrote Congressman Elihu B. Washburne, "It is the only battle field I have ever seen where a plan could be followed and from one place the whole field is within view." PUSG, IX: 490–91.
221 "weakening his center"   Ibid., 562.
221 "holding a fine lot"   SM, 393.
221 "domiciled"   Ibid.
222 "bleeding feet"   M, 246.
222 Dodge's account of Grant and Sherman's day in Nashville   Hirshson, *Dodge*, 86–87.
224 "being calculated to do injustice"   PUSG, IX: 562. Among the words stricken from Grant's report were, "I have been thus particular in noticing this matter because public notices have, unintentionally no doubt given accounts of the battle of Chattanooga, calculated to do injustice to as brave and gallant troops as fought in that battle."
224 "only be considered"   LL, 329.
224 "for their gallantry"   SM, 413.
224 "permit your name"   The exchange of letters between Burns and Grant is in PUSG, IX: 541, 542n.
225 "in the name of the people"   SG, 284.
225 "Nothing could induce me"   PUSG, IX: 542n.
225 "You occupy a position"   Ibid., 555n.
225 "the next year"   SCW, 573.
226 "pecked and pounded"   Hirshson, *Dodge*, 87.
226 "The only vote that now tells"   SCW, 564.
226 "MY DEAR MADAM"   PUSG, IX: 524.
226 "With him I am"   SCW, 580.
227 "will go to the front"   PUSG, IX: 577.
227 "camp dysentery and typhoid fever"   PUSG, X: 74.
227 Julia Grant's eyes   Ibid., 126–27.
228 "Longstreet has"   Ibid., 86.
228 "that there was much"   Ibid., 85.
228 "As it is rather desirable"   Ibid., 96.
228 "secure the entire"   Ibid., IX: 500.
229 "Somehow our cavalry"   Hirshson, *White Tecumseh*, 185.
230 "I have one of my best"   PUSG, X: 20.
230 "Enemy is scattered"   Ibid., 21.
230 "It now looks as if "   Ibid., 100.
231 "He was to go for Lee"   LL, 345.

10. GRANT AND SHERMAN BEGIN TO DEVELOP THE
    WINNING STRATEGY

232 "I was ordered"   GMS, 469.

232 Exchange of letters upon Grant's promotion to lieutenant general and general in
    chief   Grant to Sherman, March 4, 1864, PUSG, X: 186; Sherman to Grant,
    March 10, 1864, PUSG, X: 187–88n; SCW, 602.

233 Account of Grant and his son Fred's arrival in Washington, and Grant's reception
    that evening at the White House   SG, 289–90.

234 "Assuring him"   Ibid., 291.

235 "I never met"   Ibid.

235 "serve to the best" and "assured him"   GMS, 470.

236 "to the lowest number"   SG, 296.

236 "Sherman was tall"   SG, 295.

236 "On reflection I agree"   SCW, 604.

237 "You I propose to move"   PUSG, X: 274.

237 "Like yourself," "I will not," and "Enlightened War"   SCW, 617.

238 "Lee's army will be"   PUSG, X: 274.

239 "My entire headquarters"   LL, 353.

239 "I think I rank you" and "You and I"   SG, 297.

240 "We must make up"   Ibid., 301.

240 "In battle, the sphinx awoke"   Ibid., 295.

240 "the most belligerent man"   Ibid., 346.

241 "At times the wind"   Porter, *Campaigning*, 72–73.

241 Page's description   Page, *Letters*, 50.

241 "I never saw a man so agitated"   Freeman, *Lee*, III: 298n.

241 "We fought them"   McWhiney, *Battle*, 45.

242 "Who are you, my boys?"   Flood, *Lee*, 54–55.

243 "He looks as if he meant it"   Catton, *Grant Takes Command*, 159.

243 "General" through "instead of what Lee is going to do'"   Porter, *Campaigning*,
    59–70.

244 "Most of us thought"   SG, 337.

245 "Our spirits rose" and "Give way"   Ibid., 338.

246 "Wild cheers"   Porter, *Campaigning*, 79.

246 "Undismayed"   SG, 338–39.

246 "We have now ended"   PUSG, X: 422.

247 "After eight days"   McKinney, *Battle in the Wilderness*, 88.

11. SHERMAN SAVES LINCOLN'S PRESIDENTIAL CAMPAIGN

249 "He is a butcher"   Ross, *The General's Wife*, 184.

250 "a stupendous failure"   Boatner, *Civil War Dictionary*, 649.

250 "Grant is as good a leader"   SCW, 613.

251 "I begin to see it"   SG, 373.

251 "a mere question of time"   Freeman, *Lee*, III: 398. The observation was made to
    General Jubal Early.

252 Greeley's meeting at Niagara Falls   Roland, *Iliad*, 192.

252 Weed and Raymond's remarks   Ibid., 195.

252 "This morning"   SG, 383n.

253 "Who shall revive" LL, 398.

254 "Hood is a bold fighter" Wagner, *Civil War Desk Reference*, 413.

254 "so directly opposite" and "that we should force" Symonds, *Johnston*, 324.

254 "At all points" SM, 531.

255 "Hello, Johnny" LL, 368.

255 "to carry the presidential" Symonds, *Johnston*, 328.

255 "My satisfaction" JDG, 326.

256 "I expected something" LL, 387.

256 "His mouth twitched" Simpson, *Grant*, 360.

256 "the nation had" PUSG, XI: 397, 397–98n.

256 "Your progress" Ibid., 381.

256 "I was gratified" Ibid., 381n.

257 "We must try" Ibid., 392.

257 "Is there any" Ibid., 408.

257 "will be immediately" Ibid., 401.

257 "The draft must be" Ibid., 425.

258 "My withdrawel now" Ibid., 424.

258 "I have seen" SG, 382.

258 "great victory" LL, 406.

259 "I do not wish to waste lives" Ibid., 408.

259 "amid great rejoicing" PUSG, XII: 127.

259 "We want to keep" Ibid., 144.

259 "you have accomplished" Ibid., 155.

260 "I found him" Porter, *Campaigning*, 283.

260 "They would seek" Ibid., 284.

260 "Mrs. Grant, who was" Ibid., 379.

260 "And he had sent" JDG, 137.

261 "there is no chance" SCW, 685.

261 "We have Atlanta close aboard" Ibid., 671.

261 "They must understand" Ibid., 664.

262 "He was just forty-four" Porter, *Campaigning*, 290–91.

12. PROFESSIONAL JUDGMENT AND PERSONAL
    FRIENDSHIP: SAVANNAH FOR CHRISTMAS

263 "I admire" SCW, 724.

264 "Even without" Ibid., 751.

264 "the utter destruction" through "I can make the march" Ibid., 731.

265 "If there is any way" PUSG, XII: 290.

265 "On reflection" Ibid., 298.

265 "a misstep now" Ibid., 303.

265 "Do you not think" Ibid., 370.

265 "if I turn back" Ibid., 372.

265 "I do not really see" Ibid., 373.

266 "Great good fortune" Ibid., 394.

266 "the army will forage liberally" M, 506.

266 "Behind us lay Atlanta" SCW, 147.

266 "Started this morning" Hanson, *Soul of Battle*, 150.

267 "this may be the last" Ibid., 150.

267 "Grant has the bear" LL, 485.

267 "Oh, no, we have heard nothing" M, 306.

267 "on Salt Water some place" PUSG, XIII: 129.

268 "If you delay" Ibid., 107.

268 "had been in the service" Hanson, *Soul of Battle*, 153.

268 "It is a magnificent army" Ibid., 182.

268 "prowling around" M, 304.

268 "General Sherman is" Barrett, *Sherman's March*, 33.

269 "Yankee soldiers" M, 321.

270 "Dar's millions of 'em" Hanson, *Soul of Battle*, 155.

270 "the distinction between" Liddell-Hart, *Sherman*, 333–34.

270 "soldiers emerging" M, 302.

270 "My husband is a captain" Hanson, *Soul of Battle*, 175.

270 "The old lady forced it on me" Ibid., 197.

270 "The boys would stir up" Glatthaar, *March*, 72.

270 "well dressed" Ibid., 76.

271 "wild-animal stare" LL, 448.

271 "every effort" Long, *Civil War*, 599.

271 "I know that in the beginning" LL, 442.

272 "It is impossible" Hanson, *Soul of Battle*, 176.

272 "I have seen officers themselves" LL, 440.

272 "It was very touching" Ibid.

273 "The negro should be" SCW, 522. Also see p. 227.

273 "A nigger as such" Kennett, *Sherman*, 107.

273 "spare nothing" SM, 662. See also Hanson, *Soul of Battle*, 211.

273 "has settled down" Ibid., 206.

273 "Anything and Everything" Glatthaar, *March*, 79.

273 "The prevailing feeling" Hanson, *Soul of Battle*, 160.

274 "Is Fort McAllister taken?" LL, 463. Figures of Union losses at Fort McAllister vary. In his memoirs (SM, 675), Sherman puts it as "killed and wounded, ninety-two."

275 "The last letter" SCW, 785.

275 "come here by water" and "I have concluded" PUSG, XIII: 72–73.

276 "I beg to present you" SCW, 772.

276 "He's made it!" LL, 470.

276 "Our Military Santa Claus" M, 311.

276 "My Dear General Sherman" LL, 470.

276 "I congratulate you" PUSG, XIII: 129.

276 "Sherman has now demonstrated" Ibid., 149.

277 *Edinburgh Review* and London *Times* M, 311.

277 "After seeing what we have" Jones, *When Sherman Came*, 105–106.

277 Account of Allie Travis Ibid., 5–6.

277 "I feel a just pride" SCW, 788.

277 "There are some" Ibid., 792.

278 "I can hardly realize it" Ibid., 785.

279 "His conduct and deportment" Merrill, *Sherman*, 278.

280 "Where is Mary?" Thomas, *Stanton*, 35.

280 "The blood spouted up" Ibid., 41.

281 "so as to communicate"   Ibid., 342.
281 "Mr. Stanton has been here"   SCW, 538.
282 "Let 'em up easy"   Winik, *April 1865*, 208.

13. THE MARCH THROUGH THE CAROLINAS, AND
AN ADDITIONAL TEST OF FRIENDSHIP

283 "close out Lee"   PUSG, XIII: 72.
283 "I don't like to boast"   SCW, 774.
284 "I am fully aware"   Ibid., 784.
284 "How few there are"   PUSG, XIII: 203.
284 "I can not say"   Ibid., 154.
285 "I will accept no commission"   SCW, 809.
285 "I would rather have you"   PUSG, XIII: 351n.
285 "I have received"   Ibid., 350.
285 "the possibility of arriving"   Lee to Grant March 2, 1865, PUSG, XIV: 99n.
285 "The President directs"   Stanton to Grant, March 3, 1865, ibid., 91n.
286 "I was afraid"   GMS, 535.
286 "Don't forget"   Hirshson, *White Tecumseh*, 277.
286 "I almost tremble"   SCW, 776.
286 "Should you capture Charleston"   LL, 472.
286 "How shall I let you know"   Davis, *Sherman's March*, 141.
286 "If Sherman has really left"   LL, 457.
287 Experience recounted by Mrs. Alfred Proctor Aldrich   Jones, *When Sherman Came*, 114–21.
288 "Northern snow-storm"   SM, 760.
288 "the whole air"   Ibid., 767.
289 "very frequently had to"   Hirshson, *White Tecumseh*, 283.
289 "no one ordered it"   LL, 504.
289 "If I had made up my mind"   Fellman, *Sherman*, 231.
289 "Columbia!"   Merrill, *Sherman*, 289.
289 "It's the damnedest"   LL, 513.
290 "Our combinations were"   PUSG, XIV: 205n.
290 "A locomotive"   SCW, 847.
291 "Splendid legs!"   LL, 517.
291 "Sherman is simply"   Ibid., 468.
291 "it is the talk"   Glatthaar, *March*, 175.
291 "It might lead"   GMS, 712.
292 "had never thought of it"   Ibid.
292 "if I get" and "I think I see"   SCW, 828, 830.
292 "rushed around him" and "I'm going up to see"   Barrett, *Sherman's March*, 195.
292 "There is no doubt"   SCW, 836–37.

14. GRANT, SHERMAN, AND ABRAHAM LINCOLN HOLD
A COUNCIL OF WAR — AND PEACE

295 "will not let the lady"   JDG, 147.
295 "Three tiny kittens"   Porter, *Campaigning*, 410.

296 "You may tell"   SCW, 833.

297 "General Grant and"   Porter, *Campaigning*, 417.

297 "Sherman then seated"   Ibid., 418.

297 "his sandy whiskers"   M, 336.

297 "his features express"   LL, 525.

297 "I'm sorry to break up this"   Porter, *Campaigning*, 419.

298 "Did you see Mrs. Lincoln?"   Ibid., 140. A slightly varying version of this conversation concerning Mrs. Lincoln is in Hirshson, *White Tecumseh*, 301.

298 "Well, Julia"   This part of the conversation among Grant, Julia Grant, and Sherman is from Porter, *Campaigning*, 420–21.

298 "wide of the mark"   JDG, 135.

299 "a long talk of troops and movements" through "no, I can manage everything" Ibid.

300 "A crow could not fly"   Lyman, *Quotations*, 206.

300 "Men, by God"   Morris, *Sheridan*, 234.

300 "Retreat, hell!"   Ibid.

300 "turning what bid fair"   Ibid., 219.

300 "join General Grant"   Ibid., 239.

301 "or go on to Sherman"   PUSG, XIV: 183.

301 a "blind"   Morris, *Sheridan*, 241.

301 "Sheridan became a good deal"   Porter, *Campaigning*, 422.

302 "After the general compliments"   SM, 811.

302 "at that very instant"   Ibid.

302 "was strong enough"   Ibid.

302 "blood enough shed"   Ibid.

302 "one more desperate"   Ibid., 812.

302 "sat smoking"   Ibid.

303 "What was to be done"   Ibid.

303 "When at rest"   Ibid., 813.

303 "ought to clear out" through " 'unbeknown' to him"   Ibid., 812.

303 "In his mind"   Ibid., 813.

304 "wanted peace on almost any terms"   Ibid., 814–17.

304 "you are not to decide"   PUSG, XIV: 91n.

304 "Of all the men I ever met"   SM, 813.

304 "incontestably the greatest"   SG, 412.

15. "I NOW FEEL LIKE ENDING THE MATTER":
GRANT'S FINAL OFFENSIVE

305 "She bore the parting"   Porter, *Campaigning*, 425.

305 "Mr. Lincoln looked" through "I think we can send him"   Ibid., 425–26.

306 "I will haul out"   SCW, 847.

306 "the next two months"   Ibid., 849.

306 "I now feel like"   PUSG, XIV: 136.

306 "our troops have all been"   Ibid., 135.

306 "I shall . . . endeavor"   Freeman, *Lee*, IV: 21.

307 "until it is seen"   PUSG, XIV: 253.

307 "the heavy rain"   Morris, *Sheridan*, 243.

307 "pacing up and down" Porter, *Campaigning*, 429.

308 "hold on to Dinwiddie" Ibid., 246.

308 "Well, Colonel, it has happened" Freeman, *Lee*, IV: 51.

308 "I advise" Ibid., 49.

309 "All indications are" and "Rebel Armies" PUSG, XIV: 352.

309 "I am delighted" SCW, 850.

310 "I therefore request" Freeman, *Lee*, IV: 129–30.

310 "sick headache" through "I felt like anything rather" GMS, 730–35.

311 "I take it" through "And it will be a great relief" Flood, *Lee*, 10–11.

312 "Then there is nothing left" Ibid., 4.

312 "we are all one country" Ibid., 152.

312 "there was not a man" Ibid., 22.

312 "I knew" Ibid.

16. THE DAYS AFTER APPOMATTOX: JOY AND GRIEF

314 "I have this moment" SCW, 859.

314 "Lee's surrendered!" Barrett, *Sherman's March*, 207.

314 "I never heard such cheering" Glatthaar, *March*, 176.

314 "Yankee Doodle" M, 339.

314 "Glory to God" Barrett, *Sherman's March*, 207.

315 "We had a great blowout" Ibid., 207–208.

315 "No further destruction" SCW, 834.

316 "As far as the eye can reach" Bradley, *This Astounding Close*, 144.

316 "It is all over with us" Ibid.

316 "the gentlemanly bearing" Barrett, *Sherman's March*, 248.

316 "a temporary suspension" Bradley, *This Astounding Close*, 143.

316 "I undertake to abide" Ibid., 148.

317 "will be followed" through "all the details" SCW, 862.

317 "it would be the greatest" Bradley, *This Astounding Close*, 142.

318 "Messiah" Foner, *Reconstruction*, 73.

318 "Don't kneel to me" Lyman, *Quotations*, 159.

318 "If I were in your place" Winik, *April 1865*, 208.

318 "It has been intimated to me" Lincoln, *Speeches and Writings*, 694.

319 "He has a face" Betts, *Lincoln and the Poets*, 37–38.

319 "Let us convert" Hyman, *Radical Republicans*, 37.

319 Henry A. Wise Cauble, *Proceedings*, 189.

319 "very greatly rejoiced" through "I presented the question" Lincoln, *Speeches and Writings*, 696.

320 Lincoln's speech of April 11, 1865 Ibid., 697–701.

321 "just seen" Ibid., 701.

321 "to effect" through "had perhaps made a mistake" Donald, *Lincoln*, 59.

322 "Do not allow them" Lincoln, *Speeches and Writings*, 701–702.

322 "About fifty generals" Julia's account of their arrival in Washington, JDG, 153–54.

322 "Mr. Lincoln is indisposed" PUSG, XIV: 483–84.

323 "To this plan" and "This was all satisfactory" JDG, 154.

324 "the reduction of the army" Julia Grant's account of the events of April 14 begins in JDG, 154–56.

325 "We can't undertake" The account of the cabinet meeting of April 14, Donald, *Lincoln*, 590–92.

326 "This same dark, pale man" The remainder of Julia Grant's account of the events of April 14 is in JDG, 156–57.

327 "*Sic semper tyrannis*" Donald, *Lincoln*, 597.

327 "The South shall be free!" Kauffman, *American Brutus*, 7.

327 The account of the attack on Seward is based on Winik, *April 1865*, 224–26.

328 "Why didn't he shoot me!" Kauffman, *American Brutus*, 625.

329 "Mr. Lincoln cannot recover" Ibid., 34.

329 "Now he belongs to the ages" Donald, *Lincoln*, 599.

329 Empty frame in the window. Epstein, *Lincoln and Whitman*, 275.

329 "seemed stupefied" Davis, *Lincoln's Men*, 239–40.

329 "He was our best friend" Ibid.

329 "What a hold Old Abe had" Ibid., 239.

330 "The United States has lost" Lewis, *Yankee Admiral*, 167.

330 "The President stood before us" Emerson's eulogy is at http://xroads.virginia.edu/~CAP/LINCOLN/Emerson1.html.

330 "Hush'd Be the Camps To-day" Whitman, *Complete Poetry*, 468.

331 Sherman learns of Lincoln's assassination SM, 836.

331 "As soon as we were alone" through "satisfying me that" Ibid., 837–38.

332 "the great mass" and "to watch the soldiers" SM, 838.

332 "The army is crazy for" Bradley, *This Astounding Close*, 163; for a related incident, see pp. 163–65.

332 "We'll Hang Jeff Davis," the thwarted riotous march, and "Had it not been" M, 343–44.

332 "to express our utmost abhorrence" Bradley, *This Astounding Close*, 165.

333 "they all dreaded" SM, 839.

333 "escape from the country" and "If asked for" SM, 840.

333 Meeting between Sherman and Johnston Ibid.

334 "There is great danger" SCW, 863.

334 Remainder of Sherman-Johnston meeting SM, 841–42.

17. SHERMAN IN TROUBLE

337 "not to vary" and "if approved" SM, 843–44.

337 "I can see no slip" SCW, 867.

337 The full text of Sherman's terms for surrender is in Bradley, *This Astounding Close*, 268–69.

338 "I have rec'd" PUSG, XIV: 423.

339 "the greatest consternation" GMS, 756.

339 "seemed frantic" and "victorious legions" LL, 550.

340 "You will give notice" PUSG, XIV: 423–24n.

341 "The rebels know well" Ibid., 424.

341 "It is now nearly 11 O'Clock" Ibid., 428.

342 "I dread the change" JDG, 156.

343 "On to Mexico" McFeely, *Grant*, 221.

343 "For myself I would enjoy" PUSG, XIV: 405.

343 "They hope, it is said" LL, 550–51.

344 Stanton's statement to the press   *New York Times*, April 24, 1865.

345 Halleck's orders to disregard Sherman   SM, 860–61.

345 "there is some screw loose again"   Marszalek, *Commander*, 223.

345 "like the true and loyal soldier that he was"   GMS, 756.

346 "I therefore demand"   Bradley, *This Astounding Close*, 211.

346 "Grant is here"   Davis, *Sherman's March*, 273.

346 *Chicago Tribune*   LL, 553.

346 "I admit my folly"   SM, 850–51.

347 "The suffering that must exist"   PUSG, XIV: 435n.

347 "to insure a crop" and "enlightened and humane"   LL, 556.

347 Grant's written endorsement   PUSG, XIV: 435n.

348 "on the basis"   Ibid., 434.

348 "I have just returned"   Ibid., 436. This indicates that Grant was not yet aware of the furor caused by Stanton's statement to the press. In his memoirs written twenty years later, he said that he saw newspapers carrying this story when he was at Goldsboro, North Carolina, returning to Washington after meeting with Sherman.

348 Newspaper reactions   LL, 552.

348 "usurped more than" and "loyal men deplore"   *New Haven Journal*, ibid.

348 "I knew that Sherman"   GMS, 736.

349 "It is infamous"   Simpson, *Grant*, 446.

349 "like a caged lion" through "the fellows that wielded"   LL, 557.

349 "Tell General Slocum."   Liddell-Hart, *Sherman*, 399.

349 "I do think that my Rank"   PUSG, XV: 13n. This also appears, with slight variations, in SM, 861.

350 "send a copy to Mr. Stanton"   SCW, 884.

350 Sherman's calculation regarding the gold and the capture of Jefferson Davis   SM, 861.

350 "I doubt not"   Ibid., 884.

351 "I have no hesitation"   Ibid., 885.

351 "an act of Perfidy"   Sherman's Special Field Orders No. 69, SCW, 891n.

352 Special Field Orders No 69   Ibid.

352 "You have not had"   Bradley, *This Astounding Close*, 249.

353 "I cannot possibly reconcile" and "I will march my Army"   SCW, 895–96.

354 "secretary Stanton's newspaper order"   Ibid.

354 "I know of no order"   Ibid., 27.

354 "I do think a great outrage"   Ibid., 894.

355 "not know how to answer" through "made no change in my estimate"   PUSG, XV: 12.

355 "very spick and span"   LL, 566.

356 "truly charmed"   Hirshson, *White Tecumseh*, 316–17.

356 *Cincinnati Commercial* and *Louisville Journal*   M, 350.

356 "I think you have made"   Fellman, *Sherman*, 251.

356 "for a time you lost"   Ibid.

356 "act prudently"   Merrill, *Sherman*, 297.

356 "It is amusing"   Liddell-Hart, *Sherman*, 400.

356 "must expect open defiance"   Hirshson, *White Tecumseh*, 319.

357 "a set of sneaks"   SCW, 897.

357 "look out . . . or they would have"   Merrill, *Sherman*, 277.

18. GRANT, SHERMAN, AND THE RADICALS

359 "a military commander interferes" and "My terms of surrender" SG, 418.

359 "placed in that relation to the military forces" PUSG, XV: 40–41.

360 "assigning you to command" Ibid., 43.

360 "very kind" PUSG, XV: 52, 52n.

360 "would respectfully recommend" and "Citizens of the Southern States" PUSG, XV: 48.

361 Grant's testimony PUSG, XV: 45–46. This exchange also appears in U.S. Congress's *Report of the Joint Committee on the Conduct of the War*: 1524.

362 "to hell with your reveille" Glatthaar, *March*, 184. It reads as, "To h__l with your reveille."

363 "where I will move" SCW, 883.

363 "All my army" Ibid., 901.

364 "I am just in receipt" PUSG, XV: 72.

364 "the threats of Gen. Sherman" LL, 567.

365 "General Sherman, I am *very* glad" M, 353.

19. A PARADE FOR EVERYONE, AND A HEARING FOR SHERMAN

368 "Great Rush of Visitors" *New York Times*, May 23, 1865.

368 "I am informed" PUSG, XV: 115.

369 "I present this saddle" Ibid., 666.

369 "General Sherman was in no mood" LL, 569.

369 "Dear Van" through "I prefer to give" Fellman, *Sherman*, 255.

369 "Stanton wants to kill me" SCW, 896.

370 "Of course I have nothing to do with" Ibid., 795.

370 "I am to go before" Bradley, *This Astounding Close*, 251–52.

370 "Have been taking a walk" Whitman, *Prose Works*, I: 105.

371 "Please direct" PUSG, XV: 87n.

373 "I sometimes feel very nervous" Meade's remark can be found at http://history-sites.com/alcwmb/old-archive/archivefiles/6371.html.

373 Sherman's testimony U.S. Congress, *Report of the Joint Committee on the Conduct of the War*: 423.

375 "Gen. Sherman and his brother" *New York Times*, May 23, 1865.

376 "Many grizzled veterans" Flood, *Lee*, 27.

377 "What regiment are you?" Leech, *Reveille*, 414–15.

20. THE PAST AND FUTURE MARCH UP PENNSYLVANIA AVENUE

379 Sherman's family It is not clear whether Sherman's family was in the reviewing stand on May 23. Sherman said, "I had telegraphed for Mrs. Sherman, who had arrived that day, accompanied by her father, the Hon. Thomas Ewing, and my son Tom, then eight years old" (SM, 865). Ishbel Ross, in *The General's Wife*, 191, states that "Julia [Mrs. Grant] and Ellen Sherman sat together in the reviewing stand opposite the White House to watch the great victory parade."

380 "THE ONLY NATIONAL DEBT" *New York Times*, May 24, 1865.

380 "Gettysburg! Gettysburg!" LL, 572. For many details of the grand review, see Fleming, "The Big Parade."

381 "he was not reviewed at all" LL, 572. *The New York Times*, May 24, took a different view of the incident, saying that Custer had brought the horse under control and "resumed his place at the head of his division."

382 "Their muskets shone like a wall of steel" Porter, *Campaigning*, 508.

382 Tunes the bands played Leech, *Reveille*, 415.

383 "turned their eyes," "pampered and well-fed," and "I'm afraid" LL, 572–73.

383 "a Niagara of men" Garland, *Grant*, 321.

383 "Now a girlish form" Chamberlain, *Passing of the Armies*, 339.

384 "These were my men" Ibid.

384 Page's account Page, *Letters*, 392.

385 "Be careful about your intervals" LL, 696.

385 "it is mentioned" *New York Times*, May 24, 1865.

386 "Directly all sorts of colors" LL, 573.

386 "dressed up" Ibid.

386 "shining bay" Ibid., 575.

386 "opposite the northern entrance" *New York Times*, May 25, 1865.

387 "was a group of orderlies" Ibid.

387 "The Star-Spangled Banner" LL, 573.

387 "The enthusiasm to-day" and banners *New York Times*, May 25, 1865.

387 "raised their hands" Ibid.

388 "He was vociferously cheered" Ibid.

388 "there was something almost fierce" Catton, *Grant Takes Command*, 491.

388 "in his eye" M, 356.

388 Sergeant Upson Merrill, *Sherman*, 300.

388 Young private from Wisconsin. Davis, *Sherman's March*, 294.

388 "one footfall" Ibid.

388 "When I reached" SM, 865.

389 "I believe it was" LL, 575.

389 "took off my hat" SM, 865.

389 "Marching Through Georgia" Porter, *Campaigning*, 509.

389 "The acclamation" LL, 575.

389 Eyewitnesses differed in their description of Sherman's snub of Stanton. Charles A. Dana, in *Recollections of the Civil War*, says, "I sat directly behind Mr. Stanton" in the reviewing stand, and saw this: "The Secretary made no motion to offer his hand or to exchange salutations with him in any manner. As the General passed Mr. Stanton gave him merely a slight forward motion of the head, equivalent perhaps to a quarter of a bow" (250–51). Both Sherman and his aide Hitchcock said that Stanton offered his hand, and that, in Sherman's words, "I declined it publicly." See Hirshson, *White Tecumseh*, 319, and M, 356. Grant's aide Horace Porter says in *Campaigning with Grant*, that when "Stanton reached out his hand," Sherman's "whole manner changed in an instant: a cloud of anger overspread his features," and that "the general turned abruptly away" (510).

390 " 'Veteran' was written all over" M, 356.

390 "largely animal" Whitman, *Prose Works*, I: 106.

390 The two *New York Times* stories *New York Times*, May 25, 1865.

390 "moving floral gardens" Davis, *Sherman's March*, 291.

390 "talismanic banners"   LL, 577.

390 "a battalion of black pioneers"   *New York Times*, May 25, 1865.

391 "It was a most nonchalant"   Ibid.

391 References to Mrs. Herman Canfield   JDG, 99–101, 116*n*.

392 Meeting at the White House between Grant and Lee   Flood, *Lee*, 208–16.

393 "What money will pay Meade for Gettysburg?"   M, 431.

393 "inherited prejudice"   Ibid., 380.

393 "believed in the doctrine" and "The only good Indian is a dead Indian"   Ibid., 381.

394 "If nominated I will not run"   LL, 631.

394 "I don't like to give him pain" through "Yes, Mister President"   M, 385.

395 "Grant says my visits"   LL, 638.

395 Taps   Ross, *The General's Wife*, 313.

395 "It will be a thousand years"   LL, 639.

395 "His Virginia was"   Ibid.

396 "You are the only man"   Bleser, *Intimate Strategies*, 154.

396 "Wait for me Ellen"   LL, 645.

397 "Faithful and Honorable"   Hirshson, *White Tecumseh*, 386.

397 "General, please put on your hat"   LL, 652.

398 Sherman's sword on funeral train   Ibid., 652–53.

398 "The Paris I remember" and "performed some of his own"   JDG, 216, 211.

399 "I, his wife"   Ibid., 331.

L'ENVOI

401 "We were as brothers"   Ward, "We Were as Brothers," 14.

BIBLIOGRAPHY

Anderson, Bern. *By Sea and by River: The Naval History of the Civil War*. New York: Alfred A. Knopf, 1962.

Anderson, Nancy Scott, and Dwight Anderson. *The Generals: Ulysses S. Grant and Robert E. Lee*. New York: Alfred A. Knopf, 1988.

Anderson, William M. *We Are Sherman's Men: The Civil War Letters of Henry Obendorff*. Macomb: Western Illinois University, 1986. Western Illinois Monograph Series, No. 6.

Arnold, Matthew. *General Grant. With a Rejoinder by Mark Twain*. Edited by John Y. Simon. Carbondale: Southern Illinois University Press, 1966.

Barrett, John G. *Sherman's March Through the Carolinas*. Chapel Hill: University of North Carolina Press, 1956.

Belz, Herman. *Reconstructing the Union: Theory and Policy During the Civil War*. Ithaca, NY: Cornell University Press, 1969.

Betts, William W., Jr. *Lincoln and the Poets*. Pittsburgh: University of Pittsburgh Press, 1965.

Black, Robert C., III. *The Railroads of the Confederacy*. Chapel Hill: University of North Carolina Press, 1952.

Bleser, Carol K., and Lesley T. Gordon, eds. *Intimate Strategies of the Civil War: Military Commanders and Their Wives*. New York: Oxford University Press, 2001.

Boatner, Mark Mayo, III. *The Civil War Dictionary*. New York: David McKay, 1987.

Bogue, Allan G. *The Earnest Men: Republicans of the Civil War Senate*. Ithaca, NY: Cornell University Press, 1981.

Bowman, John S., ed. *The Civil War Day by Day*. Greenwich, CT: Dorset Press, 1989.

Bradley, Mark L. *This Astounding Close: The Road to Bennett Place*. Chapel Hill: University of North Carolina Press, 2000.

Brooks, William E. *Grant of Appomattox: A Study of the Man*. Westport, CT: Greenwood Press, 1971.

Burne, Alfred H. *Lee, Grant and Sherman: A Study in Leadership in the 1864–1865 Campaign*. New York: Scribner, 1939.

Casey, Emma Dent. "When Grant Went A-Courtin', By His Wife's Sister: Emma Dent Casey." Manuscript Collections, Missouri Historical Society.

Catton, Bruce. *Grant Moves South*. Boston: Little, Brown, 1960.

————. *Grant Takes Command.* Boston: Little, Brown, 1968.

Cauble, Frank P. *The Proceedings Connected with the Surrender of the Army of Northern Virginia.* Appomattox, VA: Appomattox Court House National Park, 1975.

Chamberlain, Joshua Lawrence. *The Passing of the Armies.* Dayton, OH: Morningside Bookshop, 1974.

Clarke, Dwight L. *William Tecumseh Sherman: Gold Rush Banker.* San Francisco: California Historical Society, 1969.

Conger, A. L. *The Rise of U.S. Grant.* New York: Century Company, 1931.

Coombe, Jack D. *Thunder Along the Mississippi: The River Battles That Split the Confederacy.* New York: Sarpedon, 1996.

Cramer, M. J. *Ulysses S. Grant: Conversations and Unpublished Letters.* New York: Eaton & Mains, 1897.

Croce, Paul Jerome. "Calming the Screaming Eagle: William James and His Circle Fight Their Civil War Battles." *The New England Quarterly,* March 2003, p. 5.

Dana, Charles A. *Recollections of the Civil War.* New York: Collier Books, 1963.

Davis, Burke. *Sherman's March.* New York: Random House, 1980.

Davis, William C. *Lincoln's Men: How President Lincoln Became Father to an Army and a Nation.* New York: Free Press, 1999.

Dodge, Major-General Grenville M. *Personal Recollections of President Abraham Lincoln, General Ulysses S. Grant, and General William T. Sherman.* Denver: Sage Books, 1965.

Donald, David Herbert. *Lincoln.* New York: Simon & Schuster, 1995.

————. *"We Are Lincoln Men": Abraham Lincoln and His Friends.* New York: Simon & Schuster, 2003.

Dornbusch, C. E. *Military Bibliography of the Civil War.* 4 vols. New York: New York Public Library, 1971–87.

Dowdey, Clifford. *Lee's Last Campaign: The Story of Lee and His Men Against Grant—1864.* Boston: Little, Brown, 1960.

Downey, Fairfax. *Storming of the Gateway: Chattanooga, 1863.* New York: David McKay, 1960.

Dyer, John P. *The Gallant Hood.* Indianapolis: Bobbs-Merrill, 1950.

Eckenrode, H. J., and Bryan Conrad. *George S. McClellan: The Man Who Saved the Union.* Chapel Hill: University of North Carolina Press, 1941.

Eisenhower, John S. D. *Agent of Destiny: The Life and Times of General Winfield Scott.* New York: Free Press, 1997.

Epstein, Daniel Mark. *Lincoln and Whitman: Parallel Lives in Civil War Washington.* New York: Ballantine Books, 2004.

Fellman, Michael. *Citizen Sherman: A Life of William Tecumseh Sherman.* New York: Random House, 1995.

Fleming, Thomas. "The Big Parade." *American Heritage,* March 1990, pp. 98–104.

Fleming, Thomas J. *West Point: The Men and Times of the United States Military Academy.* New York: William Morrow, 1969.

Fleming, Walter L. *General William T. Sherman as College President.* Cleveland: Arthur L. Clark Company, 1912.

Flood, Charles Bracelen. *Lee: The Last Years.* Boston: Houghton Mifflin, 1981.

Foner, Eric. *Reconstruction: America's Unfinished Revolution, 1863–1877.* New York: Harper & Row, 1988.

Foote, Shelby. *The Civil War: A Narrative.* 14 vols. Alexandria, VA: Time-Life Books, 1999.

Freeman, Douglas Southall. *Lee's Lieutenants: A Study in Command.* 3 vols. New York: Scribner, 1942–44.

———. *R. E. Lee: A Biography.* 4 vols. New York: Scribner, 1949.

Fuller, Major General J.F.C. *Grant and Lee: A Study in Personality and Generalship.* New York: Scribner, 1955.

Garland, Hamlin. *Ulysses S. Grant: His Life and Character.* New York: Macmillan, 1920.

Glatthaar, Joseph T. *Forged in Battle: The Civil War Alliance of Black Soldiers and White Officers.* New York: Meridian, 1991.

———. *The March to the Sea and Beyond: Sherman's Troops in the Savannah and Carolinas Campaigns.* New York: New York University Press, 1985.

———. *Partners in Command: The Relationships Between Leaders in the Civil War.* New York: Free Press, 1994.

Goldhurst, Richard. *Many Are the Hearts: The Agony and the Triumph of Ulysses S. Grant.* New York: Reader's Digest Press, 1975.

Govan, Gilbert E., and James W. Livingood. *A Different Valor: The Story of General Joseph E. Johnston, C.S.A.* Westport, CT: Greenwood Press, 1973.

Grant, Ulysses S. *Memoirs and Selected Letters.* New York: Library of America, 1990.

Green, Horace. *General Grant's Last Stand: A Biography.* New York: Scribner, 1936.

Groom, Winston. *Shrouds of Glory: Atlanta to Nashville: The Last Great Campaign of the Civil War.* New York: Atlantic Monthly Press, 1995.

Hanson, Victor Davis. *The Soul of Battle: From Ancient Times to the Present Day; How Three Great Liberators Vanquished Tyranny.* New York: Free Press, 1999.

Hassler, Warren W., Jr. *General George B. McClellan: Shield of the Union.* Baton Rouge: Louisiana State University Press, 1957.

Headley, J. T. *The Life and Travels of General Grant.* Philadelphia: Hubbard Brothers, 1879.

Heidler, David S., and Joanne T. Heidler. *Encyclopedia of the American Civil War: A Political, Social, and Military History.* Santa Barbara, CA: ABC-CLIO, 2000.

Hess, Earl J. *The Union Soldier in Battle: Enduring the Ordeal of Combat.* Lawrence: University of Kansas Press, 1997.

Hirshson, Stanley P. *Grenville M. Dodge: Soldier, Politician, Railroad Pioneer.* Bloomington: Indiana University Press, 1967.

———. *The White Tecumseh: Biography of General William T. Sherman.* New York: John Wiley, 1997.

Hitchcock, Henry. *Marching with Sherman.* New Haven, CT: Yale University Press, 1927.

Hoehling, A. A. *Vicksburg: 47 Days of Siege.* Englewood Cliffs, NJ: Prentice-Hall, 1969.

Horan, James. *Matthew Brady: Historian with a Camera.* New York: Bonanza Books, 1955.

Howe, Mark DeWolfe, ed. *Home Letters of General Sherman.* New York: Scribner, 1909.

http://www.history-sites.com/alcwmb/old-archives/archivefiles/6371.html (for Meade's statement concerning the Joint Committee on the Conduct of the War).

http://xroads.virginia.edu/~CAP/LINCOLN/Emerson1.html (for Emerson's eulogy of April 19, 1865).

Hyman, Harold M. *The Radical Republicans and Reconstruction, 1861–1870.* Indianapolis: Bobbs-Merrill, 1967.

Johnston, Terry A., Jr., ed. *"Him on the One Side and Me on the Other": The Civil War Letters of Alexander Campbell, 79th New York Infantry Regiment and James*

Campbell, 1st South Carolina Battalion. Columbia: University of South Carolina Press, 1999.

Jones, Katharine M. *When Sherman Came: Southern Women and the "Great March."* Indianapolis: Bobbs-Merrill, 1969.

Kaltman, Al. *Cigars, Whiskey & Winning: Leadership Lessons from General Ulysses S. Grant.* Paramus, NJ: Prentice Hall, 1998.

Kauffman, Michael W. *American Brutus: John Wilkes Booth and the Lincoln Conspiracies.* New York: Random House, 2004.

Kennett, Lee. *Marching Through Georgia: The Story of Soldiers and Civilians During Sherman's Campaign.* New York: HarperCollins, 1995.

———. *Sherman: A Soldier's Life.* New York: HarperCollins, 2001.

Kiper, Richard L. *Major General John Alexander McClernand: Politician in Uniform.* Kent, OH: Kent State University Press, 1999.

Kirschberger, Joe H. *The Civil War and Reconstruction: An Eyewitness History.* New York: Facts on File, 1991.

Korda, Michael. *Ulysses S. Grant: The Unlikely Hero.* New York: HarperCollins, 2004.

Lash, Jeffrey N. *Destroyer of the Iron Horse: General Joseph E. Johnston and Confederate Rail Transport, 1861–1865.* Kent, OH: Kent State University Press, 1991.

Lee, Richard M. *Mr. Lincoln's City: An Illustrated Guide to the Civil War Sites of Washington.* McLean, VA: EPM Publications, 1981.

Leech, Margaret. *Reveille in Washington: 1860–1865.* New York: Harper, 1941.

Lewis, Lloyd. *Captain Sam Grant.* Boston: Little, Brown, 1950.

———. *Sherman: Fighting Prophet.* New York: Harcourt, Brace, 1932.

Lewis, Paul. *Yankee Admiral: A Biography of David Dixon Porter.* New York: David McKay, 1968.

Liddell-Hart, B. H. *Sherman: Soldier, Realist, American.* New York: Frederick A. Praeger, 1958.

Lincoln, Abraham. *Speeches and Writings, 1859–1865.* New York: Library of America, 1989.

Long, E. B., with Barbara Long. *The Civil War Day by Day: An Almanac 1861–1865.* Garden City, NY: Doubleday, 1971.

Lotchin, Roger W. *San Francisco 1846–1856: From Hamlet to City.* New York: Oxford University Press, 1947.

Lowry, Thomas P. *The Civil War Bawdy Houses of Washington, D.C.* Fredericksburg, VA: Sergeant Kirkland's Museum and Historical Society, 1997.

Lucas, Marion B. "History and Memory in Late Twentieth Century Civil War Literature: The Good, the Bad, and the Ugly." *The Kentucky Review* 15 (2003), p. 41.

Lyman, Darryl. *Civil War Quotations.* Conshohocken, PA: Combined Books, 1995.

Markle, Donald E. *Spies and Spymasters of the Civil War.* New York: Hippocrene Books, 1994.

Marszalek, John F. *Commander of All Lincoln's Armies: A Life of General Henry W. Halleck.* Cambridge, MA: Harvard University Press, 2004.

———. *Sherman: A Soldier's Passion for Order.* New York: Free Press, 1993.

Marx, Karl, and Frederick Engels. *The Civil War in the United States.* New York: International Publishers, 1937.

Maurice, Major General Sir Frederick. *Statesmen and Soldiers of the Civil War: A Study of the Conduct of the War.* Boston: Little, Brown, 1926.

McCrary, Peyton. *Abraham Lincoln and Reconstruction: The Louisiana Experiment.* Princeton, NJ: Princeton University Press, 1978.

McFeely, William S. *Grant: A Biography*. New York: W. W. Norton, 1981.

McKnight, W. Mark. *Blue Bonnets O'er the Border: The 79th New York Cameron Highlanders*. Shippensburg, PA: White Mane Books, 1998.

McPherson, James M. *Drawn with the Sword: Reflections on the American Civil War*. New York: Oxford University Press, 1996.

——. *Ordeal by Fire: The Civil War and Reconstruction*. New York: Alfred A. Knopf, 1982.

McWhiney, Grady. *Battle in the Wilderness: Grant Meets Lee*. Fort Worth, TX: Ryan Place Publishers, 1995.

——, ed. *Grant, Lee, Lincoln and the Radicals*. Evanston, IL: Northwestern University Press, 1964.

Merrill, James M. *William Tecumseh Sherman*. Chicago: Rand McNally, 1971.

Miers, Earl Schenck. *The Last Campaign: Grant Saves the Union*. Philadelphia: Lippincott, 1972.

Miller, Richard F. "The Trouble with Brahmins: Class and Ethnic Tensions in Massachusetts' 'Harvard Regiment.'" *The New England Quarterly*, March 2003, p. 38.

Milligan, John D. *Gunboats Down the Mississippi*. Annapolis, MD: United States Naval Institute, 1965.

Mitgang, Herbert, ed. *Abraham Lincoln: A Press Portrait*. Athens: University of Georgia Press, 1989.

Morison, Samuel Eliot. *The Oxford History of the American People*. New York: Oxford University Press, 1965.

Morris, Roy, Jr. *Sheridan: The Life and Wars of General Philip Sheridan*. New York: Crown Publishers, 1992.

Murfin, James V. *Battlefields of the Civil War*. Godalming, Surrey, England: Colour Library Books, 1988.

Nash, Howard P., Jr. *A Naval History of the Civil War*. South Brunswick, NJ: A. S. Barnes, 1972.

Nevin, David. *The Road to Shiloh: Early Battles in the West*. Alexandria, VA: Time-Life Books, 1983.

Nichols, George Ward. *The Story of the Great March*. Williamstown, MA: Corner House Publishers, 1972.

Page, Charles A. *Letters of a War Correspondent*. Boston: L. C. Page and Company, 1899.

Perry, James M. *Touched with Fire: Five Presidents and the Civil War Battles That Made Them*. New York: Public Affairs, 2003.

Phisterer, Frederick, ed. *New York in the War of the Rebellion, 1861–1865*. 5 vols. and Index. Albany, NY: J. B. Lyon, 1912.

Porter, General Horace. *Campaigning with Grant*. Edited with introduction by Wayne C. Temple. New York: Bonanza Books, 1961.

Pratt, Fletcher. *Stanton: Lincoln's Secretary of War*. New York: W. W. Norton, 1953.

Randall, J. G., and David Donald. *The Civil War and Reconstruction*. Boston: Little, Brown, 1969.

Reed, Robert. *Old Washington, D.C., in Early Photographs, 1846–1932*. New York: Dover Publications, 1980.

Roland, Charles P. *An American Iliad: The Story of the Civil War*. Lexington: University Press of Kentucky, 1991.

Rosebault, Charles J. *When Dana Was the Sun.* New York: Robert M. McBride, 1931.

Ross, Ishbel. *The General's Wife: The Life of Mrs. Ulysses S. Grant.* New York: Dodd, Mead, 1959.

Sanborn, F. B. *The Life and Letters of John Brown: Liberator of Kansas and Martyr of Virginia.* New York: Negro Universities Press, 1969.

Sears, Stephen W. *George B. McClellan: The Young Napoleon.* New York: Ticknor & Fields, 1988.

Sefton, James E. *The United States Army and Reconstruction, 1865–1877.* Baton Rouge: Louisiana State University Press, 1967.

Sherman, William Tecumseh. *Marching Through Georgia.* Edited by Mills Lane. New York: Arno Press, 1978.

——. *Memoirs of General W. T. Sherman.* New York: Library of America, 1990.

——. *William T. Sherman Papers.* Washington, DC: Library of Congress Manuscript Collections.

Simon, John Y., ed. *The Papers of Ulysses S. Grant.* 27 vols. Carbondale: Southern Illinois University Press, 1967–2005.

——, ed. *The Personal Memoirs of Julia Dent Grant (Mrs. Ulysses S. Grant).* Carbondale: Southern Illinois University Press, 1988.

——. "Ulysses S. Grant One Hundred Years Later." Illinois State Historical Society Reprint Series #1. Reprinted from the Winter 1986 issue of *The Illinois Historical Journal.*

Simpson, Brooks D. *Let Us Have Peace: Ulysses S. Grant and the Politics of War and Reconstruction, 1861–1868.* Chapel Hill: University of North Carolina Press, 1991.

——. *Ulysses S. Grant: Triumph over Adversity, 1822–1865.* Boston: Houghton Mifflin, 2000.

Simpson, Brooks D., and Jean V. Berlin, eds. *Sherman's Civil War: Selected Correspondence of William T. Sherman, 1860–1865.* Chapel Hill: University of North Carolina Press, 1999.

Smith, Jean Edward. *Grant.* New York: Simon & Schuster, 2001.

Stanchak, John E., ed. *Leslie's Illustrated Civil War.* Jackson: University Press of Mississippi, 1992.

Steele, Janet E. *The Sun Shines for All: Journalism and Ideology in the Life of Charles A. Dana.* Syracuse, NY: Syracuse University Press, 1993.

Symonds, Craig L. *A Battlefield Atlas of the Civil War.* Annapolis, MD: Nautical and Aviation Publishing Company of America, 1983.

——. *Joseph E. Johnston: A Civil War Biography.* New York: W. W. Norton, 1992.

Taylor, Joe Gray. *Louisiana Reconstructed, 1863–1877.* Baton Rouge: Louisiana State University Press, 1974.

Thomas, Benjamin P., and Harold M. Hyman. *Stanton: The Life and Times of Lincoln's Secretary of War.* New York: Alfred A. Knopf, 1962.

Thomas, Wilbur. *General George H. Thomas: The Indomitable Warrior.* New York: Exposition Press, 1964.

Thorndike, Rachel Sherman, ed. *The Sherman Letters: Correspondence Between General Sherman and Senator Sherman from 1837 to 1891.* New York: Da Capo Press, 1969.

Todd, William. *The Seventy-Ninth Highlanders, New York Volunteers, in the War of the Rebellion, 1861–1865.* Albany, NY: Brandow, Barton & Co., 1886.

Trefousse, Hans L. *Historical Dictionary of Reconstruction*. New York: Greenwood Press, 1991.

———. *The Radical Republicans: Lincoln's Vanguard for Racial Justice*. New York: Alfred A. Knopf, 1969.

Tunnell, Ted. *Crucible of Reconstruction: War, Radicalism and Race in Louisiana, 1862–1877*. Baton Rouge: Louisiana State University Press, 1984.

Ulysses S. Grant Homepage. http://mscomm.com/~ulysses/.

U.S. Congress. Joint Committee on the Conduct of the War. *Report of the Joint Committee on the Conduct of the War*. Washington, DC: Government Printing Office, 1865.

U.S. Congress. Senate. Select Committee on the Harper's Ferry Invasion, 1860. Published as *Invasion at Harper's Ferry*. New York: Arno Press, 1969.

U.S. Congress. *Supplemental Report of the Joint Committee on the Conduct of the War*. 2 vols. Washington, DC: Government Printing Office, 1866.

Vandiver, Frank E. *Civil War Battlefields and Landmarks: A Guide to the National Park Sites*. New York: Random House, 1996.

Villard, Oswald Garrison. *John Brown, 1800–1859: A Biography Fifty Years After*. New York: Alfred A. Knopf, 1943.

Wagner, Margaret E., Gary W. Gallagher, and Paul Finkelman, eds. *Civil War Desk Reference*. New York: Simon & Schuster, 2002.

War of the Rebellion, The: A Compilation of the Official Records of the Union and Confederate Armies. Series I, Volume I. Washington, DC: Government Printing Office, 1880.

Ward, Geoffrey C. "We Were as Brothers." *American Heritage*, November 1990, p. 14.

Warner, Ezra J. *Generals in Blue: Lives of the Union Commanders*. Baton Rouge: Louisiana State University Press, 1972.

———. *Generals in Gray: Lives of the Confederate Commanders*. Baton Rouge: Louisiana State University Press, 1959.

Welles, Gideon. *Diary of Gideon Welles*. 3 vols. Boston: Houghton Mifflin, 1911.

Whitman, Walt. *Complete Poetry and Collected Prose*. New York: Library of America, 1982.

———. *Prose Works 1892*. Edited by Floyd Stovall. Volume I: *Specimen Days*, Philadelphia: David McKay, 1892.

Williams, A. Dana. *The Praise of Lincoln: An Anthology*. Indianapolis: Bobbs-Merrill, 1911.

Williams, Hermann Warner, Jr. *The Civil War: The Artist's Record*. Boston: Beacon Press, 1961.

Williams, T. Harry. *Lincoln and His Generals*. New York: Alfred A. Knopf, 1952.

———. *Lincoln and the Radicals*. Madison: University of Wisconsin Press, 1941.

———. *McClellan, Sherman and Grant*. New Brunswick, NJ: Rutgers University Press, 1962.

Wilson, James Harrison. *The Life of Charles A. Dana*. New York: Harper, 1907.

Winik, Jay. *April 1865: The Month That Saved America*. New York: HarperCollins, 2001.

Woodward, W. E. *Meet General Grant*. New York: Liveright, 1965.

Woodworth, Steven E., ed. *Grant's Lieutenants: From Cairo to Vicksburg*. Lawrence: University Press of Kansas, 2001.

Wright, General Marcus J. *General Scott*. New York: D. Appleton, 1897.

ACKNOWLEDGMENTS

The way this book came into being is a story in itself, but there would be
no story to tell were it not for the steadfast devotion and unwavering support always shown me by my wife, Katherine Burnam Flood.

At the time I finished *Lee—The Last Years*, which begins with Robert
E. Lee's surrender of his Army of Northern Virginia to Ulysses S. Grant
at Appomattox Court House, I found myself feeling that Grant was a singularly easy person to underestimate, both as a general and as a man.
Grant played only a minor role in that book, and my attention in the ensuing years shifted to subjects and individuals far from the Civil War.

In the autumn of 2002, having finished a manuscript devoted to one
of those other subjects, I began casting about for an idea for yet another
book. This led me into a number of discussions with my friend Lee Van
Orsdel, who was then Dean of Libraries at Eastern Kentucky University
in my city of Richmond, Kentucky. For some thirty years I have been
making use of Eastern's John Grant Crabbe Library. I have done research in a number of leading libraries in the United States and abroad,
and in my judgment this million-volume library is, in relation to its mission, the best in which I have worked. I have seen it grow under the leadership of my dear friend the late dean Ernest E. Weyhrauch and his
successor, Marcia Myers, and Lee Van Orsdel has brought it to new
heights in every way, including the friendly efficiency and cooperation
extended to its student and faculty users and others like myself. As a result, it has become my second home.

After Lee and I discussed possible topics ranging from the GI Bill to
Napoleon to the battles in China during 1945 to 1949, I found myself

thinking about Ulysses S. Grant. I knew that the library's holdings included a number of Civil War items, and soon found myself sitting down with Chuck Hill, university archivist. At this point I wanted to see if I could write about a less-known aspect of Grant's life, such as his marriage, which Bruce Catton described in these terms: "They shared one of the great, romantic, beautiful loves of all American history." Chuck is the proverbial quick study, and better connected in Civil War circles than I knew. Within three minutes, he had me on the phone with Professor John Y. Simon of Southern Illinois University, the editor of the monumental multivolume *Papers of Ulysses S. Grant*, one of the great works of American scholarship, which is published by the Ulysses S. Grant Association. Professor Simon told me that a study of the Grants' marriage was currently being written; during that and another call, I backed away from this idea, but Professor Simon kindly indicated his willingness to hear from me again, as I continued to think about Grant.

My search for the right idea then led me to my friend William Marshall, the director of Special Collections and Archives at the University of Kentucky. Bill is a Civil War expert who is currently the secretary of the Kentucky Civil War Round Table, the nation's largest such group. During two lengthy conversations, I found myself coming back to the idea of Grant as soldier, but in what context? His rise in four years from being a former captain in the prewar Regular Army, a man who had never commanded more than a hundred soldiers and who was forced to resign for drinking while on duty, to being the victorious commanding general of the Union Army that had a strength of a million men? Or should it be a description of his bloody battles with Lee in Northern Virginia? Bill listened, made suggestions, and spoke of what had and had not been written on Grant and the Civil War. I left him knowing that it would be Grant—but Grant, how? Grant, what?

I was ripe for a "Eureka!" moment, and it came during a long-distance telephone call with Thomas Fleming, a close friend of many years, the author of some forty books on various aspects of American history, and the man who succeeded me as president of PEN American Center. We went through a number of possibilities, and I suddenly said, "Grant and Sherman." Tom, who has been enormously helpful to me in his insightful readings of several of my manuscripts as they developed, agreed that it was an interesting, worthwhile idea: there had never been a book that focused on Grant and Sherman's military partnership and their warm, supportive friendship. (It should be noted, however, that Joseph T. Glatthaar devotes a chapter to the Grant-Sherman relationship in his excellent Civil War study, *Partners in Command: The Relationships Between Leaders in the Civil War*, where he makes this insightful defining comment: "Grant comprehended problems in all their simplicity; Sherman grasped them in all their complexity.")

From the moment of my conversation with Tom Fleming I never looked back, but there was a lot of looking ahead to do. In an exchange of letters with John Simon, he gave me important recommendations on what to begin reading, and was good enough to read the first part of my book as it evolved. During this time, I had some three hours of meetings with Charles P. Roland, alumni professor of American history at the University of Kentucky and the author of works including the admirable one-volume history, *An American Iliad: The Story of the Civil War*. His enthusiasm for the subject, combined with his suggestions for research, sent me forth with an added understanding of why he is so highly regarded by his colleagues and his graduate students. I later had the benefit of a most interesting and helpful conversation concerning the post–Civil War period with George C. Herring, alumni professor of history at the University of Kentucky, who provided me with a useful article regarding American attitudes about the nation's westward expansion and entrance into the Spanish-American War.

An important figure who soon appeared on my horizon was Professor John F. Marszalek, the W. L. Giles Distinguished Professor Emeritus of History at Mississippi State University and author of *Sherman: A Soldier's Passion for Order* and *Commander of All Lincoln's Armies: A Life of General Henry W. Halleck*. Of all the authorities I have ever approached on whose time I had no claim, John Marszalek has been the most generous in his thoughtful, careful readings of my manuscript at successive stages, and in his candid reactions to what he was seeing. He tells me that his graduate students call him "The Cheerful Assassin"; I can only say that his helpful comments are characteristically those of a tough, fair, gifted teacher who is sharing his large and significant fund of knowledge. His part in this has improved my book greatly, and its shortcomings remain mine alone.

From that point in the development of *Grant and Sherman*, I was in the library at Eastern Kentucky University virtually every day, doing research and writing the story as it evolved. Every kind of cooperation and assistance was afforded me. Eastern has a team of excellent and dedicated young reference librarians, all of whom responded most helpfully to many queries on a great array of topics. A special word of thanks is due to Linda Sizemore, Government Documents Librarian, who, often on short notice, came up with articles on various important subjects, and answers to endless questions that ranged from the population of Vicksburg, Mississippi, in early 1863, to the time that Lincoln's funeral train left Washington on the morning of April 21, 1865. Other reference librarians and assistants, working under the direction of Julie George, were unfailingly helpful. They include Karen Gilbert, Kevin Jones, Linda Klein, Victoria Koger, Brad Marcum, Leah Banks, and student workers Christine Cornell and Jennifer Mason. Rob Sica shared in his fellow reference librarians' efforts, and has directed me to materials useful for my future projects. Steve Stone, a reference librarian now at Lexington Community College, assisted me earlier in this book's evolution. I also wish to thank Carrie L. Cooper, Coordinator of Research and Instructional Services, who became the interim Dean of Libraries at Eastern when Lee Van Orsdel accepted the position of Dean of Libraries at Grand Valley State University in Allendale, Michigan.

When it came time to go beyond the library's own collection, I had the pleasure of working with Pat New, head of Interlibrary Loan, and Rene McGuire and Mia Fields, all of whom managed to produce books and other materials for me from all over the country in what seemed a remarkably short time, and to know when to renew them even before I asked them to do so. Thoughout all this, I had the almost daily assistance of the friendly workers in the Circulation Department, which operates under the able direction of Cheryle Cole-Bennett, coordinator of Retrieval Services, who has frequently shared her excellent supply of coffee with me. Headed by Webber Hamilton, this group has included Crystal Brookshire, Chandra Chaffin, Shona Green-Benge, Betty Hays, Kyle McQueen, Lana Takacs, Jeremy Turner, and Judy Warren, as well as student workers Katie Klopher, Brian McDaniel, and Tiffany Swindeman. For the past thirty years, I have had help in finding periodocals from Samira Tuel, who is now in Patron Information Services. I also appreciate the role played within the library over the years by Peggy Flaherty in developing the collection.

As this project has come along, I have had vital help in the world of word processing from my computer guru, Ward Henline of the university's Academic Computing Department. I am also particularly grateful to Carol T. Thomas for her work in bringing order out of the mass of assorted filing cards and legal pads within which were both my bibliography and chapter notes, and skillfully preparing those for publication. She continues to assist me as I study materials for future work. Jo Lane and Linda Witt have helped me keep in touch with different individuals within the large building in which this project has come to fruition. Finally, I very much appreciate the patience shown by two members of the custodial staff, Shirley Dickerson and Eleanor Land, who managed to stay even with the shifting tides of stacks of books and assorted papers that I accumulated in the work space the library most generously allotted to me. From the Dean of Libraries on down, I have been the beneficiary of a high standard of professional librarianship, as well as friendly and encouraging interest in what I do.

A different and essential part of what would be involved in the finished book was provided by my gifted friend, Professor E. Carroll Hale of the university's Department of Art and Design, whose many talents include cartography. He prepared the maps for this book, keeping his good-natured composure as I continued to think of changes I wished to make to them, right to the last moment of this book's production deadline.

I am indebted to a number of other libraries around the nation; in addition to the help I received at the University of Kentucky, I appreciate the help always given me by Barbara Power, head of Circulation at Berea College's Hutchins Library, located a few miles from my house. (It is worth noting that the fine holdings of that collection have recently been increased by a bequest of three to four thousand Civil War items, still being cataloged, from the impressive private library of the late dean Warren D. Lambert of Berea College, who was a true Civil War expert.) Other institutions whose holdings I have consulted include the Library of Congress, The New York Public Library, and the New-York Historical Society. Russell Flinchum, archivist of the Century Association in New York City, was generous with his time during my first efforts to select the illustrations for this book.

As my manuscript went through successive drafts, I became the beneficiary of a number of insightful and constructive readings and suggestions. Once again my friend Thomas Fleming read one of my efforts from start to finish, as did my sister, Mary Ellen Reese, herself an author.

Edward H. Pulliam of Alexandria, Virginia, who contributes articles to history magazines, gave me the benefit of his sound editorial instincts as well as his knowledge of Civil War events and sites in Virginia and in Washington. Dwight D. Taylor of San Francisco posed some thought-provoking questions on matters ranging from major campaigns to word usage. William Marshall of the University of Kentucky, having helped me at the outset as I discussed possible approaches to the life of Grant, subsequently read the manuscript at a late stage in its development and made a number of excellent suggestions that I incorporated. I received some exceptionally important business advice on matters related to this book's publication from Alfred Donovan and Stephen Hill, both of Chestnut Hill, Massachusetts. When I have been away from Kentucky during summers in Maine, I have relied on the computer-processing skills of Larry Gray of Bucksport, Maine, who teaches at the George Stevens Academy in Blue Hill. My daughter Lucy, a published young writer now in an M.A. program in creative writing at the University of Texas at Austin, made corrections and suggestions during a careful reading of one of my later drafts.

When it came time to submit my manuscript to publishers, this was done in masterful fashion by my literary agent, John Taylor ("Ike") Williams. His friendly and efficient assistant Hope Denekamp has been greatly helpful during every subsequent stage of the agent-author relationship. Alexis Rizzutto, then with the Kneerim and Williams agency, played an important part in the initial publishing negotiations. As a result, I signed a contract with the publishing house Farrar, Straus and Giroux, and found myself working with Eric Chinski, a truly gifted editor who has the best of editorial talents: the ability to help the writer make the most of what is already on paper at the time the author-editor relationship begins. His efforts on my behalf have been supported by a team of pleasant, able, energetic, dedicated young people. Eric Chinski's assistant, Gena Hamshaw, has played her part in bringing this manuscript forward in a friendly, highly efficient manner; she has swiftly and informatively answered every e-mail query or response from my end, on matters of every kind, and I thank her for her many hours of work on this one project. Sarah Russo, my publicist, has similarly worked hard in her role of arranging a number of appearences to widen the audience for this book, and has made many other efforts to bring attention to it. During the copyediting, Cynthia Merman gave the manuscript her careful and constructive attention, and Wah-Ming Chang of Farrar, Straus and Giroux has brought it through the overall copyediting process in a manner that I very much appreciate. I also thank Kathryn Lewis for her role as an editorial assistant when she was with the publishing house.

In closing this long list of those who have helped me, I wish to mention some individuals whose contribution to this book largely consists of the way in which they helped to form my approach and style through their reading and commenting on my earlier manuscripts. They include my friends the late Eleanor Parsons of St. Petersburg, Florida, as well as Sidney Offit of New York City; Barbara Pluff of Brunswick, Maine; Helen Poz of Melbourne, Florida; Bridget Saltonstall of Concord, Massachusetts; and Gerald Toner of Louisville, Kentucky. The author Thomas Parrish of Berea, Kentucky, has continued the help he has given me over the years by making excellent suggestions as to Civil War experts I should talk to when I began this book. A writer's work is in part a reflection of the individuals he has known, and, first to last in these acknowledgments, I have been fortunate in that reguard.

INDEX